T0227662

Energy Development in the Southwest

First published in 1980, the second volume of *Energy Development in the Southwest* ana-lyses water conditions and habitat life in the Upper Colorado River Basin, based on alternative national energy scenarios and attempts to assess some of the economic, demographic, and environmental impacts of each development scenario. The energy development scenarios considered in this book involve coal development and use, oil share production, and uranium mining and milling. This title will be of particular interest to students of Environmental Science.

Energy Development in the Southwest

Problems of Water, Fish and Wildlife in the Upper Colorado River Basin
Volume Two

Edited by
Walter O. Spofford, Jr., Alfred L. Parker, and
Allen V. Kneese

First published in 1980
by Resources for the Future, Inc.

This edition first published in 2016 by Routledge
2 Park Square, Milton Park, Abingdon, Oxon, OX14 4RN
and by Routledge
711 Third Avenue, New York, NY 10017

Routledge is an imprint of the Taylor & Francis Group, an informa business

Publisher's Note
The publisher has gone to great lengths to ensure the quality of this reprint but points out that some imperfections in the original copies may be apparent.

Disclaimer
The publisher has made every effort to trace copyright holders and welcomes correspondence from those they have been unable to contact.

A Library of Congress record exists under LC control number: 80008020

ISBN 13: 978-1-138-99889-6 (hbk)
ISBN 13: 978-1-315-65847-6 (ebk)
ISBN 13: 978-1-138-99895-7 (pbk)

Energy development in the Southwest

Problems of Water, Fish and Wildlife
in the Upper Colorado River Basin

Volume Two

WALTER O. SPOFFORD, JR., ALFRED L. PARKER,
and ALLEN V. KNEESE, editors

RESEARCH PAPER R-18

RESOURCES FOR THE FUTURE / WASHINGTON, D.C.

RESOURCES FOR THE FUTURE, INC.
1755 Massachusetts Avenue, N.W., Washington, D.C. 20036

Resources for the Future is a nonprofit organization for research and education in the development, conservation, and use of natural resources and the improvement of the quality of the environment. It was established in 1952 with the cooperation of the Ford Foundation. Grants for research are accepted from government and private sources only if they meet the conditions of a policy established by the Board of Directors of Resources for the Future. The policy states that RFF shall be solely responsible for the conduct of the research and free to make the research results available to the public. Part of the work of Resources for the Future is carried out by its resident staff; part is supported by grants to universities and other nonprofit organizations. Unless otherwise stated, interpretations and conclusions in RFF publications are those of the authors; the organization takes responsibility for the selection of significant subjects for study, the competence of the researchers, and their freedom of inquiry.

Research Papers are studies and conference reports published by Resources for the Future from the authors' typescripts. The accuracy of the material is the responsibility of the authors and the material is not given the usual editorial review by RFF. The Research Paper series is intended to provide inexpensive and prompt distribution of research that is likely to have a shorter shelf life or to reach a smaller audience than RFF books.

This is volume two of a two-volume work.

Library of Congress Catalog Card Number 80-8020

ISBN 0-8018-2495-8

Copyright © 1980 by Resources for the Future, Inc.

Distributed by The Johns Hopkins University Press,
   Baltimore, Maryland 21218

Manufactured in the United States of America

Published September 1980. $25.00 per set.

Contents of Volume II

List of Tables

PART II.   IMPACTS ON WATER QUALITY, FISH, AND WILDLIFE

Chapter 7

THE INFLUENCE OF REDUCED STREAMFLOWS ON WATER QUALITY

James R. Gosz*

Introduction

Since the quantity of water in any river system is finite, and since
the quality of water is a function of the quantity, then any consumptive
use of water can be expected to affect the quality of the remaining water.
While this conclusion seems logical, the relationship is not straightforward or simple because of the myriad of chemical and biological processes
that are involved. It is the purpose of this paper to discuss some of
the complexities involved in the relationship between reduced streamflow
and water quality, and to identify research priorities.

I would like to approach the discussion of the effects of dewatering
on water quality and stream organisms from two positions: (1) the influence of reduced stream volume and velocity; and (2) the influence of
the location in the stream system at which dewatering takes place.

The first of these deals primarily with physical and chemical factors
associated with reduced stream velocity, while the second is a result of
land-water and within-stream interactions--physical, chemical, and biological.

Before discussing the influences of dewatering, it is important to
review briefly the temporal relationship between water quantity and quality

---

*Associate Professor, Department of Biology, University of New Mexico,
Albuquerque, New Mexico.

under natural conditions. A river or stream system is a drainage system
for a definable land area occupying an elevational gradient. This elevational gradient causes significant changes to occur in all of the environmental factors, most notably in atmospheric temperature, precipitation,
and evapotranspiration. These in turn greatly affect vegetative cover,
organic matter accumulation, and soil development. Since the conditions
at a particular point in a stream are greatly influenced by the watershed
above that point, no two sections of a stream are precisely the same.
In addition, throughout the elevational gradient of the stream the various
environmental and biological factors will change in terms of their relative importance in affecting stream conditions. This must be kept in
mind when discussing the effect of modifying any one factor controlling
water quality.

A common feature of the Southwest is the logarithmic increase in
precipitation with increasing elevation. Studies in the Sangre de Cristo
Mountains of New Mexico, for example, show that annual precipitation
approximately doubles with each 1,000-meter increase in elevation (table
1). In the Colorado River region, annual precipitation varies from less
than 15 cm (6 inches) in desert areas to over 127 cm (50 inches) in the
high-elevation headwaters (Utah State University, 1975).

It is clear from this that in the Southwest, a very small portion of
a river basin, the headwater region, contributes the vast majority of the
stream volume in the drainage system. This region must also greatly influence the water quality throughout the drainage system. As a result of
the hard rock geology and high volume of streamflow commonly found at
high elevations, the total dissolved solids are low and water quality

5

Table 1. Characteristics of the Tesuque Watersheds in the Sangre de Cristo Mountains of New Mexico (1974-75 water year)

					Watershed				
	P-J	2	4	5	6	7	AW-1	8	15
Elevation (m)									
Maximum	2,560	2,850	3,383	3,444	3,520	3,490	3,231	3,648	3,734
Minimum	2,365	2,423	2,621	2,804	2,972	2,987	3,109	2,941	3,231
Vegetation (ha)									
Piñon-juniper	17	10	0	0	0	0	0	0	0
Pine	0	106	0	11	0	0	0	0	0
Mixed Conifer	0	0	100	18	0	0	0	0	0
Aspen	0	0	0	23	19	13	3.4	203	0
Spruce-fir	0	0	80	112	103	64	0	45	123
Non-timbered	0	0	0	0	0	23	0	166	40
Total	17	116	180	164	122	100	3.4	414	163
Precipitation (cm)	55.0	62.8	80.5	87.3	93.0	97.2	88.4	98.0	111.4
Runoff (cm)	1.1	2.5	9.9	13.3	28.1	41.5	50.4	42.3	58.0
Evapotranspiration (cm)	53.9	60.3	70.6	74.0	64.9	55.7	38.0	55.7	53.4
Percent	98	96	88	85	70	57	43	57	48

Notes: Watersheds: P-J--Pinon-Juniper; 2--Little Tesuque Creek Tributary #2; 4--Little Tesuque Creek Tributary #4; 5--Little Tesuque Creek Tributary #5; 6--South Fork Tesuque Creek; 7--Middle Fork Tesuque Creek; AW-1--Aspen; 8--North Fork Tesuque Creek; 15--Rio en Medio. m--meters; ha--hectares; cm--centimeters

Source: Data on elevation, precipitation, runoff, and evapotranspiration generated by the author. Data on vegetation from J. O. Carleton, J. M. Gass, and H. R. Martinez, "Soils Report for Tesuque Watersheds, Santa Fe National Forest" (Santa Fe, N.M., U.S. Department of Agriculture, Forest Service, Southwestern Region, 1972).

is high. Tables 1 and 2 show the relationship between precipitation, runoff, and several dissolved ions in streams over an elevational gradient in
the Sangre de Cristo Mountains of New Mexico. Since the geology of these
watersheds is similar, the variation in dissolved salts is primarily a
result of differences in the environmental and biological factors among
watersheds (e.g., evapotranspiration, soil development).

Figure 1 shows how the concentration of a dissolved ion varies when
all of the elevational zones shown in table 1 are integrated into a single
drainage system, in this case, the Santa Fe Municipal Watershed. While
the lower elevations show increased concentrations of a dissolved nutrient, the concentration in the mainstream remains low compared to the
tributaries because of the influence of discharge from higher elevations,
the well known dilution effect. This helps to explain the commonly observed inverse relationship between total dissolved solids and discharge
rates in most stream systems. Data for the San Juan River in New Mexico
demonstrate this relationship clearly (figure 2).

The highest discharge and lowest total dissolved solids concentrations
occur during the spring months, as a result of snowmelt at the higher elevations. During the low flow conditions of the summer months, ion concentrations normally rise for a given stream area, which suggests an evapora-
tion-concentration effect; however, quite high concentrations can also occur
during the initial increase in stream discharge following a summer storm.
This causes a great deal of variability in water quality during the snow-
free months, as can be seen in figure 2. A particular nutrient may demonstrate a significant negative correlation with the discharge rate on
an annual basis, and both significant positive and negative correlations

Table 2. Weighted (by flow volume) Average Annual Concentrations of Ions in Streams of the Tesuque Watersheds (1974-75 water year)

(mg/1)

	Watershed								
	P-J	2	4	5	6	7	AW-1	8	15
Ca^{++}	78.70	22.87	5.61	3.81	2.97	2.97	2.78	2.40	2.65
Mg^{++}	38.22	9.20	2.80	1.63	1.16	1.01	0.91	0.79	0.60
Na^{+}	17.57	9.88	4.35	3.04	2.49	2.07	2.21	2.02	1.69
K^{+}	1.70	0.95	1.22	0.92	0.62	0.89	0.64	0.62	0.51
NH_4^{+}	0.16	0.12	0.12	0.09	0.09	0.09	0.07	0.07	0.07
NO_3^{-}	0.37	0.02	0.07	0.04	0.03	0.06	0.03	0.03	0.09

Notes: For characteristics and names of the watersheds, see table 1.

Source: Data generated by the author.

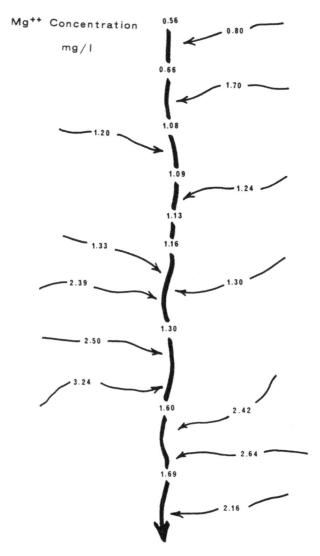

Mg⁺⁺ Concentration

mg / l

Note: The portion of the river shown covers an elevational range of 2,377 to 2,536 meters and forest communities from pinon-juniper to spruce-fir.

Figure 1. Stream Water Concentrations of Mg^{++} (mg/l) in the Mainstream and Tributaries of the Santa Fe River on 3 June 1972

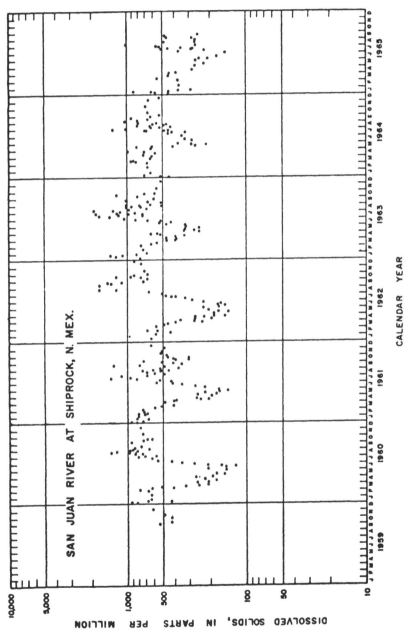

Figure 2. Temporal Variation in Dissolved Solids in the
San Juan River at Shiprock, New Mexico

Source: J. W. Hernandez, Proposed Water Quality Standards for the San Juan River, LaPlata River, and Animas River in New Mexico (Reston, Va., U.S. Geological Survey; Santa Fe, N.M., New Mexico Department of Public Health, 1967).

with discharge during different time intervals following a rain storm. The variability of stream water quality parameters during short periods (i.e., summer months) is often a result of the local watershed environmental and biological factors. Larger streams tend to show less short-term variability, since they integrate the effects of a large land area and are less affected by local occurrences.

It is tempting to use the temporal variations in discharge rate and water quality to predict the effects of a dewatering operation; however, reduced stream volume by a diversion operation or by reduced precipition at higher elevations (or on some small area of a large watershed) represent distinctly different cases and must be evaluated separately.

Reduced Stream Discharge

Any reduction in stream discharge will cause a reduction in stream velocity, since velocity and discharge are mathematically related (Hynes, 1970; Leopold and coauthors, 1964). The effect of velocity is primarily a physical one, and a change in stream velocity can be expected to affect temperature, properties of water, sediment-transport capacity, erosion potential, streambed characteristics, and the composition of the stream community.

The influence on stream temperature is reasonably straightforward in that any dewatering causes an increase in temperature below the dewatering point. This is a result of two interacting factors: (1) a decrease in the volume of cooler water from higher elevations; and (2) decreasing the depth, velocity and volume of water allowing it to warm more rapidly. An increase in the temperature regime of a stream will cause predictable changes in the physical properties of the water itself.

Increasing water temperature causes a decrease in density, surface tension, coefficient of friction, viscosity, heat of vaporization, specific heat, and an increase in thermal conductivity (Hynes, 1970; Slatyer, 1967). Thus, although the overall water velocity will decrease after the dewatering point, the subsequent warming of the remaining water will cause it to begin to flow faster (0.5 percent faster for each 1° C rise between 4° C and 20° C). The decreasing viscosity at warmer temperatures also reduces the silt-carrying capacity of the water, causing an unusual situation in which a given volume of water can increase in velocity but decrease its silt-carrying capacity. Warmer water also will produce a thinner boundary layer on the bottom (Ruttner, 1963), evaporate more easily, heat more rapidly, and distribute that heat more rapidly in the water body. The variable properties of water have seldom been considered in biological studies in either terrestrial or aquatic ecosystems, thus the significance of these changes is difficult to assess.

A major effect of a dewatering operation would be a significant decrease in the suspended solids load below the dewatering point because of the reduction in discharge. It can be shown that the suspended solids load, which is an indicator of the erosive power of the river, varies as some power (usually between 2 and 3) of the discharge. This decrease in discharge may cause some alteration of the streambed below the dewatering point, as both stream depth and width are mathematically related to discharge (Leopold and coauthors, 1964). It can be expected that the streambed will be reduced in size because of the deposition of sediment and the invasion of rooted vegetation in the shallower and slower stream areas, which further enhances the trapping and accumulation of silt.

Whether the reduction of the streambed becomes permanent depends upon the efficiency of the dewatering operation at high discharge, since it seems that the form of a channel is determined and maintained by discharges that would just about fill it (Hynes, 1970). In many rivers and streams this discharge is reached about every one and one-half years on the average (Leopold and coauthors, 1964).

All of the above discussion is rather qualitative because of the variations which can occur as a result of specific site factors. The size and location of the stream channel, the magnitude of the dewatering operation, and the quality and quantity of water in the natural stream at specific times of the year, all will influence the actual changes that follow dewatering. For example, Baxter (1961) found that most of the beds of Scottish salmon rivers were still covered by water at even one-eighth of the mean annual discharge. However, small, stony streams began to contract at one-half of mean discharge, and at one-eighth, only one-third of the bed remained wet. These differences between large and small streams can be expected to have important biological consequences.

The physical factors associated with a dewatering operation are expected to produce significant changes in the stream biota. The effects will be as varied as the combinations of physical factors themselves and at this point we can only identify trends.

Velocity is the most significant characteristic of running water and it is not surprising that organisms demonstrate strong adaptations to it as well as to its indirect effects. Decreasing velocity following dewatering is expected to influence organisms through a combination of:

13

1. decreased concentrations of dissolved gas

2. increased temperature

3. more pools, fewer riffles

4. finer textured substrate

5. decreased nutrient availability

6. decreased turbidity

Decreased concentrations of gases (particularly oxygen and carbon
dioxide) occur primarily as a result of decreased turbulence at lower
discharge, increased temperature which lowers the solubility of these
gases in water (Crockford and Knight, 1959), increased physiological
requirement for oxygen at warmer temperatures (Hynes, 1970), and increased
oxidation of detritus in the additional pools formed as a result of lower
discharge and warmer temperatures.

These factors are likely to be most severe for species such as trout
in which the rate of metabolism rises rapidly with temperature. Any conditions which even slightly decrease oxygen tensions in warm water are
deleterious for these species, quite apart from their intolerance of warm
temperatures. Many other cold-water headstream species are probably
limited in this way (Hynes, 1970). A great many stream animals are steno-
thermic (that is, restricted to a narrow, or in this case cold, temperature range). Pomeisl (1961), for example, states that most species of
stonefly in Austria are confined to waters where the summer temperature
does not exceed 15° C, and very few occur in waters which regularly exceed
25° C. It seems probable, however, that the intolerance of higher

temperatures is at least as much concerned with respiration and the lower availability of oxygen as it is with temperature itself.

A similar relationship holds for bryophytes, particularly mosses, and their need for dissolved CO_2. In addition to their need for a firm unsilted object to which to attach themselves (a result of relatively high discharge), mosses are unable to photosynthesize bicarbonate ions (Ruttner, 1963). In contrast to most algae and the flowering plants, they are confined to turbulent water where there is an adequate supply of carbon dioxide. This means they are favored in spring areas (water emerging from the ground is normally rich in CO_2) and "white water" areas of rapids and falls because this increases the absorption of atmospheric carbon dioxide to a saturated or supersaturated condition. When water becomes less turbulent, carbon dioxide is lost to the atmosphere and other plants and algae take over.

The reduced flow which accompanies a dewatering operation further exacerbates the inability of organisms to get sufficient levels of dissolved O_2 or CO_2. The concentration of oxygen which limits a given stream organism is inversely related to the flow rate. At low flow rates, the density of dissolved oxygen molecules must be relatively high in order for the organism to obtain a sufficient quantity. Flow makes water "physiologically richer" because of its constant renewal of materials in solution near the surfaces of organisms (Hynes, 1970; Ruttner, 1963). This may well be the explanation for the demand for flow by many species or for the demand for increased flow in warm weather when dissolved gases are low. This is no doubt true for other essential nutrients as well since flow increases the effective physiological concentration in the water, and running water is a richer environment for animals, as well as plants, than still water of the same

chemical concentration. A most difficult aspect is quantifying the nutrients or dissolved gases which come in contact with stream organisms,
since measured stream velocities are those of the main mass of water
and not the velocities that many bottom invertebrates are exposed to.
Many invertebrates are sheltered under stones, in moss, or in the boundary layer on the bottom where water velocities are very low. The effect
on availability of nutrients following a reduction in the boundary layer
(by increasing water temperature) is not known.

A great deal of literature is available on the change in stream
species from riffles (clean, stony runs) to pools. There is an innate
flow demand in some species (plants, invertebrates, and fish), and
different species become dominant at different velocities. This not
only causes community differences from location to location in a stream,
but also at one location as discharge varies throughout the year.

One feature of the ecology of algae which makes them difficult to
study is their rapid change. During periods of low flow and clear water,
considerable populations are built up and every solid object in the water
may become thickly covered with a brown carpet of diatoms and trailing
streamers of filamentous green algae. In a natural stream these populations rise and fall in relation to the occurrence of summer storms. A
dewatering operation can be expected to reduce the flow in the stream
which would favor the development of low-flow communities. It can also
be expected to reduce the frequency and intensity of flushes from summer
storms which would tend to remove these communities. The overall effect
would probably be a decrease in the number of species in the stream community since, in general, the fauna of riffles is richer than that of

silty reaches and pools both in number of species and in total biomass
(Hynes, 1970). Holden and Stalnaker (1975) reported that fish species
diversity was found to be greater in the more heterogeneous, high gradient, rocky bottom habitats than in low gradient, sandy bottom areas of
the Colorado River. In one sense a dewatering operation which reduces
discharge fluctuation may be of benefit since the instability of the flow
regime in many rivers makes them difficult habitats for some species of
fish (e.g., trout). On the other hand, many species inhabiting large
rivers depend upon high discharge periods for spawning. Johnson (1963)
reported that the bigmouth buffalo (Ictiobus cyprinellus) spawns successfully only at times of high discharge and is definitely controlled by the
lack of high flows in some years.

Another important aspect of fluctuating discharge is the occasional
or annual drying out of streams, leaving isolated pools or, perhaps, no
surface water at all. A dewatering operation is expected to increase the
severity and/or frequency of this phenomenon. Although numerous studies
have been made on the biota which survive these drying spells under natural conditions (or leave and reinvade when water levels increase), little
information is available on changes in community composition as a result
of more frequent or more severe drying periods following dewatering. It
would appear that the species most affected would be those that are in-
tolerant to the higher temperatures and lower oxygen concentrations of
stagnant pools, especially if large amounts of organic matter enter the
stream (leaf fall) before normal flows are restored. Hynes (1970) reports that the ability to withstand drought is by no means universal
among stream invertebrates, and that many take few steps to avoid its

consequences. The exceptions are those species which endure drought,
even in an active phase, by moving down into the hyporheic habitat
(streambed) where they find sufficient moisture to survive, e.g.,
flatworms, oligochaetes, Elminthidae and their larvae, some chironomid
larvae, and Hydracarina (Hynes, 1970). Mann (1959) reports that a few
species, such as the leech, Dina lineata, are found only in places which
dry up more or less regularly. It is possible that only in such places
can they compete with species which do not have the ability to avoid
the consequences of drought.

Although the major controlling factor in the distribution of invertebrates is often cited as the nature of the river bed (Hynes, 1960), it
must be remembered that many factors are correlated with the type of substrate. A high velocity, or eroding stream condition, will be associated with a gravelly or rocky substrate, as well as a high oxygen
content and low temperature. These streams tend to be at the headwaters of river basins or near springs, and they also tend to be low
in dissolved salts. The decreased velocities found at lower elevations are conditions favoring the deposition of material (fine textured
substrate), warmer temperatures, higher concentrations of dissolved salts,
and at times oxygen deficiencies.

Stream Dewatering Location

The material discussed so far, as well as a great deal of additional
literature, document the longitudinal variation that occurs in the stream
of any drainage basin. Since no two stream sections along the longitudinal gradient are the same, it follows that the location of a dewatering
operation will make a difference in the water quality parameters below

that point. Referring again to figure 1, it can be seen that a water
quality parameter in the Santa Fe River, in this case Mg^{++} concentration,
changes continuously from within-stream factors and differences among the
tributaries, which are the result of the land-water interaction. The relatively large volume of water from high elevations reduces the effect (concentration) of the tributaries and it should be apparent that a dewatering
operation on the main stream will increase the ability of the tributaries
to modify stream chemistry below that point. The magnitude of the modification caused by the lower elevation tributaries is dependent on the location of the dewatering operation on the main stream.

It is now appropriate to discuss the factors that influence the
variation in chemical parameters of a stream and its tributaries in more
detail, since this is the type of information which may play a role in
the decision for the location of a dewatering operation. The various
factors can be discussed in terms of reactions within the stream, reactions within the terrestrial portion of the watershed, and the interaction between the aquatic and terrestrial portions of the watershed (via
land-water linkages). The land-water interactions are the least understood of these three categories. This is unfortunate since the available
literature demonstrates that water quality is greatly affected by these
interactions (Likens and Bormann, 1974, 1975). The geologic output of
water, particulate matter, dissolved nutrients, and other chemicals from
terrestrial systems, often represents the major output for these systems
and, at the same time, represents the major input for most aquatic systems. This relationship has been used to an advantage by using changes
in stream chemistry to quantify the effects of land management activities.

The results of these studies have shown that undisturbed or natural ter-
restrial ecosystems minimize the geologic output of inorganic material
(Likens and Bormann, 1972, 1974; Hobbie and Likens, 1973; Cooper, 1969;
Gosz, 1975) and that different types of natural ecosystems differ in
their ability to minimize outputs (i.e., minimize inputs to streams
(Gosz, 1975)).

Table 3 demonstrates this pattern for the geologic output of soluble
Ca^{++} and Mg^{++} from terrestrial ecosystems in New Mexico ranging from pinon-
juniper to spruce-fir and alpine tundra. While there may be large differences from year to year as a result of different precipitation and
stream discharge quantities, the pattern of ability to minimize geological output remains the same among the different communities. It is of
value to discuss the factors which may be responsible for these patterns,
since they should hold for similar communities throughout the Upper Colorado River Basin.

Precipitation-Evapotranspiration-Runoff

The change in the hydrologic regime over the environmental gradient
is of obvious importance. It has already been mentioned that annual precipitation approximately doubles for a 1,000-meter increase in elevation
in northern New Mexico. Annual stream volumes, however, increase between
twenty and thirty times for a 1,000-meter increase in elevation. This
difference is primarily the result of the influence of temperature on
evapotranspiration. High elevations accumulate significant quantities
of winter precipitation in the snowpack which runs off largely during
spring snowmelt. The warmer winter temperatures at low elevations allow
much more evaporation of winter precipitation, or more frequent melting

Table 3. Stream Output of Soluble Ions from the Tesuque Watersheds in the Sangre de Cristo
Mountains of New Mexico During Three Water Years--1972-73, 1973-74, and 1974-75

(kg/ha)

	Watershed								
	P-J	2	4	5	6	7	AW-1	8	15
Ca++									
1972-73	54.8	17.8	17.3	14.3	14.5	24.0	-	19.6	24.5
1973-74	3.4	2.4	1.5	2.2	3.8	4.6	5.6	4.9	8.0
1974-75	8.7	5.7	5.6	5.1	8.3	12.3	13.6	10.2	15.4
Mg++									
1972-73	26.7	6.8	7.3	5.3	5.3	8.2	-	5.9	5.0
1973-74	1.5	0.9	0.7	0.9	1.5	1.5	2.0	1.6	1.9
1974-75	4.2	2.3	2.6	2.2	3.3	4.2	4.6	3.4	3.5

Note: For characteristics and names of the watersheds, see table 1.

Source: Data generated by the author.

conditions, which permit water movement into the soil where it is subject to transpiration by plants, even during the winter months (Swanson, 1967). Summer precipitation is also more subject to evapotranspiration at lower elevations because of warmer temperatures and dryer soil conditions which can hold a greater proportion of the precipitation for subsequent evapotranspiration. Soils at higher elevations (under forest vegetation) typically have a higher average moisture content, and a larger proportion of summer precipitation ends up in the drainage system.

These patterns explain the decrease in the percentage of annual precipitation which is evaporated with increasing elevation (table 1). When evapotranspiration is expressed in centimeters of water, however, an interesting pattern develops, showing the greatest evaporation occurring on watersheds at intermediate elevations and lower amounts occurring on watersheds at both high and low elevations. This pattern is inversely related to the pattern of output of soluble ions from the watersheds and is consistent year after year in spite of large variations in annual precipitation and seasonal distributions (see tables 1 and 3). It is a result of a complex of factors including temperature, precipitation, and the resulting biological community (table 1).

Biological Community

The greatest evaporation should occur at elevations with the most favorable combination of temperature and moisture. High elevations have abundant moisture but low temperatures and low elevations in the semiarid Southwest have a scarcity of moisture. The most favorable combination of temperature and moisture can also be expected to produce the greatest amount of plant growth. Not only are temperature and moisture important in the physiology of plants but the ability to transpire water

(evapotranspiration) also has been identified as important for normal
growth. Factors which inhibit transpiration also inhibit growth (Kramer
and Kozlowski, 1960; Odum and Pigeon, 1970) and the net primary production of a variety of communities is significantly correlated with actual
evapotranspiration (Rosenzweig, 1968). The primary productivity of a
community is a tremendously important factor in determining the organic
biomass, both living and dead, the diversity of species, soil formation
and depth, assimilative capacity (for toxins and pollutants), and nutrient
uptake and retention properties (Odum, 1971).

Of all the communities along the elevational gradient, the mixed
conifer association at the intermediate elevations demonstrates the greatest development of these characteristics which no doubt explains the ability of this community to minimize geological outputs of inorganic material
(tables 1 and 3). The large output of dissolved nutrients at high elevations
is a function of the very high precipitation and discharge volumes at those
elevations; however, the large output of dissolved material at low elevations is a function of low organic biomass, both living and dead, which
is important in intercepting and evaporating water and reducing erosional
loss (Reynolds and Knight, 1973); of thin soils (low storage capacity);
and of salt accumulations in the soil (caliche).

The effect of a large organic biomass and evapotranspiration ability
on resultant stream water quality can best be shown by looking at their
influence on peak stream discharge. High discharges, especially those
following summer storm activity, are responsible for high sediment yields
(Bormann and coauthors, 1969), high nitrogen concentrations (Gosz,
1978), increased organic levels (Gosz and Barr, 1974), and high

23

total outputs of dissolved nutrients (Likens and Bormann, 1972). Figure 3 shows peak instantaneous discharge rates and precipitation for watershed communities along an elevational gradient. It is readily apparent that the communities of the intermediate elevations greatly dampen peak discharge which helps to explain the low total losses of inorganic material. This does not necessarily hold for soluble organic matter (Gosz and Barr, 1974).

Geology

One other major factor which can influence stream water quality is the geology of the area. The results presented so far are for communities on a similar bedrock which allows us to see the differences caused by the separate communities. Where the geology changes significantly, such as from the igneous rock of high elevations to sedimentary rock or alluvium at low elevations, inorganic water quality differences would be dominated by those geological differences (Hem, 1970), and the effects of individual communities would be difficult to observe. In Montana, the influence of geology (sedimentary vs. igneous) on some aspects of inorganic water quality has been shown to be greater than the clearcutting of forests (Forcier, personal communication).[1] The geology of the Upper Colorado River Basin is very complicated and variable, ranging from hard, igneous rock to easily erodible, salt-yielding shales. This spatial variability can cause significant changes in the water quality indicators in relatively short distances (Williams, 1975). Over half of the total salt load in the Colorado River is contributed by the natural land and water system (Utah State University, 1975).

---

[1]Personal communication with Larry Forcier, School of Forestry, University of Montana, December 1974.

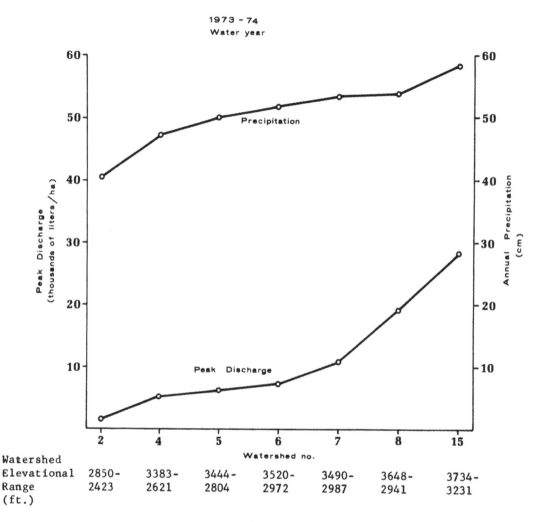

Note: Increased interception and evapotranspiration at the intermediate elevations causes a reduction in the peak discharge rate.

Figure 3. Weighted (by area) Annual Precipitation and Peak
Discharge Rate for the Tesuque Watershed During
the 1973-74 Water Year

Organic Quality Parameters

The organic aspects of water quality is a subject that has been
treated only superficially in relation to its potential significance.
The tremendous variety of organic compounds which may be present, their
normally low concentrations, the difficulty and expense of analysis, and
the lack of good procedures all have contributed to a paucity of basic
knowledge about this water quality parameter. Even less is known about
the organic requirements of aquatic organisms or the regulatory role of
these compounds. Adams and coauthors (1975) report that more than 300
naturally occurring organic compounds have been identified so far from
natural water systems including antibiotics, vitamins, common metabolites, and cellular compounds. Many of these are known to produce effects in laboratory bioassays; however, little is known of their effects
in a natural aquatic environment. McGauhey (1968) reports, for example,
that the lethal dose of phenol for rainbow trout is 6 mg/l with an exposure time of three hours. What are the effects of more natural levels
(0.005 to 5 mg/l, depending on the type of compound and analysis) on trout
or any other aquatic organism (Gosz and Barr, 1974)? Some fish, for
example, are able to detect natural substances at concentrations as low
as 1×10^{-11} mg/l (Warren, 1971). An eel, by conditioned response train-
ing, was able to detect B-phenylethyl alcohol in concentrations of
3×10^{-18} parts (Hasler, 1970). Are there levels at which certain
organic compounds are stimulatory or even required? At what levels are
there behavioral changes or changes in community composition resulting
from species moving out or in?

A number of reports are available demonstrating that fish utilize
very low levels of organic substances to guide themselves in low visi-
bility water, to home in on spawning grounds, for predator avoidance,
in sex, and in care of young (Hasler, 1970). What other organic or inorganic compounds react with a particular organic compound to cause
synergistic or antagonistic effects? How different is the effect of one
type of phenol from another? How do reactions within terrestrial communities, streams, and land-water interactions alter the levels and types
of organic compounds and how will the location of a dewatering operation
change those conditions? As a result of the soils and vegetation, a
drainage basin lends a fragrance to water which is distinctive for every
stream (Hasler, 1970). A stream can be expected to change in fragrance
over an elevational gradient because of the changing vegetation and soils.
How important are these compounds in the normal activity of aquatic organisms? Answers to these questions represent a tremendous amount of
research yet to be performed.

Adams and coauthors (1975) illustrate some possible ecosystem roles
and interactions of organic compounds which are influenced by or which
control populations of microorganisms. A modification of their scheme
is shown in figure 4. The figure identifies with the fact that all
aquatic organisms release organic compounds into their surroundings. The
release of organics by algae is well known, releasing perhaps as much as
10 percent of the total carbon fixed as organic carbon. The release may
be a complex toxin, a waste product, an internally used compound that
leaks out of the cell, or the loss of cell coatings or other extracellu-
lar compounds. These organic compounds may affect ecosystem function,

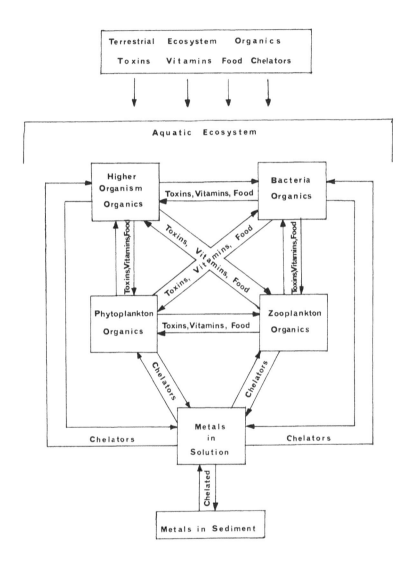

Note: Organic compounds serve as toxins, vitamins, food sources, and chelators. These organic compounds may inhibit or stimulate growth or attract or repel other organisms as interchange arrows indicate. The source may be the terrestrial ecosystem or the aquatic organisms themselves.

Figure 4.   Potential Roles of Organic Compounds
in an Aquatic Ecosystem

community composition or succession, and at times make an entire lake useless for many beneficial uses of society (Adams and coauthors, 1975). For example, toxins of the blue-green algae have been known to cause the death of higher life forms including fish, birds, cattle, and sheep (Fitch and coauthors, 1934; Gorham, 1960; Ingram and Prescott, 1954). They are also known to affect simpler life forms, such as bacteria, by producing antibiotic substances (Nielsen, 1955).

Figure 4 shows that organic compounds, either from the surrounding watershed or produced by aquatic organisms, may affect other organisms by inhibiting or stimulating growth acting as a toxin, vitamin, food source, or transporting chelator. Chelators may act to transport needed metals into the cells or lower the concentration of metals around the organism to nonlethal levels. That these organics are a natural and necessary part of the aquatic environment is evidenced by the history of development of artificial sea water media for dinoflagellates (Martin and Martin, 1973). Few dinoflagellates can be grown in completely in-organic media and most seem to require some organic growth factors. Ini-tially, soil extracts were used, and these had growth-promoting actions that were ascribed to the chelating action of the humus component, pre-sumably through a reduction in toxicity or enhancement of the availabil-ity of trace elements. Soil extracts, however, should also be a source of vitamins, auxins, growth hormones, etc. River water also was used as an additive to provide growth enhancement and these were eventually and effectively replaced by vitamins, iron, and a chelating agent.

Additional evidence for the significant influence of organic com-pounds comes from studies of red tide outbreaks. Prakash and Rashid (1968) noted, for example, that substantial amounts of humic substances enter the Bay of Fundy (where blooms of _Gonyaulax tamerensis_ are prominent) just before the onset of dinoflagellate maximum. Ingle (1965) suggested the involvement of organic substances (tannic and humic acids) in out-breaks of _Gymnodinium breve_ in Florida coastal waters. These materials were presumed to aid transport of iron from bogs and swamps to potential outbreak areas, and rivers have been shown to have increased iron concen-trations following periods of heavy rainfall (Martin and Martin, 1973). Carr and Whitton (1973) report that fulvic acid, a natural chelator in soils, caused a marked increase in cell numbers for a number of freshwater plankton. The nature of the organic compound also is important as evidenced by the fact that some plankton did not show a similar increase when either Ethylenedi-nitrilotetraacetate (EDTA) or citric acid were used as the chelating agent.

The organics in freshwater systems also have been shown to play a significant role in the transport of heavy metals, their solubility, sorp-tion on suspended and bottom sediments, oxidation states (thus affecting their toxicity) and uptake by biota. Naturally occurring organics like fulvic acid can complex heavy metals and concentrate at the air-water in-terface where they have been shown to enrich the concentrations of such trace elements as lead, iron, copper, and nickel (Duce and coauthors, 1972). Organic compounds in water and soil-water systems can affect the oxidation state of trace elements, thereby exerting a strong influence on their be-havior, such as their solubility and ability to be absorbed and ion-exchanged. For example, the reducing environment created by organic pol-lution (natural or man-caused) may cause the reduction of chromium-VI to chromium-III, thereby affecting its transport in water since sediments and

biomass have different capacities for adsorbing or absorbing these two
species (Gloyna and coauthors, 1971).

The variability of concentrations of trace elements in natural fresh
waters is great, both on a macro- and micro-geographic scale, as well as
with time at a given location. In most instances their concentrations
do not correlate with such characteristics as streamflow, inorganic ion
concentrations, or lithologic environment. Knowledge is limited and adequate models for the movement of trace elements into and through these
waters have yet to be developed (Andelman, 1973). No doubt some of the
difficulty in modeling trace elements results from the variability of
organic compounds in terrestrial and aquatic systems. Not only do quantities vary between and at given locations but the properties of the organics
differ as a function of their environment. Humic acids of lake sediments
may be unlike those in soils because of the difference in source material
and the microorganisms involved in decomposition (Andelman, 1973). Rashid
and coauthors (1972) compared humic acids of terrestrial and marine origin
and found that, compared to their properties in soils, humic acids in
ocean sediments have higher molecular weight, lower cation exchange capacity, conspicuously low phenolic hydroxyl or high carbonyl and quinone content, lower metal-holding capacity, and similar concentrations of amino
groups.

Consider now the variation that occurs along the elevational gradient
of a stream: temperature, quantity, and type of bottom sediments; quantity
and chemical composition of organic detritus (e.g., aspen leaves vs. pine
or spruce needles); oxygen content; and changes in organisms (including
microorganisms). With this knowledge, we could predict differences

in stream organics throughout the course of the stream; that is, differences in the toxins, vitamins, food sources, and chelators, as well as their functional importance. At this time, however, that is about all we can predict. We do not have information to predict what the magnitude of the change will be, when it will occur, or what it will do.

The same haunting questions occur concerning the effects of a dewatering operation. Are there particular terrestrial vegetational communities which yield specific organic compounds required by downstream organisms (i.e., vitamins, or precursors of vitamins)? If so, the location of a dewatering operation is critical. If a certain terrestrial community produces toxic material, for example, the high levels of phenols associated with aspen vegetation (Gosz and Barr, 1974; Dormaar, 1971), the location of a dewatering operation may be used to an advantage. This requires that we know not only the types and quantities of compounds coming from terrestrial ecosystems, but also the subsequent changes which occur in the stream. For organics this is virtually unknown. Changes in the inorganic nutrient status of the stream have been studied and the results show that the concentrations of some nutrients (e.g., Ca^{++}, Mg^{++}) tend to increase in amount, while others decrease (e.g., NO_3^-, $PO_4^=$). The controlling factors are:

1. Increased concentrations of nutrients coming from lower elevation tributaries (see figure 1). As a result of higher evapotranspiration on lower elevation land areas, the drainage water has higher salt concentrations. The reduced plant biomass at the lower and drier elevations also allows more erosional loss during storms.

2. Increased evaporation from the stream. As a typical stream or

river drops in elevation, it slows (less steep gradient), widens, warms, and becomes more illuminated. This can cause more evaporation and an increased salt concentration.

3. Uptake by stream biota. In a uniform reach of stream with no additions along its length, ions, such as nitrate and phosphate, may decline in amount. This is undoubtedly due to uptake by plants (Neel, 1951) and/or conversion to other compounds by microorganisms followed by uptake (e.g., NO_3^- to NH_4^+) (Adams and coauthors, 1975). The concentration factors of evapotranspiration are also acting on these nutrients; however, the requirements of stream organisms dominate the pattern. Whether a particular nutrient increases or decreases depends on the biological demand in relation to the abundance. Although there is a requirement for Ca^{++} and Mg^{++}, it is small in relation to the amount normally present, and concentration factors dominate.

These same factors should be important for organic compounds with the addition of the significant changes that can occur by photooxidation (Owen, 1971; Rosen, 1971) and the production of additional organic compounds by stream organisms (Hynes, 1970).

Biological Indices

The above discussions clearly identify the fact that changes occur continuously in stream systems both through space and time. It is clear that changes in land or stream use practices will cause additional changes in water quality parameters which will be reflected by the biota. One of the important questions then becomes, "what is a significant change in water quality parameters in terms of their impact on stream biota?" Secondly, what significant water quality changes represent natural types of changes versus unnatural? That is, a change from one natural water quality condition to another natural water quality condition must be evaluated differently from a change to a new aquatic environment which organisms have not yet experienced. Finally, there is the question of indicator species. What types of species, if any, best reflect those changes in community composition to be expected from a change in the stream environment?

In a regional assessment study of the Colorado River prepared for the National Commission on Water Quality, it was reported that salinity is the major water quality factor in the basin and it has interactions which outweigh all other considerations except the availability of the water resource (Utah State University, 1975). The study also reports that it was not possible to demonstrate deleterious effects of salinity on aquatic organisms for the 100-1,000 mg/l range of total dissolved solids typically found in the river proper. Changes in the aquatic community obviously have occurred; however, it was not possible to relate the disappearance of species, productivity changes, etc., to salinity increases because of confounding by instabilities in flow and substrate, sedimentation effects, flow pattern changes, time lag in biological response, and introduced species. These statements do not mean that salinity changes do not have an effect, only that in most situations the natural ecosystem is too complex to be able to show a clear, direct relationship with any one parameter.

Only when the level of a given parameter reaches extreme conditions (e.g., high toxin levels, deoxygenation, siltation) may it dominate the myriad of chemical, physical, and biological interactions and demonstrate a clear effect. Typically, this is the kind of situation reported in studies of how an organism responds to changes in a particular water quality parameter. The more subtle effects are often masked by antagonistic or synergistic reactions with other components, plus behavioral (e.g., migration), competitive, or predator-prey patterns.

A rather simple example reported in Hynes (1970) illustrates this. Increased levels of carbon dioxide are known to increase the susceptibility of fishes to low levels of dissolved oxygen, and low levels of oxygen increase the toxicity of ammonia. Increasing the level of carbon dioxide, however, reduces the toxicity of ammonia by lowering the pH causing the formation of ammonium ions which are much less toxic. While increasing the carbon dioxide level increases the protection against ammonia poisoning, the higher level of carbon dioxide may be toxic in itself. Here, then, is an example of three water quality parameters which could act in either a synergistic or antagonistic way. The effect of a particular combination of these three factors on an organism can be expected to be influenced by many of the other parameters in an increasingly complex way.

There are other examples of interest, too, especially those dealing with heavy metal toxicity. Rarely is a river system affected by just one potential toxin, and a number of studies have shown that a combination of toxins may increase or decrease the toxic effect of a single toxin (see Hynes, 1960). For example, the presence of a small concentration of lead nitrate reduces the toxicity of copper nitrate but when more lead is added to the solution the toxicity of the mixture increases above the level attributed to copper alone. These interactions may also act in complex ways involving the sensory perception of organisms and their ability to avoid toxic levels of a substance. The ten-spined stickleback avoids copper sulphate only when the salt is fairly concentrated. However, relatively low concentrations of this salt destroy the organism's sensitivity to alcohol, chloroform, formalin, mercuric chloride, and unknown others which it normally avoids (Hynes, 1960).

Fish have been studied in the greatest detail concerning their ability to avoid natural and man-introduced substances in streams (see Jones, 1964; Höglund, 1961). This ability is often modified under laboratory conditions, sometimes to the point where they will not avoid conditions that could prove rapidly fatal to them. Unfortunately, most of our knowledge of fish behavior or avoidance patterns is largely based on laboratory findings. There are significant problems in trying to extrapolate these types of data to a natural aquatic ecosystem. Most of the heavy metal studies are laboratory studies using simple metal salt solutions; however, it is highly improbable that any organism would be subject to free heavy metal ions in solution in a natural aquatic environment. A high proportion of the heavy metal content of stream water is a component of the particulate load (Andelman, 1973; Gosz, 1977). Even for the ions in the filtrate, the vast majority, if not all, of the heavy metal ions are complexed by soluble organics (Morel and coauthors, 1973). Because the complexing or precipitating organics (ligands) are usually in large excess of the heavy metal ions in natural stream systems, organics play a major role by complexing trace metal ions, keeping them in solution, and mediating interactions among them. Most studies of heavy metal toxicity have not considered the role of organics; therefore, the majority of the literature on this subject would seem to be of questionable value for use in setting water quality standards for aquatic organisms.

During the evolutionary process, the problems faced by all species in surviving (reproducing, growing, and moving) have been solved in a variety of ways. Each species now, however, can be successful only within a limited range of environmental conditions. As a result of a species' morphology, physiology, and behavior, it is sensitively adapted to the environment in which it has evolved. The stream environment has been shown to be quite variable in space and time; therefore, the physiology and behavior of stream organisms have come to be sufficiently plastic to permit their success. If a change in environmental conditions is much beyond the range within which a species has evolved, individuals of that species may fail to adapt and their population may not persist. Within the overall range of conditions in which a species can exist, there may be a number of different ranges for various behavioral responses (territorial, reproductive, feeding, migratory, avoidance) and physiological responses. The presence of a very low level of a given toxic substance could quite conceivably change only one aspect of a species' behavior and thereby decrease its survival chances and lead to the loss of the population. Even though behavior is as much a part of animals and as important to their persistence as are their morphological structures, the study of behavior has been largely neglected (Warren, 1971).

Since it has been asserted that it is difficult to predict the effects of a given alteration of water quality on a species, it also follows that it is extremely difficult to say whether that alteration results in an unnatural water quality condition. A point that needs to be stressed is that although an aquatic community may be altered drastically following a change, for example, after the introduction of organic pollution, none of the animals found in the polluted water are confined to it (Hynes, 1960). All may be normal inhabitants of natural habitats such as river muds and ponds, and they merely happen to be favored by the conditions associated with organic pollution. Even natural streams experience various degrees of what would be identified as pollution. In the densely forested headwater regions, autumn leaf-fall may add so much organic matter to water that fish are asphyxiated (Schneller, 1955). The water may be murky or discolored, foul smelling, and have decaying organic matter covered with fungi. These conditions near a human establishment would be cited as gross negligence. It is indeed fortunate that "naturally" polluted situations have occurred and continue to occur for they allowed the evolution and maintenance of a biota which can survive and, indeed, aid in cleaning up human-caused organic pollution.

One may wish to consider an increase in the level of potentially toxic compounds (i.e., those which can be lethal at relatively low concentrations), such as heavy metals and some organic compounds, as an indication of an unnatural water quality condition. With the possible exception of the synthetic biocides, however, most potentially toxic compounds exist in the natural aquatic environment at varying levels, and we can expect aquatic organisms to be tolerant of them to some degree. This feature also may cause laboratory experiments to be misleading. An animal's recent experience with a lethal factor is often particularly important in determining its tolerance. Previous exposure to a toxic substance that did not kill the animal but caused some stress usually increases the tolerance for that factor (Warren, 1971). Under stress conditions, adaptive responses usually occur rendering the animal more tolerant not only of those conditions but also of more critical ones. This is the process of acclimation. Fish and most animals can acclimate to a wide variety of factors and conditions in their environments. While the information is rather sparse, fish have been shown to acclimate to phenol, synthetic detergents, excessive hydrogen ions (low pH), ammonia, cyanide, zinc, and other toxic substances (Warren, 1971).

These discussions do not present a very favorable picture for predicting or recognizing water quality changes which will begin to affect the success of a certain population. Laboratory studies have been shown to both overestimate and underestimate the effect of a certain water quality parameter while in situ studies have difficulty because of the complexity of a natural ecosystem. This is not to say that we cannot predict the effect of a gross insult to an environment which results in a rapid change (e.g., oil spill, sewage effluent), but it does not appear that we can predict slow and creeping changes arising from modest changes in water quality. These modest changes in water quality are probably the types of change to be expected from a dewatering operation, a change from one natural water quality condition to another. No new noxious or toxic materials are anticipated from dewatering, only a change in concentrations of dissolved or suspended material and a change in the physical characteristics of the stream.

If a survey were made of chemists, biologists, and ecologists concerning useful indices of stream change, each would identify his specialty as very valuable. This should not be surprising since a change in the stream environment should be reflected in all of the physical, chemical, and biological features. Biological indicators should provide good sensitivity to change because they respond to an integration of relatively minor changes. That benefit, however, has a weakness in that biologists are seldom able to demonstrate that indicators are specific for a change in a certain parameter (Hooper, 1969). Fish (Larkin and Northcote, 1969), zooplankton (Brooks, 1969), bottom fauna (Jónasson, 1969), phytoplankton (Lund, 1969), and bacteria (McCoy and Sarles, 1969) all have been identified as useful indices of eutrophication; however, each group also demonstrates some disadvantages. For example, fish being at the top of the production pyramid are the last affected. Having growth and survival rates that fluctuate widely, being flexible in their food habits and able to exploit alternatives, and being associated with other species in a complex interaction on food organisms, they pose a degree of variability which is difficult to correlate with minor changes. Still, we know more about fish than any other aquatic group and they can be of use.

Bottom dwellers are among the most common of freshwater animals and can be greatly influenced by stream changes. The techniques for sampling, sieving, and sorting, however, are defective and inefficient, and the number of bottom fauna is usually underestimated (Jónasson, 1969).

Bacteria are prominent and important members of the aquatic community but it is difficult to describe the bacterial population. Our understanding of the equilibrium determining the numbers and kinds is imperfect because of the limitation of the counting methods, lack of identification of the types of bacteria, and lack of knowledge of the ecological relations (McCoy and Sarles, 1969).

In spite of the arguments in favor of a wide variety of groups of organisms, the list of sensitive indicator organisms is small. For an individual species to be useful as an indicator organism, it must have a rather narrow range of suitable environmental conditions that are known. Few are the species that satisfy these qualifications, and those that do are likely to be rare in their numbers of individuals for these very reasons, hardly an ideal condition.

Research Needs

In any listing of suggested future research on aquatic ecosystems, it would be difficult to leave out anything. There are areas of knowledge that are relatively weak compared to others, however, and these would be logical candidates for increased emphasis.

The physical and inorganic chemical aspects of lotic environments are perhaps the best understood parameters while their influence on stream biota is at best poorly understood. The influence of stream parameters on population dynamics needs a great amount of attention. There are abundant studies of tolerance limits under laboratory conditions; however, in the author's opinion many of these studies only "muddy the waters" with respect to our understanding of natural systems. There is a need for tolerance level studies but they must be performed under real-world conditions; that is, they must be performed under the influence of behavioral changes (territorial, feeding, migratory, avoidance), natural competition, and trophic level interactions, and they must be measured in terms of the success of the population. Difficult? Exceedingly so, and it will take a great deal of ingenuity, not only on the part of the biologists, but perhaps also of sociologists, psychologists, and economists. The methodology developed in a number of other disciplines may provide the answers or breakthroughs which we need. It is not enough to simply use information from other disciplines (multidisciplinary approach); these disciplines must be encouraged to join with biologists to design experiments and interpret results (interdisciplinary approach).

Along the same lines, the vast majority of biologists are not able to handle the data they are capable of gathering. As a result of the use of statistics, researchers can be described as "operating on a 95 percent confidence interval." This means nearly all of the variation must be accounted for before a statistically valid conclusion can be generated. It would seem that only in relatively few situations when one, or at most several, factors dominate the natural environment could we hope to explain nearly all of the variation. This is exactly why biologists have reverted to laboratory studies where all of the factors except those of interest are fixed, allowing them to satisfactorily explain the variation. It seems like a case of letting the tail wag the dog. In order to be able to explain the variation in natural communities, much more sophisticated data analysis procedures must be used. The procedures available currently may be inadequate, which means we need to attract the interest of more mathematicians and statisticians in order to develop sophisticated multivariate analyses which will work for the type of data collected. A good example of the interaction needed is the International Symposium on Statistical Ecology which brings together ecologists who use mathematical and statistical methods, and statisticians whose work has ecological applications (Patil and coauthors, 1971).

A great deal of interest and expertise has appeared in recent years in mathematical modeling (e.g., Patton, 1971, 1972, 1975). This discipline can be of tremendous value in helping to understand the complexity of ecosystems. The formulation of mathematical models of real-world phenomena requires one to operationally define the important variables and state the hypothetical relationships among those variables. The model then simulates a real-world condition and represents a complex hypothesis which can generate predictions. To be accepted, however, the model and its predictions must be tested against additional real-world measurements (Kowal, 1971). This requires the statistical analysis of those measurements as well as a statistical comparison with the predictions of the model. The statistical tests, which use an arbitrary confidence level, may show that several different models may be accepted as explaining the same set of real-world data. This does not reflect a philosophical conflict since a model is an abstract concept arbitrarily identified with the real world (Kowal, 1971); however, it does reinforce a previous point--the development of additional or more refined statistical analyses.

Finally, there is one aspect of water quality which is very poorly understood--the levels and influence of natural organics. Largely as a result of the difficulty and expense of analysis, we do not have a knowledge of the quality, quantity, and variability of the tremendous array of organic compounds which exist in the natural environment. We have even less knowledge of their regulatory role on the behavior, competition, and reproductive success of a population. It seems again as though we need the help of outside disciplines (e.g., organic chemists, electrical engineers) to develop the methodology and equipment required for our needs. This is a common thread which runs through all of the above discussions, and it does not appear as though significant progress can be made until a number of disciplines can be infused into our studies of the biological world. This conclusion should be obvious if one thinks of the number of disciplines required to understand the human population.

References

Adams, V. D., R. R. Renk, P. A. Cowan, and D. B. Porcella. 1975. _Naturally Occurring Organic Compounds and Algal Growth in an Eutrophic Lake_ (Logan, Ut., Utah State University, Utah Water Research Laboratory, PRWG 137-1).

Andelman, J. B. 1973. "Incidence, Variability and Controlling Factors for Trace Elements in Natural, Fresh Waters," in P. C. Singer, ed., _Trace Metals and Metal-Organic Interactions in Natural Waters_ (Ann Arbor, Mich., Ann Arbor Science Publications, Inc.).

Baxter, G. 1961. "River Utilisation and the Preservation of Migratory Fish Life," _Minutes of the Proceedings of the Institution of Civil Engineers_ vol. 18, pp. 225-244.

Bormann, F. H., G. E. Likens, and J. S. Eaton. 1969. "Biotic Regulation of Particulate and Solution Losses from a Forest Ecosystem," _BioScience_ vol. 19, pp. 600-610.

Brooks, J. L. 1969. "Eutrophication and Changes in the Composition of the Zooplankton," in National Academy of Sciences, _Eutrophication: Causes, Consequences, Correctives_ (Washington, D.C., NAS).

Carr, N. G., and B. A. Whitton. 1973. _The Biology of Blue-Green Algae_, Botanical Monographs, Vol. 9 (Berkeley, Calif., University of California Press).

Cooper, C. F. 1969. "Nutrient Output from Managed Forests," in National Academy of Sciences, _Eutrophication: Causes, Consequences, Correctives_ (Washington, D.C., NAS).

Crockford, H. D., and S. B. Knight. 1959. _Fundamentals of Physical Chemistry_ (New York, John Wiley & Sons).

Dormaar, J. F. 1971. "Prolonged Leaching of Orthic Black Ah Material with Water and Aqueous Extracts of _Populus tremuloides_ and _P. balsamifera_ Leaves," _Journal of Soil Science_ vol. 22, pp. 350-358.

Duce, R. A., J. G. Quinn, C. E. Olney, S. R. Piotrowicz, B. J. Ray, and T. L. Wade. 1972. "Enrichment of Heavy Metals and Organic Compounds in the Surface Microlayer of Narragansett Bay, Rhode Island," _Science_ vol. 176, p. 161.

Fitch, C. P., L. M. Bishop, and W. L. Boyd. 1934. "Water Bloom as a Cause of Poisoning in Domestic Animals," _Cornell Veterinarian_ vol. 24, p. 30.

Gloyna, E. F., Y. A. Yousef, and T. J. Padden. 1971. "Nonequilibrium Systems in Natural Water Chemistry," _Advances in Chemistry Series_ vol. 106 (Washington, D.C., American Chemical Society).

44

Gorham, P. J. 1960. "Toxic Waterblooms of Blue-Green Algae," The Canadian Veterinary Journal vol. 1, pp. 235-245.

Gosz, J. R. 1975. "Nutrient Budgets for Undisturbed Ecosystems Along an Elevational Gradient in New Mexico," in F. G. Howell, J. B. Gentry, and M. H. Smith, eds., Mineral Cycling in Southeastern Ecosystems, ERDA Symposium Series CONF-740513 (Washington, D.C., U.S. Energy Research and Development Administration).

_____. 1977. Influence of Road Salting on the Nutrient and Heavy Metal Levels in Stream Water (Las Cruces, N.M., New Mexico Water Resources Research Institute, Technical Completion Report Project No. 3109-68, December).

_____. 1978. "Nitrogen Inputs to Stream Water from Forests Along an Elevational Gradient in New Mexico," Water Research (in press).

Gosz, J. R., and M. L. Barr. 1974. Stream Organics to Evaluate Land Management (Las Cruces, N.M., New Mexico Water Resources Research Institute, Technical Completion Report Project No. 3109-146, December).

Hasler, A. D. 1970. "Chemical Ecology of Fish," in E. Sondheimer and J. B. Simeone, eds., Chemical Ecology (New York, Academic Press).

Hem, J. D. 1970. Study and Interpretation of the Chemical Characteristics of Natural Water, Geological Survey Water-Supply Paper 1473 (2nd ed., Reston, Va., U.S. Geological Survey).

Hernandez, J. W. 1967. Proposed Water Quality Standards for the San Juan River, La Plata River, and Animas River in New Mexico (Reston, Va., U.S. Geological Survey; Santa Fe, N.M., New Mexico Department of Public Health).

Hobbie, J. E., and G. E. Likens. 1973. "The Output of Phosphorus, Dissolved Organic Carbon and Fine Particulate Carbon from Hubbard Brook Watersheds," Limnology and Oceanography vol. 18, pp. 734-742.

Höglund, L. B. 1961. The Reactions of Fish in Concentration Gradients (Drottningholm, Fishery Board of Sweden, Institute of Fresh-Water Research Report 43).

Holden, P. B., and C. B. Stalnaker. 1975. "Distribution and Abundance of Mainstream Fishes of the Middle and Upper Colorado River Basins, 1967-1973," Transactions of the American Fisheries Society vol. 104, pp. 217-231.

Hooper, F. F. 1969. "Eutrophication Indices and Their Relation to Other Indices of Ecosystem Change," in National Academy of Sciences, Eutrophication: Causes, Consequences, Correctives (Washington, D.C. NAS).

Hynes, H. B. N. 1960. The Biology of Polluted Waters (Liverpool, Liverpool University Press).

_____. 1970. The Ecology of Running Waters (Toronto, University of Toronto Press).

Ingle, R. M. 1965. "Red-Tide Research at the Florida State Laboratory," in J. E. Sykes, ed., Bureau of Commercial Fisheries Symposium on Red Tide, Special Science Report, Fisheries No. 521 (Washington, D.C., U.S. Department of Commerce, National Marine Fisheries Service).

Ingram, W. M., and G. W. Prescott. 1954. "Toxic Fresh-Water Algae," The American Midland Naturalist vol. 52, no. 1.

Johnson, R. P. 1963. "Studies on the Life History and Ecology of the Bigmouth Buffalo Ictiobus cyprinellus (Valenciennes)," Journal of the Fisheries Research Board of Canada vol. 20, pp. 1397-1429.

Jónasson, P. M. 1969. "Bottom Fauna and Eutrophication," in National Academy of Sciences, Eutrophication: Causes, Consequences, Correctives (Washington, D.C., NAS).

Jones, J. R. E. 1964. Fish and River Pollution (London, Butterworth).

Kowal, N. E. 1971. "A Rationale for Modeling Dynamic Ecological Systems," in B. C. Patten, ed., Systems Analysis and Simulation in Ecology Vol. I (New York, Academic Press) pp. 123-194.

Kramer, P. J., and T. T. Kozlowski. 1960. Physiology of Trees (New York, McGraw-Hill).

Larkin, P. A., and T. G. Northcote. 1969. "Fish as Indices of Eutrophication," in National Academy of Sciences, Eutrophication: Causes, Consequences, Correctives (Washington, D.C., NAS).

Leopold, L. B., M. G. Wolman, and J. P. Miller. 1964. Fluvial Processes in Geomorphology (San Francisco, W. H. Freeman).

Likens, G. E., and F. H. Bormann. 1972. "Nutrient Cycling in Ecosystems," in J. Weins, ed., Ecosystems: Structure and Function (Corvallis, Oreg., Oregon State University Press).

_____. 1974. "Linkages Between Terrestrial and Aquatic Ecosystems," BioScience vol. 24, pp. 447-456.

_____. 1975. "Nutrient-Hydrologic Interactions," in A. Hasler, ed., Coupling of Land and Water Systems, Ecological Studies, Analysis and Synthesis, Vol. 10 (New York, Springer-Verlag).

Lund, J. W. G. 1969. "Phytoplankton," in National Academy of Sciences, Eutrophication: Causes, Consequences, Correctives (Washington, D.C., NAS).

Mann, K. H. 1959. "On Trocheta bykowskii Gedroyc, 1913, A Leech New to the British Fauna, with Notes on the Taxonomy and Ecology of Other Erpobdellidae," Proceedings of the Zoological Society of London vol. 132, pp. 369-379.

Martin, D. F., and B. B. Martin. 1973. "Implications of Metal-Organic Compounds in Red Tide Outbreaks," in P. C. Singer, ed. Trace Metals and Metal-Organic Interactions in Natural Waters (Ann Arbor, Mich., Ann Arbor Science Publications, Inc.).

McCoy, E., and W. B. Sarles. 1969. "Bacteria in Lakes: Populations and Functional Relations," in National Academy of Sciences, Eutrophication: Causes, Consequences, Correctives (Washington, D.C., NAS).

McGauhey, P. H. 1968. Engineering Management of Water Quality (New York, McGraw-Hill).

Morel, F., R. E. McDuff, and J. J. Morgan. 1973. "Interactions and Chemostasis in Aquatic Chemical Systems: Role of pH, pE, Solubility, and Complexation," in P. C. Singer, ed., Trace Metals and Metal-Organic Interactions in Natural Waters (Ann Arbor, Mich., Ann Arbor Science Publications, Inc.).

Neel, J. K. 1951. "Interrelations of Certain Physical and Chemical Features in a Head-Water Limestone Stream," Ecologist vol. 32, pp. 368-391.

Nielsen, E. S. 1955. "An Effect of Antibiotics Produced by Plankton Algae," Nature vol. 176.

Odum, E. P. 1971. Fundamentals of Ecology (3rd ed., Philadelphia, W. B. Saunders).

Odum, H. T., and R. F. Pigeon, eds. 1970. A Tropical Rain Forest: A Study of Irradiation and Ecology at El Verde, Puerto Rico (Washington, D.C., U.S. Atomic Energy Commission).

Owen, E. D. 1971. "Principles of Photochemical Reactions in Aqueous Solution," in S. J. Faust and J. V. Hunter, eds., Organic Compounds in Aquatic Environments (New York, Marcel Dekker).

Patil, G. P., E. C. Pielou, and W. E. Waters. 1971. Statistical Ecology (University Park, Pa., Pennsylvania State University Press).

Patten, B. C. 1971. Systems Analysis and Simulation in Ecology Vol. I (New York, Academic Press).

_____. 1972. Systems Analysis and Simulation in Ecology Vol. II (New York, Academic Press).

_____. 1975. Systems Analysis and Simulation in Ecology Vol. III (New York, Academic Press).

Pomeisl, E. 1961. Okologische-biologische Untersuchungen an Plecopteren in Gebiet der niederosterreichischen Kalkalpen und des diesem vor-gelagerten Flyschgebietes ("Ecological-Biological Studies of Plecoptera in the Vicinity of the Lower Austrian Limestone Alps and the Flanking Flysch Zone") Verhandlungen der Internationalen Vereinigung fur Theoretische und Angewandte Limnologie (Transactions of the Society for Theoretical and Applied Limnology) vol. 14, pp. 351-354.

Prakash, A., and M. A. Rashid. 1968. "Influence of Humic Substances on the Growth of Marine Phytoplankton, Dinoflagellates," Limnology and Oceanography vol. 13, p. 598.

Rashid, M. A., D. E. Buckley, and K. R. Robertson. 1972. "Interactions of a Marine Humic Acid with Clay Minerals and a Natural Sediment," Geoderma vol. 8, pp. 11-27.

Reynolds, J. F., and D. H. Knight. 1973. "The Magnitude of Snowmelt and Rainfall Interception by Litter in Lodgepole Pine and Spruce-Fir Forests in Wyoming," Northwest Scientist, vol. 47, pp. 50-60.

Rosen, J. D. 1971. "Photodecomposition of Organic Pesticides," in S. J. Faust and J. V. Hunter, eds. Organic Compounds in Aquatic Environments (New York, Marcel Dekker).

Rosenzweig, M. L. 1968. "Net Primary Productivity of Terrestrial Communities: Prediction from Climatological Data," The American Naturalist, vol. 102, pp. 67-74.

Ruttner, F. 1963. Fundamentals of Limnology. Translated by D. G. Frey and F. E. J. Fry (Toronto, University of Toronto Press).

Schneller, M. V. 1955. "Oxygen Depletion in Salt Creek, Indiana," Investi-gations of Indiana Lakes and Streams vol. 4, pp. 163-175.

Slatyer, R. O. 1967. Plant-Water Relationships (New York, Academic Press).

Swanson, R. H. 1967. "Seasonal Course of Transpiration of Lodgepole Pine and Engelmann Spruce," in W. E. Sopper and H. W. Lull, eds., Forest Hydrology (Elmsford, N.Y., Pergamon Press).

Utah State University. 1975. Colorado River Regional Assessment Study. Part Three: Area-Specific Water Quality Analysis and Environmental Assessment (Logan, Ut., Utah State University, Utah Water Research Laboratory).

Warren, C. E.   1971.   Biology and Water Pollution Control (Philadelphia, W. B. Saunders).

Williams, J. S.   1975.   The Natural Salinity of the Colorado River (Logan, Ut., Utah State University, Utah Water Research Laboratory Occasional Paper 7).

THE INFLUENCE OF REDUCED STREAMFLOWS ON WATER QUALITY: A DISCUSSION

Gene E. Likens*

I believe that James Gosz has done an excellent job of outlining and discussing some of the major problems associated with the interactions between land use, streamflow, and water quality. These are critical relationships for management of landscapes, and the problem of reduced streamflow in the Southwestern United States is of particular importance. Dr. Gosz has carefully documented the complexities that are involved in these ecological relationships, and has shown the importance of understanding the role of the terrestrial ecosystem in evaluating and predicting changes in streamflow and water quality. Data such as those presented for the Tesuque Watersheds of New Mexico are of fundamental importance in the development of sound land management proposals. In my discussion of Dr. Gosz's paper, I would like to reemphasize some of the points he made, provide some elaboration on the importance of an in-depth understanding of the linkages between aquatic and terrestrial ecosystems, and briefly touch on the costs associated with this type of ecosystem research.

It is frequently stated that "a stream reflects its surroundings." A stream is an especially open ecosystem. As such, it may be thought that a stream has very little structure of its own because its characteristics appear to be ephemeral; its obvious features are transient and always changing. You cannot go back to the same spot in a flowing stream and

---

*Professor of Ecology, Division of Biological Sciences, Section of Ecology and Systematics, Cornell University, Ithaca, New York.

find the same parcel of water--it's gone! Therefore, it may be argued that the entire drainage area or landscape is the functional unit, with the aquatic ecosystem (the stream) only reflecting these surroundings. This may be particularly so in headwater streams. For example, it is here that the basic chemical characteristics of the entire stream are established, largely by amount of precipitation and geologic substrate. These chemical characteristics persist with relatively little modification (within an order of magnitude, or so) all the way to the mouth unless man intervenes. For example, Gibbs (1967) found that about 85 percent of the total amount of dissolved salts and suspended solids carried by the Amazon River originated from the Andean headwaters. Thus, a stream ecosystem indeed cannot be separated from its drainage area in attempting to understand its structure, its functions and, particularly, its management. This should be an axiom. On the other hand, a stream is a legitimate ecosystem in its own right. It has definite and definable biotic and abiotic components that interact and exchange both energy and nutrients. As such, it functions as an ecosystem and modifies, regulates, and processes the outputs from the terrestrial drainage area that flow through it.

Some General Characteristics of Forested
Landscapes in the North Temperate Zone

The quantity and quality of rainwater and snowmelt water are altered as the water passes through a forested ecosystem. Chemicals obtained in precipitation or released by weathering are cycled by living vegetation, animals, and microbes accumulated in living and dead organic matter and the soil, and are then lost in drainage waters. However, outputs from undisturbed forested ecosystems are generally highly regulated. For example, the outputs of water as streamflow are controlled in both amount and timing by the high infiltration capacity and storage of the soil in undisturbed forest ecosystems, and most importantly, by evapotranspiration of water vapor. With evapotranspiration, liquid water is converted to water vapor, and dissolved substances are concentrated or precipitated in plant tissues and soils. Water lost by evapotranspiration cannot transport dissolved substances or particulate matter out of the ecosystem in streamflow. This is probably the most important way in which an undisturbed forested ecosystem regulates the potential degrading effects of moving water in the landscape. As a result, losses of dissolved substances, erosion, and losses of particulate matter are relatively small and controlled. Thus, the undisturbed forested ecosystem has a very strong control over the amount and types of materials that enter the drainage streams (potentially available for loss in drainage waters from the watershed).

Terrestrial outputs are the inputs for stream ecosystems that drain them (figure 1). Stream ecosystems alter, store, and process the dissolved substances and particulate matter that pass through. In contrast to terrestrial ecosystems, there have been very few material balance studies for stream ecosystems. In my opinion, this subject deserves careful attention and much could be learned from comparative studies of such balances in stream ecosystems for elements like carbon, nitrogen, phosphorus, and potassium in particular. From the few studies that have been done in headwater streams, it can be shown that streams do operate as functional ecosystems.

Within the Hubbard Brook Experimental Forest in New Hampshire, we have done some studies of material balance for the Bear Brook ecosystem. Bear Brook is a small, headwater stream which drains a portion of the deciduous

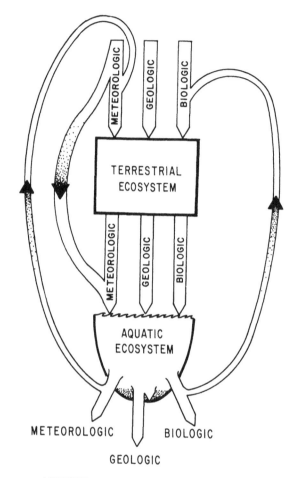

NUTRIENT LINKAGES BETWEEN
TERRESTRIAL AND AQUATIC ECOSYTEMS

Note: Vectors may be meteorologic, geologic, or biologic components moving nutrients or energy along the pathway shown.

Figure 1. Diagrammatic Model of the Functional
Linkages Between Terrestrial and
Aquatic Ecosystems

Source: G. E. Likens and F. H. Bormann, "Linkages Between Terrestrial and Aquatic Ecosystems," BioScience vol. 24, no. 8, 1974.

forest. It is densely shaded and more than 99 percent of the energy, as organic carbon, originates from allochthonous sources (Fisher and Likens, 1973). Thirty-four percent of these organic carbon inputs are respired to CO_2 and 66 percent are exported. Moreover, the ratio of dissolved-to-particulate organic matter entering Bear Brook from all sources was 0.86, whereas the ratio in export was 2.3. The stream ecosystem degraded about 63 percent (by weight) of the incoming particulate organic matter to dissolved organic matter or to CO_2. Thus, the stream ecosystem functioned as an efficient decomposer.

One of my graduate students, Judy Meyer, is doing a similar study on phosphorus in Bear Brook. She finds that losses nearly balance inputs for the stream ecosystem but much processing occurs; for example, large particulate organic matter is converted to fine particulate organic matter. The major export of phosphorus occurs as fine particulate organic matter, again showing the functional role of the stream ecosystem (Meyer, 1976).

Disturbance from Long Range

Man can alter these linkages between terrestrial and associated stream ecosystems from within drainage areas as well as from great distances. Air-borne substances may be transported throughout air sheds spanning vast areas, and affect a variety of terrestrial and aquatic ecosystems within those areas. A good example is the problem of acid precipitation in the northeastern United States. It is thought that acid precipitation falling on remote areas of northern New England stems from airborne pollutants which were emitted from industrial and urbanized regions to the west and south of these areas. The strong acids (sulfuric and nitric) which are produced from gaseous by-products in the combustion of fossil fuels have lowered the average annual pH of precipitation to between 4.0 and 4.5 over much

of the eastern United States (figure 2). During the last two decades, the

distribution of this acid precipitation has intensified and increased in

geographic extent, particularly in the northeastern United States (figures

3 and 4). Cragin, Herron, and Langway (1975) have shown that concentrations

of lead and sulfur in deposits of snow in Greenland after 1955 have in-

creased, relative to pre-1840 values, by about ten- and five-fold, respec-

tively. This is attributed to the increased combustion of fossil fuels

and smelting following the industrial revolution in the Northern Hemi-

sphere. Therefore, long-range transport of air pollutants within air sheds

is affecting widespread terrestrial and aquatic ecosystems, including remote

areas.

The effects of acid precipitation and other airborne pollutants on

organisms is often very severe. The serious decline or extinction of fish

populations in widespread aquatic habitats as a consequence of acid precipi-

tation has been documented in Sweden, Norway, the United States, and Canada

(Bolin, 1971; Dickson, 1975; Braekke, 1976; Beamish and Harvey, 1972;

Schofield, 1976a, 1976b). Serious ecological disruptions have been observed

at all levels of the food web in acidified lakes and streams, and ecosystem

function has been seriously impaired (Braekke, 1976). Aquatic ecosystems

on granitic substrates are most sensitive to acid precipitation, since their

buffering capacity is low. Effects on natural forest vegetation are equiv-

ocal at present, but laboratory and field studies show that serious damage

can occur to vegetation in the pH range frequently observed (pH < 3.5) dur-

ing the summertime in the northeastern United States (see Likens, 1976).

As shown in figure 2, rain and snow in the southwestern United States are

currently not acid (pH < 5.6) although there are few data. However, with the

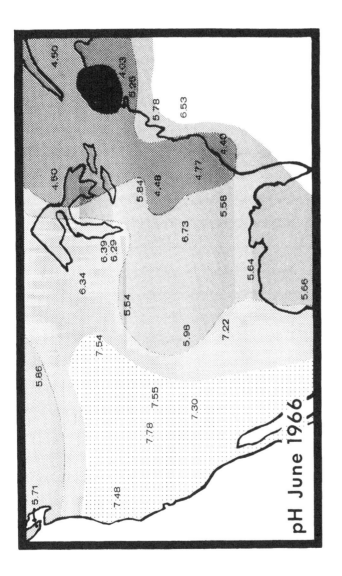

Figure 2. The pH of Precipitation over the United States During June 1966

<u>Source:</u> Data courtesy of the National Center for Atmospheric Research, Boulder, Colorado (A. L. Lazrus, personal communication).

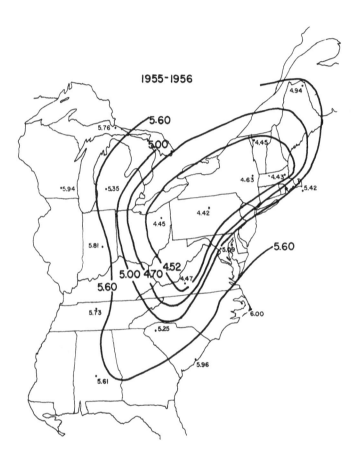

Figure 3. Calculated pH of Precipitation over the Eastern
United States During the Period 1955-1956

Source: C. V. Cogbill and G. E. Likens, "Acid Precipitation in the
Northeastern United States," Water Resources Research vol. 10, no. 6, 1974.

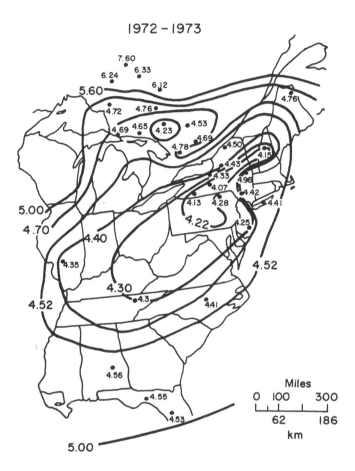

1972 – 1973

Figure 4. Observed pH of Precipitation over the Eastern
United States During the Period 1972-1973

Source: Modified from G. E. Likens, "Acid Precipitation,"
Chemical and Engineering News (November 1976).

increased development and use of power plants and smelters burning fossil fuels and emitting oxides of sulfur, nitrogen, heavy metals, etc., the potential for deterioration of air quality and for the production of acid precipitation in this area is a distinct possibility in the next decade or so. In my opinion, it is critical to set up monitoring stations now, particularly in forested headwater areas, so that trends in precipitation chemistry can be evaluated in the future and control measures applied intelligently.

A special problem in headwater streams of mountainous regions where snow accumulates is in the quality of meltwater from the snowpack. In this case, pollutants accumulated in a snowpack over the entire winter may flush through streams and lakes at high concentrations. Because of the ion separation that occurs when snow or ice freezes and melts, the first part of the meltwater may be appreciably elevated in acidity, heavy metals, pesticides, and so forth (Odén and Ahl, 1970; Braekke, 1976; Hagen and Langeland, 1973). This problem of episodic impacts has been shown to have serious consequences on the aquatic ecosystems in Sweden and Norway (Braekke, 1976) and deserves very careful study. Because of the large orographic influence on the amount of precipitation in the southwestern United States, one would expect that headwater streams in this area might be most vulnerable, and the potential for acidic episodes following snowmelt is a distinct possibility. At lower elevations, acidic aerosols might adversely affect vegetation and thus indirectly influence streamflow quantity and quality through effects on the terrestrial ecosystem.

Analysis of precipitation chemistry in the eastern United States indicates clearly that concentrations of nitrate in precipitation have increased dramatically during the past ten years or so (figures 5 and 6).

Presumably this increase is related to increased air pollution. Nitrogen,

along with phosphorus, is a critical nutrient for aquatic ecosystems, and the

effect of this increased input on cultural (man-induced) eutrophication and

long-term productivity is essentially unknown at present. In aquatic eco-

systems within remote forested areas, the majority of the nitrogen and phos-

phorus input comes from precipitation falling directly on the lake's surface

(Likens and Bormann, 1974). Therefore, any increase in nitrate (or phos-

phorus) in rain and snow takes on added ecological and management signifi-

cance. The potential for stimulation of biological growth from nutrient

inputs in precipitation may offset or mask the toxic effects of airborne

pollutants transported long distances. These long-term impacts of simulta-

neously "good and bad" inputs from precipitation deserve careful attention.

A forested ecosystem serves as an important "filter" for airborne or

waterborne pollutants, at least up to a point. These ecosystems normally

produce high-quality air and water, in spite of the various insults to

which they may have been subjected. As such, undisturbed forested eco-

systems produce high-quality air and stream water "for free." When the

watershed is disturbed, man has to replace these natural functions of the

ecosystem with air conditioners, water treatment plants, flood control

dams, etc., all at high cost. These benefits (or costs) are usually ig-

nored in economic analyses of environmental impacts.

Disturbance from Within

Man's alteration of terrestrial ecosystems from within, such as de-

forestation, road building, urbanization, and so forth, can have a major

destabilizing influence on those characteristic functions of the

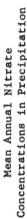

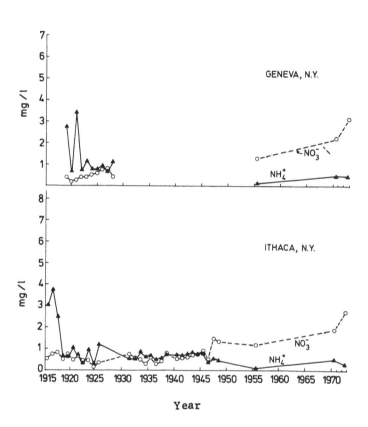

Figure 5. Average Annual Nitrate and Ammonium Concentrations
in Precipitation for Ithaca and Geneva, New York

Source: Modified from G. E. Likens and F. H. Bormann,
"Acid Rain: A Serious Regional Environmental Problem," Science
vol. 184, no. 4142, 1974.

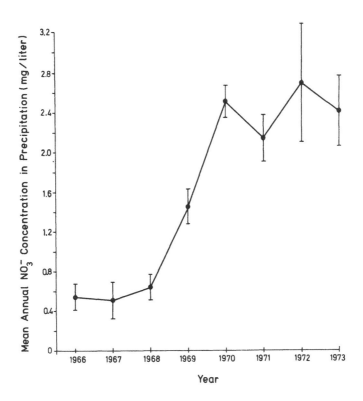

Note: The vertical bars represent one standard
error of the mean.

Figure 6. Average Annual Nitrate Concentrations in
Precipitation for Nine U.S. Geological
Survey Stations in New York State and
Pennsylvania

undisturbed forested ecosystem, such as regulation of discharge, control

of erosion, filtration of air and water, and regulation of nutrient loss.

Disturbance of normal ecosystem function affects both quality and quantity

of water draining from the system. These effects on streams are varied

and important. As Dr. Gosz mentioned, in our studies in New Hampshire

when forest ecosystems are clearcut, a variety of complex ecosystem ef-

fects were observed (figure 7). Some of these results were expected, many

were not. Detailed studies of entire ecosystems, often over long periods

of time (more than 10 years), are required to fully evaluate the ecologi-

cal impact of disturbance in natural ecosystems.

Finally, there are differences in the quality of the outputs from

headwater ecosystems in relation to changes in the hydrologic cycle. In

New Hampshire watersheds, we found that concentrations of dissolved sub-

stances were generally independent of discharge, whereas concentrations

of particulate matter increased exponentially with increased discharge

(figure 8). This is a particularly important consideration in evaluating

the loss for a nutrient like phosphorus, which is exported from streams

primarily in fine particulate matter. Since disturbance of forested eco-

systems usually results in elevated discharges of water, management of

disturbed ecosystems must consider such relationships.

Unfortunately, ecosystem research is usually complex, long-term and

expensive (see Battelle Columbus Laboratories, 1975). More than forty

scientists were engaged and approximately two million dollars were allo-

cated for interdisciplinary research on effects of acid precipitation on

forests and fish in Norway during 1976. As Dr. Gosz pointed out, many

disciplines are required but most environmental studies to date have been

multidisciplinary rather than interdisciplinary. We need much better

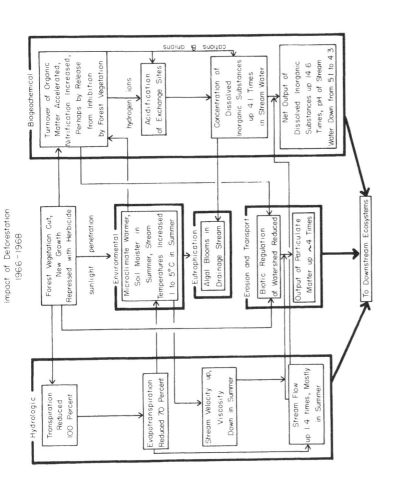

Note: Based upon data obtained during 1966-1968.

Figure 7. A Summary of Some of the Ecological Effects of Deforestation of Watershed 2 in the Hubbard Brook Experimental Forest

Source: G. E. Likens and F. H. Bormann, "Biogeochemical Cycles," The Science Teacher vol. 39, no. 4, 1972.

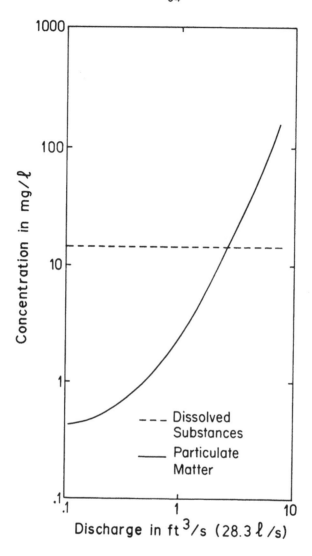

Figure 8. General Relationships Between the Concentration
of Dissolved Substances and Particulate Matter
and Stream Discharge in a Northern Hardwood Forest
Ecosystem

Source: After F. H. Bormann, G. E. Likens, and J. S. Eaton, "Biotic
Regulation of Particulate and Solution Losses from a Forest Ecosystem,"
BioScience vol. 19, no. 7, 1969.

interaction and communication between individuals and between disciplines in such studies. The end result in terms of management and protection of our natural resources would be well worth the effort and money.

Some Summary Thoughts

1. Streams reflect their surroundings, but are indeed ecosystems in their own right. They receive, alter, and respond to the effluents from terrestrial systems. The alteration of these effluents, within the stream ecosystem, then determines overall what the outputs from the entire drainage system will be.

2. Stream ecosystems are highly dependent upon events. A flood that occurs on a 1-in-100-year frequency may produce more erosion and transport of particulate matter than the total of all of the smaller events during the intervening years. This flooding, although often considered by man to be harmful, is in many ways the critical maintenance factor for stream ecosystems. For example, flooding is one of the few ways that nutrients are returned "uphill" from the aquatic system to the terrestrial system. A good example of this is the fertile agricultural areas that depend upon this type of nutrient replenishment in floodplains. Also, nutrient-rich, flooded areas along stream channels are vital nursery grounds for fish and invertebrates that man depends upon in a variety of ways, as was pointed out by Dr. Gosz.

3. The structure and function of stream ecosystems reflect the successional state and condition, that is, disturbance, of the entire drainage area. Both will change with time.

4. Stream ecosystems may be affected by man's activities from long range (via airsheds) and from disturbance within the watershed.

5. Headwater ecosystems (terrestrial and aquatic) should be protected from disturbance to ensure a supply of high-quality water.

6. Management proposals that consider streams in isolation from their drainage basins are sheer folly.

References

Battelle Columbus Laboratories. 1975. Evaluation of Three of the Biome Studies Programs. Final report to the National Science Foundation, International Biological Program (IBP), Contract NSF-C879; Research Report (Columbus, Ohio).

Beamish, R. J., and H. H. Harvey. 1972. "Acidification of the La Cloche Mountain Lakes, Ontario, and Resulting Fish Mortalities," Journal of Fisheries Research Board of Canada vol. 29, no. 8, pp. 1131-1143.

Bolin, B., ed. 1971. Air Pollution Across National Boundaries: The Impact on the Environment of Sulfur in Air and Precipitation. Report of the Swedish Preparatory Committee for the U.S. Conference on Human Environment (Stockholm, Kungl. Boktryckeriet P.A. Norstedt et Soner).

Braekke, F. H., ed. 1976. Impact of Acid Precipitation on Forest and Freshwater Ecosystems in Norway (Oslo, Sur nedbørs virkning på skog og fisk (SNSF) Project, Research Report FR 6/76).

Cragin, J. H., M. Herron, and C. C. Langway, Jr. 1975. The Chemistry of 700 Years of Precipitation at Dye-3, Greenland (Hanover, N.H., U.S. Army Cold Regions Research and Engineering Laboratory, Research Report 341).

Dickson, W. 1975. The Acidification of Swedish Lakes (Stockholm, Fisheries Board of Sweden, Institute of Freshwater Research, Report No. 54).

Fisher, S. G., and G. E. Likens. 1973. "Energy Flow in Bear Brook, New Hampshire: An Integrative Approach to Stream Ecosystem Metabolism," Ecological Monographs vol. 43, no. 4, pp 421-439.

Gibbs, R. J. 1967. "Amazon River: Environmental Factors that Control Its Dissolved and Suspended Load," Science vol. 156, pp. 1734-1737.

Hagen, A., and A. Langeland. 1973. "Polluted Snow in Southern Norway and Its Effects of Meltwater on Freshwater and Aquatic Organisms," Environmental Pollution vol. 5, pp. 45-57.

Likens, G. E., and F. H. Bormann. 1974. "Linkages Between Terrestrial and Aquatic Ecosystems," BioScience vol. 24, no. 8, pp. 447-456.

Meyer, J. 1976. "Phosphorus Dynamics in a Forest Stream Ecosystem: A Mass Balance Approach," Abstract for the 39th Annual Meeting of the American Society of Limnology and Oceanography (Savannah, Ga.).

Odén, S., and T. Ahl. 1970. Summary: The Acidification of Scandinavian Lakes and Rivers (Uppsala, Sweden, Särtryck ur Ymer, Årsbok) pp. 103-122.

Schofield, C. L. 1976a. "Lake Acidification in the Adirondack Mountains of New York: Causes and Consequences," in L. S. Dochinger and T. G. Siccama, eds., Proceedings of the First International Symposium on Acid Precipitation and the Forest Ecosystem (Columbus, Ohio, U.S. Department of Agriculture, Forest Service).

_____. 1976b. "Acid Precipitation: Effects on Fish," Ambio vol. 5, no. 5-6, pp. 228-230.

THE INFLUENCE OF REDUCED STREAMFLOWS
ON WATER QUALITY: A DISCUSSION

Lee Ischinger*

I would like to compliment Dr. Gosz on his valiant attempt at summariz-
ing a very difficult and complex topic. However, I am hopeful that the
picture he has created regarding the scientific community's ability to
determine the influence of reduced streamflows on water quality is not
quite as dismal as he has suggested.

The author initiated his discussion of the subject with a lengthy
and rather elementary presentation of the physical, geologic, and
hydrological events typically associated with stream dewatering in
western environs and some of the more classical effects on water
quality and aquatic biota. The discussion proceeded in textbook
fashion and, as such, contains many sweeping generalizations, e.g.,
"any consumptive use of water can be expected to affect the quality
of the remaining water"; poorly defined terms, e.g., "water quality
is high"; and esoteric jargon, e.g., "effective physiological concentration."

The discussion of the location in the watershed of stream dewatering structures brought together the climatic, hydrologic, and biological
elements of the stream ecosystem and described the mechanism by which

*Aquatic Ecologist, U.S. Fish and Wildlife Service, Fort Collins,
Colorado.

these elements interact and affect downstream quality. The section provided some positive insight for resource managers with reference to the
importance and ecological implications of diversion structure location
and the associated biological and chemical effects on the watershed.

I was disappointed, however, by the absence of any discussion of
water quality effects on the riparian habitat in the various vegetative
life zones. Water quality changes brought about by stream dewatering,
particularly increased total dissolved solids concentrations, may produce significant long-term successional changes in riparian vegetation
in the Upper and Lower Sonoran life zones. The successional change in
riparian vegetation resulting from invasion by hypersaline-tolerant species, such as Salt Cedar, and the resulting effect on wildlife habitat
is a significant factor in the Lower Colorado River and its tributaries.
It is one which is worthy of considerably more recognition and research
effort similar to studies conducted by Omart at Arizona State University.

In his discussion of organic quality parameters, the author placed
major emphasis on a myriad of poorly known "natural" organic substances
such as antibiotics, vitamins, common metabolites, etc., and their direct
and indirect effects on water quality and aquatic biota. The author
pointed out that many of these poorly understood organic compounds may
play an important role in affecting an organism's ability to withstand,
adapt, or avoid deteriorations in water quality or toxic materials, or
that they may serve as chemical chelating agents and thereby eliminate,
reduce, or change the toxicity of certain water quality constituents,
e.g., heavy metals. Since many of these organic compounds are terrestrial in origin, and arise from particular types of terrestrial vegetative communities, the author suggested that the location of a dewatering

structure in the watershed may have significant ramifications with reference to the stream's ability to assimilate waste products or toxic materials. I would concur with the author that many of these trace organic
compounds may play a significant role in the aquatic ecosystem's ability
to assimilate toxins. However, I must hasten to point out that the relationship between the location of a dewatering structure, the input of
trace organic compounds, and significant changes in water quality or the
biological community is not well documented in the literature. If, as
the author suggested, these changes in natural organic constituents have
significant impact on water quality and aquatic life, they should be
measurable. If the changes are too subtle or long-term to be measurable
with present methodologies and approaches, I would challenge the author
to develop new and more useful approaches and techniques to accomplish
this task.

In the discussion of biological indicators of water quality, the
author indicated as a major stumbling block the difficulty in defining
what constitutes a significant change. He posed the question, "What is
a significant change in water quality parameters in terms of stream biota?"
I feel that, given the current status of technology, a significant change in
water quality would have to be one which elicits a measurable response from
the stream biota. That is to say, in order for a water quality change to be
significant to the aquatic biota, it must be related to some measurable change
in order to be detected. Furthermore, in the real world, it may be significant only to the degree that it affects a change in some segment of the biota
or the water quality itself that is considered valuable, such as a particularly desirable fishery or drinking water supply. Similarly, the author's
second question dealing with the distinction between natural changes vs.
unnatural or man-made changes in water quality may be significant from the

standpoint of a regulatory agency, but is moot from the standpoint of the
resource itself. That is to say, if a valuable resource is threatened
with destruction, it is immaterial whether it is from a natural or man-
induced source. At this point the question becomes one of advocacy and
value judgments made within existing socio-economic and legal/institu-
tional frameworks which will be discussed at length by others at this
Forum.

The question here is whether subtle changes in water quality produce
an effect which can be measured by means of biological indicators. The
author presented a litany of well-known pitfalls and shortcomings of
many of the more traditional approaches to the biological assessment of
water quality, but failed to offer any positive substitutes other than
some ill-defined, massive modeling effort. To my mind this is a "head in
the sand" approach. Biologists in the resource management field cannot
afford to wait for a "mathematical messiah" to come down the road with
all the answers. In this regard, a recent study by the National Science
Foundation (Fromm and coauthors, 1974) of federally supported mathematical models clearly indicated that "up to two-thirds of the models failed
to meet their avowed purposes...." I believe that mathematical modeling
of complex ecosystems may provide many of the answers we need at some
point in the future. However, in the interim, resource managers are
forced to make decisions within the confines of existing data and analytical
methodology.

The problems of assessing or predicting subtle changes in water quality and the resulting ecosystem effects is by no means unique to stream
dewatering. Identical problems are encountered universally in the governmental and private sectors as a result of the National Environmental
Policy Act of 1969 and the Water Pollution Control Act Amendments of 1972.

These two pieces of federal legislation have articulated public policy with reference to environmental quality in a manner that places great emphasis on the capabilities of the scientific communities, particularly of ecologists, to assess or predict the impacts of various types of resource development or exploitation. In most cases, the bottom line of environmental impact assessments speaks to the biological changes or perturbations that are likely to occur as a result of man's activities. The net result of the above-mentioned legislation has been the production of countless volumes of largely uninterpreted and often unintelligible biological data variously titled as environmental impact statements, environmental assessment reports, detailed development plans, etc. These data-gathering exercises probably account for from 50 percent to 80 percent of the time, money, and manpower commitments of some federal and state agencies. In the private sector, these two pieces of legislation have come to be known as the National Environmental Consultants' Welfare Acts.

The point here is that the Congress, in drafting these pieces of legislation, may have overestimated the scientific communities' capability to assess or predict subtle environmental changes. Judging from what one sees in the typical environmental impact statement, this point can be easily argued. It seems that ecologists have a penchant for looking at trees without ever really seeing the forest, and tend to become rather irate when someone suggests that their particular or favorite tree may not be meaningful or relevant to what happens to the forest. The ecologist's typical reaction to this type of criticism is to double his efforts by devising all sorts of ingenious and innovative approaches and techniques to further refine his ability to describe his particular tree, only to find out that his tree and the rest of the forest has been clearcut. The point of this discussion is to emphasize that more of the same old traditional data collecting will not suffice to meet the environmental challenges of today and the future.

In terms of biological indicators of water quality, we need to move away from the component-level approach summarized by Dr. Gosz to an ecosystems approach. Instead of measuring dissolved oxygen, we should look at diurnal oxygen metabolism. Instead of compiling long lists of species, it may be more productive to determine indices of species diversity. That is, only through an understanding of those elements that measure cumulative ecosystems interaction will we be in a position to assess or predict effects of subtle changes in water quality such as those associated with a stream dewatering. In a paper presented at a Symposium on Environmental Impact Assessment, Odum and Cooley (1976) suggested several relatively simple and straightforward approaches to systems ecology. The systems approach to environmental assessment does not so much embody new methodology as it does a different focus or level of resolution on the same ecosystems components. In my view, a simplified systems approach offers one of the few viable alternatives in the short term for meaningful assessment of subtle water quality changes.

Changing land-use practices attendant to stream dewatering, and the associated effects on water quality, is another significant topic which should have warranted discussion in this paper. Rapid changes in land use associated with urban development, energy and mineral exploitation, electric power production, and more intensive agricultural uses of water will probably play the major role in reducing water quality and quantity in the western states.

Increasing demands on limited water supplies is probably the most pressing problem facing federal, state, and local resource managers and planners. It seems that the environmental costs of a continued high rate of development and exploitation will soon outstrip the economic benefits derived from that activity. It would seem the clear mission of this Forum to come to grips with these questions in an atmosphere of reason. Regulatory and resource management agencies need clear and positive recommendations of measures that could be taken to ameliorate the influence of stream dewatering on water quality. We are in need of innovative approaches to upgrade our technical capabilities and of sound economic descriptions of what the environmental costs of development really are. Until resource managers are able to successfully argue their case in the marketplace with other competitors for the limited resource, the actuarial prospectus for sustaining high water quality and the associated high fish and wildlife values is quite dismal.

In conclusion, I felt that Dr. Gosz's presentation was largely reiterative of a recent work by Stalnaker and Arnette (1976), lacked an **in-depth** perception of the problem, and proposed little in the way of meaningful recommendations for new or innovative approaches that could be utilized by resource managers.

76

References

Fromm, Gary, William L. Hamilton, and Diane E. Hamilton, 1974. Federally Supported Mathematical Models: Survey and Analysis (Washington, D.C., National Science Foundation, RANN Division, June).

Odum, Eugene, and James Cooley. 1976. "Ecosystem Profile Analysis and Performance Curves as Tools for Assessing Environmental Impact," in Council on Environmental Quality, A Symposium on Environmental Impact Assessment (Washington, D.C.).

Stalnaker, C. B., and J. L. Arnette, eds. 1976. Methodologies for the Determination of Stream Resource Flow Requirements: An Assessment (Washington, D.C., U.S. Department of the Interior, Fish and Wildlife Service, Office of Biological Services, Western Water Allocation).

Chapter 8

POTENTIAL IMPACTS OF ALTERATIONS IN STREAMFLOW
AND WATER QUALITY ON FISH AND MACROINVERTEBRATES
IN THE UPPER COLORADO RIVER BASIN

Richard S. Wydoski*

Introduction

"The Colorado is probably the most utilized, controlled, and fought over river in the world. It flows through lands of incomparable beauty and includes nearly seven percent of the nation's contiguous land mass, including parts of seven states. From the time of the early settlers to the present, the water of the Colorado River has been the key to development of the arid region."

The above preamble, as quoted from Crawford and Peterson (1974), describes concisely the history and importance of the Colorado River Basin. This river flows through 1,440 miles of arid to semiarid land, but the basin produces only 60 acre-feet of water per square mile annually (Utah Water Research Laboratory, 1975), which is less than that of any other major river in the United States. The water from this river serves 15 million people by supplying water for cities, irrigation, industry, mining, and energy production.

---

*Utah Cooperative Fishery Research Unit, Utah State University, Logan, Utah.

The Utah Cooperative Fishery Research Unit is jointly sponsored by the Utah State University, Utah Division of Wildlife Resources, and U.S. Fish and Wildlife Service.

In this region, rich with scenery and aesthetic values, competition
for water has become an environmental problem that is especially important to the people living in this area (Bishop, 1971; Whittlesay, 1976).
Agencies that act as custodians of natural resources are now being chal-
lenged to weigh partisan views such as environmental concerns against
economic needs, or preservation against use (McCloskey, 1973). Economic
growth and environmental priorities can be compatible if the social and
economic needs of the people are balanced with the environment (Goldman,
1973; Reynolds and Biswas, 1969). Historically, many decisions involving
the planning of water use at all levels--local, regional, and national--
have been made by special interest groups with little public involvement.
Ecological systems and public values are often damaged or destroyed
through alterations of the aquatic environment. Recent and major changes
in the attitudes and opinions of the public will require reordering of
priorities and decision-making policies (Biswas and Durie, 1971; Fitzsimmons and Salama, 1973; Luce, 1973; Pierce, 1976; Shira, 1976). Cost-
benefit analyses have been grossly miscalculated. Recreational benefits,
some of which are intangible, have been inadequately represented in the
past. These considerations are particularly important if the consequences
of water development or use are irreversible.

It is anticipated that the urban population of the Colorado River
Basin will increase 253 percent between 1960 and 2000 and require a three-
fold increase in water for municipal use (Crawford and Peterson, 1974).
In addition, some of the largest sources of natural energy fuels are found
under the scenic rocks, plains, and mountains of this region (Crandall, 1974).
Crandall stated that the need for natural energy fuels is growing faster

than the need for scenery, wilderness, and open spaces. He also called

attention to the large quantities of water that will be needed for

steam-generating plants, development of coal through gasification, and

extraction of other minerals such as trona. In addition, the oil shale

deposits in the Colorado River Basin are among the most extensive in the

world. Some predictions indicate that a million-barrel-per-day oil

shale industry would require about 150,000 acre-feet of water annually,

and coal development in the Four Corners region of Arizona, Colorado,

New Mexico, and Utah is expected to require a similar amount (Crawford

and Peterson, 1974).

The Joint Committee on Atomic Energy (1973) has summarized the past

uses of energy in the United States by comparing energy type, form of

use, and efficiency of use in their report, "Understanding the National

Energy Dilemma." Some of the energy problems that are facing the nation

are a result of a long history of neglect and oversight (Energy Policy

Project, 1974). Some reports have stressed the need for comprehensive

river basin planning (Everett, 1970; Kindswater, 1964; Mihursky and

Cronin, 1973; Hickman, 1975; Wydoski and coauthors, 1976; Simons, 1976).

Others have emphasized the importance of aesthetics to environmental

planning or discussed the relationship between social-psychological sys-

tems and water resource development (Bagley, Kroll, and Clark, 1973;

Fitzsimmons and Salama, 1973).

Various reports have summarized the supply and demand for water, or

considered choices in the use of water, from the Colorado River (U.S. Water

Resources Council, 1968; National Research Council, 1968). Beattie and

coauthors (1971) reviewed the economic consequences of transferring water

between river basins, a problem in the Upper Colorado River Basin today.
For example, the Central Utah Project was designed to increase irrigation in the Great Basin of Utah with water from the Uinta Mountains.
This same water, however, may be needed to develop the energy resources
in the Upper Colorado River Basin (Bishop and coauthors, 1975; U.S. Department of the Interior, 1974). For a number of reasons, including the
demand for energy resources, a comprehensive study of the Upper Colorado
River was made that provided information on a broad variety of topics
(U.S. Water Resources Council, 1971). The resulting report is intended
to be used as a guide for decisions on the best use of this water and related land resources. This report provides details on the supply and
distribution of water, projects future requirements, defines problems,
and provides a framework for alternative choices in the future. Another
more recent report summarizes the Western U.S. Water Plan (known as the
Westwide Study), which was undertaken under authority of the Colorado River
Basin Project Act (PL 90-537) of 1968 (U.S. Department of the Interior,
Bureau of Reclamation, 1975a). The main goal of the plan was to develop
adequate information on which to base future decisions on water and related resources in the eleven western states. McCloskey (1973) pointed
out that environmental concern in the United States has had a great impact in planning water projects, particularly in the West. However, the
appropriate knowledge that will allow water development to be compatible
with the production of fish and wildlife is still lacking.

Weatherford and Jacoby (1975), who have concisely summarized the
future demands and allocations of Colorado River water, stated: "In
broad terms, the problem of managing the Colorado River is the problem

of allocating a flow resource in such a way as to satisfy legally preferred current demands without foreclosing the satisfaction of a different set of configuration of demands in the future. When so viewed, it
is clear that there will be no single or final solution to the problems
of allocation and management in the Colorado River Basin. The time for
seriously addressing the emerging generation of problems, however, is
now."

This paper summarizes the present knowledge of the effects of various management practices on fish and macroinvertebrates in the Upper
Colorado River, attempts to identify future effects of alterations in
streamflow and water quality on these animals, and offers recommendations for future research.

Fish Fauna of the Upper Colorado River Basin

The Colorado River Wildlife Council listed fifty species and four
subspecies of fish in the Upper Colorado River (Richardson, 1976).
Twenty of these species (40 percent) and the four subspecies are native
to one or more states in the upper basin, but some have been introduced
into other states. Thirty species (60 percent) have been introduced at
various times into the river system. In a recent study by the Utah
Cooperative Fishery Research Unit, twenty-nine species of fish were recorded in the major tributaries to the Upper Colorado River (Holden and
Stalnaker, 1975). Ten of these species (34.5 percent) were native to
the river and nineteen were introduced. Other investigators have reported a high percentage of introduced species in the lower basin as
well (Minckley and Deacon, 1968).

The fish fauna of this basin is unique in that 74 percent of the native species are endemic only to the Colorado River system (Miller, 1959). Four of these endemic fishes are known to be rare and declining in abundance in the river system: the Colorado squawfish (Ptychocheilus lucius), humpback chub (Gila cypha), bonytail chub (Gila elegans), and humpback sucker (Xyrauchen texanus). The first two fish have been officially designated as "endangered" by the U.S. Department of the Interior, Fish and Wildlife Service (1976) and the other two were suggested for listing as "endangered" and "threatened," respectively, by the Desert Fishes Council in 1975.

These endemic species are threatened by several factors including the reservoirs and cold tailwaters resulting from the construction of high dams, and competition with introduced fish species. Dams have been constructed in canyon areas that were once the habitat for these fish and, in addition, have altered flow patterns, lowered mean water temperatures, and reduced turbidity. Although much remains to be learned about the habitat requirements of these endemic fish, the potential impacts of alterations to the river ecosystem must be evaluated if these unique species are to be protected from becoming extinct--as outlined by the Endangered Species Act (1973), PL 93-205.

In addition to the large numbers of introduced species competing with native fish, disease or parasitism may also be a factor in their decline. For example, an increased incidence of parasitism in the roundtail chub (Gila robusta) followed the establishment of various exotic fishes in the Moapa River, Nevada (Deason and Bradley, 1972). These investigators offered parasitism as a factor in the decline of this native chub. No data are available to evaluate parasitism as a cause for the decline of native species in the upper basin, but this possibility cannot be discounted.

If fish populations decline until the species is considered threatened or endangered, genetic diversity of the species may be changed to the degree that genetic heterozygosity is decreased or lost. This genetic heterozygosity allows the fish to adapt to various environmental conditions and, if lost, may accelerate the extinction of the species. Recently, Smith and coauthors (1976) have pointed out the importance of population genetics in the management of all fish and wildlife populations, and their concern is especially applicable to threatened or endangered species that are found in the Upper Colorado River Basin.

The expected distribution of fish by generalized ecotypes in the Colorado River Basin has been outlined by Bishop and Porcella (in this volume). This distribution is divided into high-elevation trout waters (>5,000 ft.), transition reaches that contain trout and other species, and meandering stretches with lower velocities and warmer water temperatures that contain the endemic species and introduced species such as channel catfish, Ictalurus punctatus; carp, Cyprinus carpio; and creek chub, Semotilus atromaculatus.

Changes Caused by Impoundments

The cold, clear, headwater streams of the Upper Colorado River originate in the Rocky and Uinta mountains. The main tributaries of the river drop into a dry desert, where spectacular canyons have been carved out over the years. Here the river has been historically warm and turbid, but the volume sometimes changed drastically and suddenly during flash floods. For example, before Flaming Gorge Dam was completed in 1963, the Green River characteristically had high spring flows and low winter flows (Vanicek, Kramer, and Franklin, 1970). These flows have been replaced by relatively stabilized flows. However, monthly and daily flows have varied as a result of water releases at the dam, which depend on power and downstream water demands.

Impoundments have changed the natural warm, turbid river into cold, clear tailwaters with a controlled flow. Below Flaming Gorge Dam, introduced rainbow trout (<u>Salmo</u> <u>gairdneri</u>) have replaced the endemic species. Between 1962 and 1966-67, while the impoundment was filling, the tailwater fishery for rainbow trout was excellent. Since 1967, however, water has been drawn from the hypolimnion of the reservoir, resulting in reduced water temperatures (table 1), which, in turn, have reduced trout production and the potential for recreational fishing. The Upper Colorado Region of the U.S. Bureau of Reclamation (BR), in cooperation with the state agencies of Colorado, Utah, and Wyoming, has proposed modifications of the penstock intake at Flaming Gorge Dam to increase the water temperature in the tailwater (U.S. Department of the Interior, Bureau of Reclamation, 1975b). This modification is expected to result in temperatures between 50° to $55^{\circ}F$ for about five months, based on a BR mathematical model of river temperature. These temperatures are within the range required for good trout growth. These temperatures are also within the range for good growth and production of various trout food organisms such as caddisflies (Tricoptera) and stoneflies (Plecoptera) (U.S. Department of the Interior, Bureau of Reclamation, 1975b). The endemic fish that virtually disappeared from the Green River between

Table 1. Average Monthly Water Temperature ($^{\circ}$F) of the Green River Below
          Flaming Gorge Dam at Greendale, Utah, Before and After Impoundment

Month	Before impoundment 1957-59	After impoundment 1964-65	After impoundment 1966-72 [a]
January	33	41.0	42.0
February	33	37.5	39.5
March	36	38.5	39.0
April	45	40.5	39.5
May	52	41.5	40.0
June	60	44.0	40.5
July	70	47.0	42.5
August	68	48.5	44.5
September	60	49.5	45.5
October	50	53.5	47.0
November	35	53.0	48.5
December	33	47.5	46.0
Annual average	47.9	45.2	42.9

Source: U.S. Department of the Interior, Bureau of Reclamation, Environmental Assessment of Proposed Penstock Intake Modifications, Flaming Gorge Dam, Utah (Salt Lake City, Ut., Bureau of Reclamation, Upper Colorado Region, 1975).

[a]From October 1967 through 1972, the temperature was recorded in degrees Centigrade and converted to degrees Fahrenheit.

the Flaming Gorge Dam and the mouth of the Yampa River might benefit from these discharges of warmer water. These endemic fish are currently found in the Green River below the confluence with the Yampa River (Holden and Stalnaker, 1975).

The agricultural economy of the West, as well as municipal and industrial needs, depend upon storage of part of the annual runoff for use at another time during the year. As a result, a number of reservoirs have been built on the Upper Colorado River, such as Lake Powell and Fontenelle and Flaming Gorge reservoirs, which offer exceptional fisheries where little sport fishing existed previously (Mullan, 1974; Wiley and Mullan, 1975; Mullan and coauthors, 1976).

Problems with Supply and Demand for Sport Fishing

Although it is difficult to predict demand for recreation because of changes in numerous factors over time, participation in recreational fishing has been increasing rapidly and the increase is expected to continue. The number of anglers in the Mountain states increased 37 percent between 1955 and 1970 (U.S. Department of the Interior, Bureau of Sport Fisheries and Wildlife, 1972). Because people of all ages fish and because the American public centers much of its recreation on water resources, the demand for this sport will increase even more in the future. Projection of the demand for hunting and fishing in the Upper Colorado Basin indicates an increase of 43.6 percent between 1965 and 2020 (U.S. Water Resources Council, 1971; table 2). Nearly 74 percent of the participation in 1965 was by fishermen and this percentage is expected to increase in the future (table 2). In the Mountain states (Arizona, Colorado, Idaho, Montana, Nevada, New Mexico, Utah, and Wyoming), the

Table 2. Projected Demand for Fishing and Hunting in the Upper Colorado
River Basin

| Year | Thousands of user-days | Use for (percent):[a] | |
		Fishing	Hunting
1965	4,815	73.7	26.3
1980	5,982	75.6	24.4
2000	8,405	77.4	22.6
2020	11,040	78.5	21.5

Source: U.S. Water Resources Council, Upper Colorado Region
State-Federal Inter-Agency Group/Pacific Southwest Inter-Agency
Committee, Upper Colorado Region Comprehensive Framework Study, Main
Report (Washington, D.C., 1971).

[a]As percentages of total user-days for a given year.

88

percentage of total fishing that is done in streams is somewhat higher
than the average for the United States but the percentage done in reservoirs is far higher in the Mountain states (46.6 percent) than elsewhere in the United States (average 28.1 percent) (table 3).

As water resources development increases in the Upper Colorado
River, more streams will be impounded and more fishing will become
available in man-made reservoirs or in the tailwaters. In a small
percentage (4.7 percent) of the fluctuating reservoirs in the Mountain region of western United States, wild trout make up as much as
20 percent of the total catch (table 4). This percentage is somewhat
higher (6.1 percent) for reservoirs in some states of the Upper Colorado River Basin (Colorado, Utah, and Wyoming). Nevertheless, the
fishery in most of these reservoirs must be sustained by stocking because self-sustaining trout populations have not become established.
The main factors that appeared to limit trout production in 105 Idaho
reservoirs were a paucity of spawning or nursery areas (69.5 percent
of the reservoirs), drawdown (62.9 percent), competition (51.9 percent),
temperature (31.4 percent), food supply (19.0 percent), and dissolved
oxygen (15.2 percent) (Gebhards, 1975). Gebhards reported that a poll
of the eleven western states indicated that the limiting factors were
similar in all;[1] they were identified as lack of spawning or nursery
areas, excessive fishing pressure, drawdown, competition, and unfavorable
water temperature.

------

[1]The eleven western states include: Arizona, California, Colorado,
Idaho, Montana, Nevada, New Mexico, Oregon, Utah, Washington, and Wyoming.

Table 3. Types of Waters Fished Most Often by Freshwater Anglers Twelve Years
Old and Older in the United States in 1970

Region	Thousands of fishermen	Type of freshwater fishing (percent)[a]				
		Reservoirs	Ponds	Natural lakes and ponds	Rivers and streams	Total
United States	29,363	28.1	13.3	27.2	31.4	100.0
Mountain states[b]	1,752	46.6	3.1	15.9	34.4	100.0

Source: U.S. Department of the Interior, Bureau of Sport Fisheries and
Wildlife, 1970 National Survey of Fishing and Hunting, Resource Publication 95
(Washington, D.C., 1972).

[a]As percentages of total fishermen.

[b]The Mountain states include Arizona, Colorado, Idaho, Montana, Nevada,
New Mexico, Utah, and Wyoming.

Table 4. Fluctuating Reservoirs in the Mountain Region of the Western
United States with Wild Trout Fisheries

State	Total number of reservoirs	Reservoirs in which wild trout composed 20% or more of the total catch
Wyoming	330	2
Colorado	250	30
Idaho	105	14
Utah	99	0
Nevada	61	3
Montana	31	3
Arizona	28	0
New Mexico	12	4
Total	916	56

Source: S. Gebhards, "Wild Trout: Not by a Damsite," in W. King,
ed., Wild Trout Management (Denver, Colo., Trout Unlimited, Inc., 1975).

Most reservoirs in the Mountain states are large; an average of 90.4 percent are larger than 16 hectares (39.5 acres) (table 5). This percentage is somewhat higher (91.4 percent) for some states of the Upper Colorado River Basin (Colorado, Utah, and Wyoming). As indicated in the preceding paragraph, the general management of these reservoirs consists of stocking--usually with fingerling rainbow trout. The stocked fingerlings feed primarily on zooplankton, benthic organisms, and sometimes terrestrial insects that fall into the water (Varley, Regenthal, and Wiley, 1971). Usually, these fish do not become piscivorous, and other fish species invariably proliferate in new impoundments and are considered to be serious competitors with trout (Gebhards, 1975). The management of undesirable, nongame species such as carp and Utah chub (Gila atraria), as well as various species of suckers and shiners, in reservoirs is one of the principal problems confronting state fishery management agencies in this region. Techniques are needed for the manipulation of these populations of undesirable species, since eradication is hardly possible in the large reservoirs such as those found in the Upper Colorado River Basin.

Management problems are compounded when large fish are caught in these reservoirs, such as the world-record brown trout (Salmo trutta) from Flaming Gorge Reservoir. Should the agencies responsible for the management of these reservoirs direct their efforts toward the management of smaller rainbow trout that are sought by most anglers or toward the management of fewer, but larger, brown trout or lake trout (Salvelinus namaycush) that will be caught by fewer anglers? Should agencies attempt to control the nongame species such as Utah chubs that compete with the rainbow trout, or should they leave them as forage for large brown trout and lake trout?

Table 5. Area of Water Surface in the Mountain States

State	Total surface area (in thousands) Hectares	Acres	Percentage of total water surface less than 16 hectares (39.5 acres) in area	Percentage of total water surface greater than 16 hectares (39.5 acres) in area
Arizona	91.9	227.0	5.7	94.3
Colorado	132.7	327.8	21.9	78.1
Idaho	242.8	599.8	6.7	93.3
Montana	454.9	1,123.6	17.7	82.3
Nevada	174.8	431.8	3.5	96.5
New Mexico	61.9	152.9	1.3	98.7
Utah	671.4	1,658.4	2.3	97.7
Wyoming	227.0	560.7	19.4	80.6
Total or average	2,057.4	5,082.0	9.6	90.4

Source: U.S. Department of the Interior, Bureau of Reclamation, Westwide Study Report on Critical Water Problems Facing the Eleven Western States (Washington, D.C., 1975).

Ecology of Fish and Instream Flows

Flows of streams in the Colorado River Basin are being rapidly altered as a result of domestic, industrial, and irrigation uses, and further alterations may develop in relation to proposed energy development schemes. All water in this basin is considered to be in use at the present time, but reallocations still take place regularly with the result that water flows in natural streams are continually being reduced (Neuhold, Herrick, and Patten, 1975). State agencies have been asked to provide information on an acceptable range of water flows to sustain a viable fishery. However, this range is difficult to determine because a stream is a complex ecosystem and the only proven way of predicting how fish will respond to changes in streamflow is to measure their microhabitat requirements by life stage. Although much published material is available on stream ecology (e.g., Hynes (1970) cited over 1,200 references), information on the response of fish to changes in flow has been scattered throughout the published literature. A number of reviews have been completed on streamflow requirements of fish, particularly salmonids (Fraser, 1972b; Giger, 1973; Hooper, 1973; Pacific Northwest River Basins Commission, 1972; Wesche, 1973). In addition, workshops are being held to further define this problem. For example, a workshop was held in September 1975 to review, synthesize, and prepare a working document on methodologies for determining streamflow needs (Stalnaker and Arnette, 1976), and another was held on instream flow needs in May 1976 that considered this subject in greater scope and depth (Orsborn and Allman, 1976).

Because recent activity in the evaluation of streamflow requirements of fish has been extensive, I cannot synthesize the existing knowledge in a few pages. However, it is clear that current methodologies do not take into account the wide range of environmental factors on which information will be needed to ensure sound decisions on streamflow recommendations. In fact, several of the literature reviews have uncovered large gaps in existing knowledge of the basic ecology of fish (e.g., the Pacific Northwest River Basins Commission (1972) suggested research on requirements for rearing young salmonids, and Hooper (1973) suggested research on the effects of water flows on the drift rates of invertebrate organisms). Giger (1973) concluded that "improved understanding of the stream ecology of juvenile salmonids is seen as a more important immediate goal, however, than efforts to develop procedures for making streamflow recommendations."

Salmonids

Although literature reviews have shown important gaps in available information on stream-dwelling salmonids, the behavior, ecology, and life history of this group of fish has been intensively studied (Allen, 1969) and is now fairly well understood. Stream-dwelling salmonids are known to be territorial in their behavior (Allen, 1969; Edmundson, Everest, and Chapman, 1968; Giger, 1973; Hooper, 1973). The locations of the territories are influenced by the availability of drifting food (Chapman, 1966; Waters, 1969), with the more dominant fish occupying the area with the greatest drift. Coupled with the availability of food is the spatial requirement of salmonids (Allen, 1969; Chapman, 1966). Giger (1973) reported that his review of the literature demonstrated a positive, fairly linear relationship between fish size and average area of streambed per fish. These food and spatial requirements are influenced by water velocity and cover (Chapman, 1966). Furthermore, Chapman and other investigators have shown that territoriality is not exhibited, or at least is reduced, in pool environments, demonstrating that water velocity is important in determining the distribution and carrying capacity of stream-dwelling salmonids. Fish size also influences habitat selection and, in general, large fish have large territories and occupy areas with a greater water velocity than those occupied by small fish. In addition, there are differences in habitat selection by species; for example, rainbow trout usually are found in faster water than cutthroat trout, and brown trout generally require more cover than cutthroat trout.

The importance of overhanging cover and pools as habitat for trout has been reported by Weber (1959) and others. He reported that stream-flows below Grandby Dam on the Colorado River are only about 11 percent of average annual historical flows and that overhanging cover is lacking during low streamflows. Weber reported that trout could be captured by electro-fishing in shallow pools (6 inches deep) at low flows if these pools were near the water line with brush cover. He also was able to demonstrate that trout would occupy reaches of the stream if cover was provided. He placed five brush piles where fish had not been previously collected and, in approximately one month, all of these piles contained trout. Numerous papers in the literature support the importance of cover for trout populations even more strongly than this example from the Colorado River. However, it is quite clear that cover is directly related to streamflows because, at the extremes, streambank cover can be eroded by excessive flows and the water may recede from cover at low flows.

Lewis (1969) demonstrated that water velocity accounted for 66 percent of the variation in numbers of brown trout and that total cover accounted for 59 percent of the variation in numbers of rainbow trout.

Lewis used a multiple linear regression in his analysis, with the number of trout per pool as the dependent variable, and pool surface area, mean depth, mean water velocity, percent cover, and total cover as the independent variables. Pool surface area, volume, and depth accounted for little of the variation in trout numbers in that study. Lewis suggested that greater food drift, accompanied by increased water velocities, might support greater densities of trout. This concept is supported by the work of other investigators (e.g. Chapman, 1966; Waters, 1969).

Some studies have focused on the microhabitat of salmonids which includes physical features in addition to water velocity and depth (Baldes and Vincent, 1969). The brown trout used in this experiment remained near structures on the bottom or along the edges of the shaded side of the experimental flume. The water velocity near the substrate was less than that in the water column and, as the velocity increased, the trout moved nearer to the substrate and became more dependent on irregularities on the flume bottom. Differences appear to exist among species, and generalizations are difficult to make concerning microhabitat of a single species because it may vary with the size of the fish, and also seasonally with behavioral adjustments to water level or to other physical features of the stream.

Because the habitat requirements of salmonids vary considerably with the factors that have already been mentioned, much emphasis has been placed on spawning; consequently, this aspect of the life history is best known for native trout species (Hooper, 1973; Smith, 1973; table 6). Although the data summarized in table 6 have been obtained from various locations, they are probably applicable to the species of trout found in the headwater streams of the Uinta and Rocky mountains.

Table 6. Water Velocity, Substrate Size, and Temperature Preferred for Spawning by Resident Trout

Species of trout	Time of spawning	Preferred water velocity[a] (cm/sec)	(feet/sec)	Preferred substrate size[a] (cm)	(inches)	Optimum spawning temperature (°F)[a]
Brown	Oct.-Dec.	40-52 (30-91)	1.3-1.7 (1.0-3.0)	0.63-3.8 (0.63-7.6)	0.25-1.5 (0.25-3.0)	44-48 (43-55)
Cutthroat	Apr.-early July	-- (30-91)	-- (1.0-3.0)	-- (0.63-7.6, to 10.0+ for large fish)	-- (0.25-3.0, to 4.0+ for large fish)	53 (45-55)
Brook	Sept.-Dec.	-- (6-91)	-- (0.2-3.0)	-- (coarse sand to 7.6 cm gravel)	-- (coarse sand to 3.0 in. gravel)	43-46 (37-50)
Rainbow	Feb.-May	about 61 (43-82)	about 2.0 (1.4-2.7)	0.63-3.8 (0.63-7.6)	0.25-1.5 (0.25-3.0)	52 (45-56)

Source: D. R. Hooper, Evaluation of the Effects of Flows on Trout Stream Ecology (Emeryville, Calif., Pacific Gas and Electric Co., Department of Engineering Research, 1973).

[a]Ranges are provided in parentheses.

This brief and incomplete account of trout behavior and ecology indicates the importance of water velocity, and the complexity of the relationship of fish ecology to streamflows. Unfortunately, the inter-relations among and between species add additional variables that make interpretations even more difficult.

Other Fish

Few data are available on streamflow requirements of most fish species inhabiting the larger main tributaries of the Colorado River. Ongoing research by the Utah Cooperative Fishery Research Unit may pro-vide data on the spawning habitats of Colorado squawfish, humpback sucker, flannelmouth sucker (Catostomus latipinnis) and bluehead sucker (Catostomus discolobus). In general, the suckers appear to require a gravel substrate and a moderate water flow for spawning. Except during spawning, they may be found in eddies or at the interface between the eddy and the main river current. Adult Colorado squawfish and round-tail chub are generally found in pools, riffles, or eddies. Ripe round-tail chubs were collected in shallow pools and eddies over rubble or boulder stream bottoms covered with silt (Vanicek and Kramer, 1969). However, because it has not been possible to effectively sample fish in fast current in the main rivers, uncertainty exists about habitat use. In addition, passive gear such as gill nets and trammel nets have been used to collect fish in these rivers. Because these nets depend on fish movement for capture, and because the fish may have been moving between habitats when captured, there is no way of knowing what part of the stream the fish were inhabiting. Also, high runoff during the spring, and ice during the winter, preclude effective sampling of these species so that habitat(s) occupied during spring and winter cannot be defined precisely.

Some information is available on the water depth and velocity used for spawning by some warm-water and cool-water fish (Bovee, 1974; table 7). Several species listed in table 7 have been introduced and have become established in the Colorado River, such as the white sucker (Catostomus commersoni), largemouth bass (Micropterus salmoides), and walleye (Stizostedion vitreum). The creek chub was reported as native to Utah (Richardson, 1976) but is now found in Colorado and Wyoming as well. Inferences can be drawn from table 7 concerning spawning requirements of related species in the Upper Colorado River, but further generalizations beyond this are risky.

In the Lower Colorado River, backwater marsh habitat was never extensive but was extremely important to various species of wildlife (Ohmart, Deason, and Freeland, 1975). This habitat is also important to the endemic fish in the Upper Colorado River. Vanicek and Kramer (1969) reported that young Colorado squawfish and chubs were commonly taken in quiet water or shallow pools over silt, sand, and occasionally gravel bottoms. These backwater habitats may become filled with accumulations of silt and organic materials if the streamflow is reduced, and the natural flushing action of the spring runoff is necessary to keep these habitats open or to create new eddies or backwaters. Reduction of the annual peak flows by storage reservoirs on the Upper Colorado River may reduce these critical nursery areas for the endemic fish species.

Another problem concerning the streamflow of the Colorado River is the continual reduction of the flow by depletions. Most (62 percent)

Table 7. Water Depth and Velocity Used for Spawning by Warm-Water and Cool-Water Fish

Species	Depth (meters)	Depth (feet)	Velocity (cm/sec)	Velocity (feet/sec)
Paddlefish	variable	variable	49-91	1.6-3.0
Shovelnose sturgeon	0.3-0.9	1.0-3.0	75-150	2.5-4.9
Creek chub	--	--	49-91	1.6-3.0
Longnose dace	0.03-0.3	0.1-0.7	15-45	0.5-1.5
Longnose sucker	0.2-0.3	0.7-1.0	31-45	1.0-1.5
White sucker	0.2-0.3	0.7-1.0	31-45	1.0-1.5
Shorthead redhorse	0.3-0.9	1.0-3.0	31-61	1.0-2.0
Smallmouth bass	0.9-1.8	3.0-5.9	11	0.36
Largemouth bass	0.3-1.8	1.0-5.9	0	0
Walleye	1.2-1.5	3.9-4.9	0-50	0-1.6
Sauger	1.2-1.5	3.9-4.9	0-50	0-1.6

Source: K. D. Bovee, The Determination, Assessment and Design of "In-Stream Value" Studies for the Northern Great Plains Region. Final Report, EPA Contract No. 68-01-2413 (Missoula, Mont., University of Montana, 1974).

of these depletions are from irrigation and associated uses (U.S. Water
Resources Council, 1971). These depletions often occur in rather small
amounts but cumulatively can have a large impact overall. Further reductions are anticipated for the system in the near future. For example,
the U.S. Water Resources Council estimated that the water exported
from the Green River will increase from 0.12 to 0.65 million acre-feet
and on-site depletions will increase from 0.87 to 1.67 million acre-feet
between 1965 and 2020. The Council estimated similar depletions for the
upper main stem of the Colorado River--from 0.43 to 0.88 million acre-
feet by export and from 0.97 to 1.35 million acre-feet by on-site depletions for the same time period.

At the present time, it is not possible to make many predictions on
the responses of fish to changes in streamflow because the requirements
of many species or the life stage of a particular species may not be known.
It is particularly important to emphasize that quantitative data be obtained on the streamflow requirements **by species** and **for each life stage**
if predictions of fish responses to changes are to be made with accuracy.

Responses of Macroinvertebrates to Streamflow

Water flows can greatly affect the invertebrates that are a major
source of food for fish. Reduced water flows can limit the abundance
and diversity of swift-water invertebrates, and excessive flows can
scour the substrate and become detrimental (Giger, 1973; Ward, 1976).
A summary of several studies comparing the relationship between water
velocity in riffles and the numbers of bottom organisms shows that the
numbers of organisms can differ considerably in different streams for a
given water velocity (table 8). Large differences can also exist in

Table 8. Numbers of Bottom Organisms per Square Foot Collected at
Different Water Velocities in Stream Riffles

Water velocity[a]		Source		
(cm/sec)	(feet/sec)	Pearson and coauthors	Surber	Kennedy
15-31	0.5-1.0	53	99	444
34-46	1.1-1.5	90	$\underline{148}$[b]	$\underline{881}$[b]
49-61	1.6-2.0	$\underline{120}$[b]	115	484
64-76	2.1-2.5	89	$\underline{152}$[b]	289
79-91	2.6-3.0	105	125	171
95-107	3.1-3.5	65	339[b]	--
110-122	3.6-4.0	62	--	--

Source: H. D. Kennedy, Seasonal Abundance of Aquatic Invertebrates
and Their Utilization by Hatchery-Reared Rainbow Trout, Bureau of Sport
Fisheries and Wildlife, Technical Paper 12 (Washington, D.C., U.S. Depart-
ment of the Interior, 1967); L. S. Pearson, K. R. Conover, and R. E. Sams,
"Factors Affecting the Natural Rearing of Juvenile Coho Salmon During the
Summer Low Flow Season." Unpublished manuscript (Portland, Oreg., Fish
Commission of Oregon, 1970); E. W. Surber, "Bottom Fauna and Temperature
Conditions in Relation to Trout Management in St. Mary's River, Augusta
County, Virginia," Virginia Journal of Science vol. 2 (1951) pp. 190-202;
as reported in R. D. Giger, Streamflow Requirements of Salmonids, Final
Report, Anadromous Fish Project AFS-62-1 (Portland, Oreg., Oregon Wildlife
Commission, 1973).

[a]Surber recorded surface velocity; depth of measurement not specified
in other studies.

[b]Underlined data indicate the mode in distribution of bottom organisms
as related to water velocity.

[c]Inadequate sample size.

various reaches of the same stream. Part of this variation is due to the diversity of species, their life histories, and their habitat requirements. In addition, ecological factors such as water temperature or chemical components that affect productivity of the stream vary and affect abundance (Knight, 1965). Riffle habitats are important in producing bottom organisms (Hill, 1965; table 9). However, different groups of organisms are found in slower water. The substrate composition can also be important in the distribution of bottom organisms (Sprules, 1947; table 10). It is not possible to review fully the literature on the ecological requirements of bottom organisms. However, the preceding examples were used to provide some insight into the factors affecting distribution and abundance of macroinvertebrates.

Drift of bottom organisms has been shown to be important to the ecology of stream-dwelling fish (Waters, 1969). This drift is another complex ecological phenomenon about which generalizations are difficult to make. However, bottom organisms may exhibit different patterns in diel and seasonal periodicity in natural drift that varies by species, streamflow, water temperature, and perhaps other ecological variables. The numbers of organisms in the drift and on the bottom may differ by species in the same location at the same time (Pearson, 1967; Pearson and Franklin, 1968; figure 1). In addition, comparisons of the same groups of bottom organisms between locations demonstrate that generalizations about the relationship between organisms on the bottom and organisms in the drift are risky (figures 1 and 2).

Changes in the flow and water temperature in the Green River after the closure of Flaming Gorge Dam also changed the distribution and species composition of macroinvertebrates. Pearson, Kramer, and Franklin (1968)

Table 9. Percentage of Total Volume Contributed by Different Bottom
Organisms in Four Categories of Stream Habitat of the North
Fork of the White River, Colorado, 1963

	Stream habitat[a]				Number of samples in which the group of invertebrates occurred
	Deep fast	Deep slow	Shallow slow	Riffle	
Plecoptera	0.0	12.3	1.2	86.5	33
Ephemeroptera	2.3	6.4	1.6	89.7	89
Trichoptera	>1.7	>8.1	2.6	>87.5	88
Diptera	1.9	6.5	2.5	89.1	89
Coleoptera	4.3	10.6	2.2	82.9	71
Hymenoptera	0.0	0.0	0.0	100.0	1
Platyhelminthes	0.0	0.0	0.0	100.0	3
Hydracarina	0.0	24.4	4.7	70.9	40
Oligochaeta	0.0	14.3	0.0	85.7	7
Pelecypoda	0.0	0.0	50.0	50.0	1
Gastropoda	0.0	33.3	16.7	50.0	4
Nematoda	0.0	28.6	0.0	71.4	7

Source: Combined data for 89 samples from R. R. Hill, White River Survey
(Denver, Colo., Colorado Department of Game, Fish, and Parks, Fisheries Research
Division, Federal Aid Report F-26-R-2, Job 2, 1965).

[a]Deep refers to water more than 1.5 feet deep; shallow, to water less than
1.5 feet deep. Fast refers to a water velocity greater than 1 foot/second;
slow to velocity less than 1 foot/second. A riffle has a depth to 1.5 feet with
water velocity over 1 foot/second.

Table 10. Relative Numbers and Volumes of Emerging Insects from Various
Substrates in Algonquin Park, Ontario, Canada

Substrate type	Relative numbers[a]	Relative volume[a]
Rubble (rapid current)	4.6	20.7
Rubble (pool)	3.3	--
Gravel	2.1	3.6
Muck	1.8	1.5
Sand	1.0	1.0

Source: D. R. Hooper, Evaluation of the Effects of Flows on Trout
Stream Ecology (Emeryville, Calif., Pacific Gas and Electric Co.,
Department of Engineering Research, 1973) citing W. M. Sprules, An Eco-
logical Investigation of Stream Insects in Algonquin Park, Ontario,
Biology Series No. 56 (Toronto, University of Toronto, 1947).

[a]All numbers and volumes per unit of area are expressed in terms of
the sand substrate which was assigned a value of 1.0. Therefore, these
values refer to a multiple factor in numbers or volume of insects over
the sand substrate.

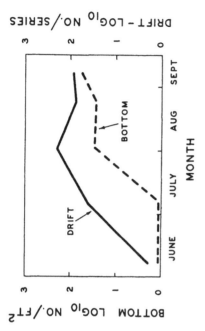

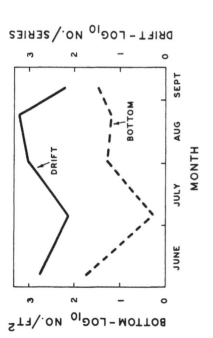

Figure 1. Comparison of the Densities of <u>Baetis</u> sp. I Nymphs (left) with Simuliidae larvae (right) on the Bottom and in the Drift at Carr Ranch, Green River, 1965

<u>Source:</u> W. D. Pearson, "Distribution of Macro-Invertebrates in the Green River Below Flaming Gorge Dam, 1963-1965" (M.S. thesis, Utah State University, Logan, 1967).

107

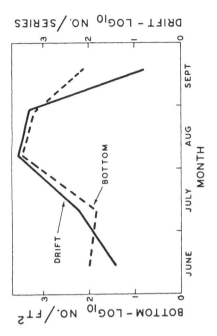

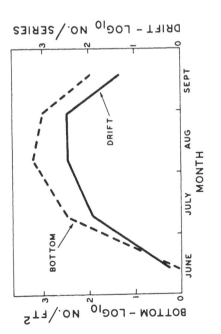

Figure 2. Comparison of the Densities of Baetis sp. I Nymphs (left)
with Simuliidae larvae (right) on the Bottom and in the
Drift at Little Hole, Green River, 1965

Source: W. D. Pearson and D. R. Franklin, "Some Factors Affecting Drift
Rates of Baetis and Simuliidae in a Large River," Ecology vol. 49, no. 1, 1968.

reported that nine forms of invertebrates found in the Green River at
Little Hole, just downstream from the Flaming Gorge Dam, before the
closure of the dam were not found after the closure. On the other hand,
ten groups of invertebrates that had not been found previously were found
after the dam was completed. Downstream, at Carr Ranch and Echo Park,
the river environment showed signs of transition resembling the pre-
impoundment condition. Still further downstream, at Island Park, the
river appeared the same in 1964-65 as it did before the dam was completed
(1963). This study and others have shown that disruption of an ecosystem
can cause changes in the species composition and abundance of macroinvertebrates.

Pronounced water level fluctuations can also have a great impact on
macroinvertebrates. For example, an estimated 25 percent of the streambed that had been submerged for the previous five months was exposed by a
drop of more than 1 foot in the water level of the Green River at Island
Park on 7 September 1964 (Pearson and Franklin, 1968). As the water level
dropped, the drift rates of Baetis sp., other mayflies, caddisflies, and
stoneflies reached the highest ever recorded at this station, especially
after sunset (figure 3). Pearson and Franklin reported that they observed nymphs of Baetis sp. moving (crawling and swimming) toward deeper
water as the river level dropped. As the water level returned to normal
two days later, the drift of macroinvertebrates decreased.

The wide fluctuations in streamflows released from the Green Moun
tain Reservoir on the Blue River, a tributary to the Colorado River, were
also shown to have a great impact on invertebrate populations (Powell,
1958). Powell reported that the average weight per unit of area was six

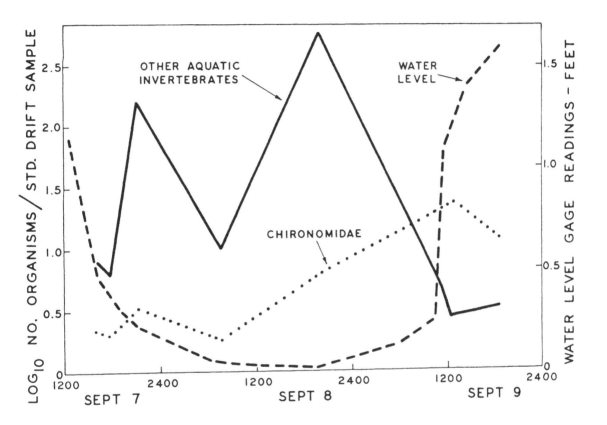

Figure 3. Drift Rates of Chironomidae and Other Aquatic Invertebrates
During a 1-1/2 ft. Drop and Rise in Water Level, Island Park,
Green River, 7-9 September 1964.

Source: W. D. Pearson and D. R. Franklin, "Some Factors Affecting Drift
Rates of Baetis and Simuliidae in a Large River," Ecology vol. 49, no. 1, 1968.

times as great above the reservoir as below the reservoir, and attributed these differences to daily and seasonal streamflows. He showed
that daily vertical fluctuations of four feet commonly occur immediately
below the reservoir and stated that a maximum fluctuation of 1,850 cubic
feet per second may occur in less than one minute. The impact on invertebrates is extremely detrimental especially when over 80 percent of the
stream environment may be dry at low streamflows and severe ice cover
occurs in winter.

Catastrophic drift of chironomid larvae was recorded in the Green
River on 11 June 1965 (Pearson, 1967; figure 4). A sudden increase in
water level of nearly three feet caused an increase in the drift rate of
chironomid larvae. The drift decreased to nearly pre-flood rates after
about 3 hours, however, even though the water level remained high, indicating that the organisms had been dislodged forcibly by the high water
velocity.

These examples demonstrate that changes in reservoir discharges to
a large stream such as the Green River can have a marked effect on bottom
macroinvertebrates in the river. The fine review of streamflow patterns
below large dams on stream benthos by Ward (1976) provided the following
main points for consideration: (1) each dam must be considered individ-
ually in establishing flow criteria, (2) benthic species composition is
considerably modified by impoundments, (3) the flow regime may enhance
or reduce benthic organisms, and (4) the responses of benthic organisms
to smaller, sublethal effects related to streamflow are presently unknown.

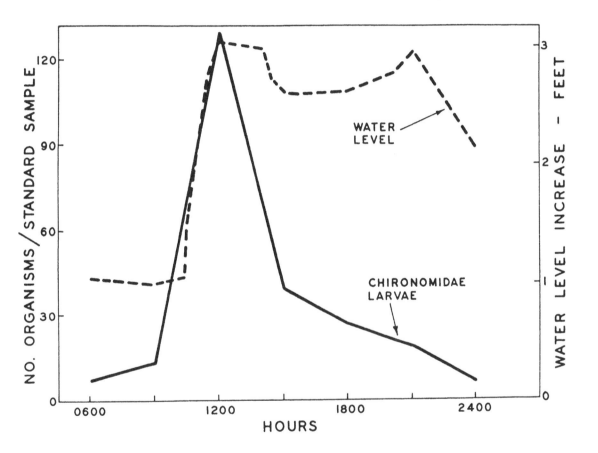

Figure 4. Increased Drift Rates of Chironomidae larvae after a Sudden
Increase in Flow, Little Hole, Green River, 11 June 1965

Source: W. D. Pearson and D. R. Franklin, "Some Factors Affecting Drift
Rates of Baetis and Simuliidae in a Large River," Ecology vol. 49, no. 1, 1968.

Stream Channelization

Stream channels are altered for various reasons including highway
construction, land reclamation, navigation, flood control, and others.
These alterations are often solutions of single purpose problems which
frequently result in the creation of other problems. Stream channelization is a form of alteration that involves straightening and reducing
the length of the original channel, thereby increasing the hydraulic
gradient and thus the water velocity. In addition, cover along the
streambank is removed and pools are often converted into riffles or runs
(Elser, 1968). Changes of this kind affect the habitat of both aquatic
and terrestrial organisms. After channelization, streambeds are known
to remain unstable, sometimes for years. Substrate instability is be-
lieved to be the most significant factor related to changes in fish and
invertebrate populations after channelization (Etnier, 1972).

Downstream areas are often subjected to increased flooding and
eutrophication as a result of these alterations. Flooding has been
shown to drastically reduce trout populations, particularly young-of-
the-year, in various parts of the United States (Hoopes, 1975; Seegrist
and Gard, 1972). Shelford's ecological law of tolerance can be applied
in this situation whereby a water velocity that is extremely high may be
as detrimental as when the streamflow is too low.

In the Intermountain West, most stream channelization studies have
been conducted in Montana and Idaho. Altered stream channels in thirteen Montana streams produced only one-fifth the number and one-seventh
the weight of game fish produced by unaltered channels (Peters and Alvord,
1964). Undisturbed sections of forty-five Idaho streams contained an

average of eight times (range, 1.5-112 times) the weight of game fish

found in channelized sections. Invertebrates are similarly affected by

altered sections of streams. For example, the standing crop of inver-

tebrates was eight times greater in unaltered than in altered sections of

the Missouri River.

The basic problem is that little consideration is given to the hy-

draulic characteristics of a stream that are favorable to fish--the chan-

nels are simply deepened and straightened to improve flow. In addition,

the streambed gravel and rubble are used to build levees along the banks,

and these levees destroy existing vegetation and thus cover. This method

is a poor approach to controlling high flows, since it increases down-

stream flooding. The existing hydraulic gradient should be preserved

when stream channels are relocated. This is especially important with

regard to sediment loads and transport (Platts and Megahan, 1975). Land-

use planning that involves the entire river system should be considered

so that a natural equilibrium of the hydraulic gradient can be maintained

(Goldman, 1973). As quantitative information becomes available for

assessing the effects of stream alterations, and for determining what

alternatives are available, managers will be better able to make intelli-

gent decisions toward solving their problems.

Most of the changes in streamflow due to stream channelization in

the Upper Colorado River would occur in mountain streams because road

construction in the narrow canyons requires channeling, or in the flood-

plain of streams because of private reclamation or flood control projects.

Nevertheless, channelization continues to affect water flows and to dis-

rupt stream ecosystems in this region because of the many, although some-

times small, alterations made particularly by private landowners.

Water Quality

Sedimentation is considered by the U.S. Geological Survey to be one
of the more important pollutants in streams. These sediments result from
land uses that denude soils. Logging (Gibbons and Salo, 1973; Hall and
Lantz, 1969), stream channelization (Elser, 1968), improper watershed
management (Platts and Megahan, 1975), and grazing (Gunderson, 1968;
Lusby, 1970; Platts, 1958) all contribute to increases in sedimentation.
Increases in fine sediments in spawning gravel related to logging practices decreased the survival of salmonid eggs in an Oregon stream (Hall
and Lantz, 1969). In the Strawberry Reservoir drainage of north-central
Utah, mortality of the eggs of the cutthroat trout (Salmo clarki) was
45 percent in the pre-eyed stage and 30 percent in the eyed stage from a
stream where heavy grazing occurred versus 29 percent and 4.5 percent for
respective stages in a section of the same stream where grazing was less
intense (Platts, 1958). The experimental addition of sand to a Michigan
stream increased the gradient and width of the stream and reduced the
depth and pools (Hansen and Alexander, 1976). Such physical changes in
a stream can have pronounced impacts on the biota of the stream (Elser,
1968; Etnier, 1972; Lewis, 1969; Ward, 1976). In the arid and semiarid
area of the Upper Colorado River Basin, turbidity caused by sediments is
common but the effect of this turbidity on the fish inhabiting this river
system is not known. Some studies have indicated that turbidity can
affect the behavior of fish (Heimstra, Kamkot, and Benson, 1969).

Sedimentation is particularly important in reservoirs (Dendy, 1968)
where suspended materials settle because of decreased velocities. The
net result is that reservoirs become filled, limiting their life spans.

Neel (1963) has pointed out that reservoirs may have no reasonably predictable pattern of limnology. In western reservoirs, erosion of silt, clay and sand fills the reservoir at a fast rate (figure 5). Workman and Keith (1974) concluded that erosion control or management in the Upper Colorado River Basin would be ineffective from the physical and biological standpoint and would also be unsound from an economic standpoint. Although the building and regulation of reservoirs provides greater quantities of water for use, reservoirs also reduce total runoff by exposing more surface for evaporation (Neel, 1963). In addition, the porous sandstone of the Lake Powell Basin absorbs much water and further reduces the amount that flows downstream in the Colorado River. Symons, Weibel, and Robeck (1964) reviewed the influence of impoundments on water quality and provided recommendations for research. In addition, the effect of water level fluctuations on fish and other aquatic organisms is covered by the bibliography of Fraser (1972a). Other investigators reported the effects of new impoundments, such as those on the Green River and its tributaries, on aquatic life (e.g., Funk and Gaufin, 1971; Varley, Regenthal, and Wiley, 1971).

A water quality assessment has been recently made for the Colorado River (Utah Water Research Laboratory, 1975). This comprehensive study (over 1,200 pages) is too detailed to summarize here; it is noteworthy, however, that the study has demonstrated that salinity is the major water quality factor in the basin, and its interactions outweigh all other considerations except the availability of the water resource. Average total salinity of the Colorado River system in 1972 was less than 50 mg/1 in the headwaters but increased to 847 mg/1 at Imperial Dam in California (Bessler, 1975). Several pieces of federal legislation are directed at

116

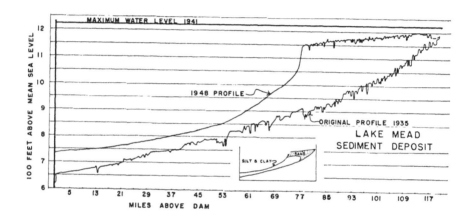

Figure 5.   Sediment Deposits in Lake Mead from the Colorado River
            Between 1935 and 1948

Source: H. R. Gould, _Sedimentology. Lake Mead Comprehensive Survey, 1948-49_ (Reston, Va., U.S. Geological Survey, 1954) from J. K. Neel, "Impact of Reservoirs," in D. G. Frey, ed., _Limnology in North America_ (Madison, Wisc., University of Wisconsin Press, 1963) p. 583.

the salinity problem (1972 Federal Water Pollution Control Act Amendments, PL 92-500; 1974 Colorado River Basin Salinity Control Act, PL 93-320). Recently, the U.S. Environmental Protection Agency (1971) reviewed the water quality problem caused by minerals and outlined some management aspects for the Colorado River. It has been estimated that the salinity results from two-thirds natural causes and one-third man-made causes. The salinity from diffuse sources, both natural and man-made, accounts for 84 percent, and that from point sources accounts for 16 percent--6 percent from natural mineral springs and 10 percent from artificial drainage (Utah Water Research Laboratory, 1975; Williams, 1975). Since only a small portion can be identified from point sources, it appears that only a small quantity of salt can be reduced. Salinity has been increasing in the Colorado River system for various reasons with about 10 million tons going downstream annually. If the Environmental Protection Agency guidelines for this river are to be met, then about 2.5 million tons per year will need to be removed. However, implementation of all salinity control measures on the Colorado River will provide for a maximum reduction of about 1.6 million tons annually or about 150 mg/l at Imperial Dam (Bessler, 1975). Current damages to agricultural crops are estimated to be between $110,000 and $230,000 for each mg/l increase in salinity at Imperial Dam, California.

The relationship between streamflow and salinity must be understood to visualize the problems confronting the upstream and downstream users of water (U.S. Environmental Protection Agency (1971); Utah Water Research Laboratory, 1975). For example, salinity of the water may decrease as the flow increases, but the total amount of salt passing a particular point may increase. The potential effects of increased salinity on fisheries

is unknown. However, it is known that larval and juvenile fish sometimes do not survive environmental changes that have little or no effect
on adult fish. The ionic regulation, circulation, and respiration of
small fish can be impaired rather easily by changing the chemistry of
their environments (Hoar and Randall, 1969, 1970). Since curtailed
reproduction of the endemic fish species has been established (Vanicek
and Kramer, 1969; Holden and Stalnaker, 1975), it appears desirable to
establish the tolerance limits of the species considered to be threatened or endangered (Colorado Squawfish, humpback chub, and bonytail
chub), as well as of the endemic fish that appear to be stable in numbers (e.g., the bluehead sucker) or perhaps even increasing (e.g.,
the flannelmouth sucker).

Bishop and Porcella (in this volume) indicated that significant concentrations of heavy metals may occur in several tributaries of the Colorado River that would seriously affect water quality. Mining operations
in Idaho destroyed anadromous runs of salmon and trout when high metal
concentrations (copper and iron), decreased pH, and other adverse factors
caused fish kills or an avoidance of the stream by the fish (McKim and
coauthors, 1975). McKim and coauthors summarized the recent published
literature on the effects of pollution on freshwater fish. The effects
of surface mining on fish and wildlife in Appalachia were reviewed by
Boccardy and Spaulding (1968). Although such a review is not available
for the Colorado River, the effects of mining on this river system can
also be drastic (Tsivoglov and coauthors, 1959; figure 6). Bottom macroinvertebrates were scarce for about 28 miles downstream from a uranium-
vanadium mine on the Animas River that drains into the Colorado River
(figure 6). This scarcity of organisms was attributed to toxicity from

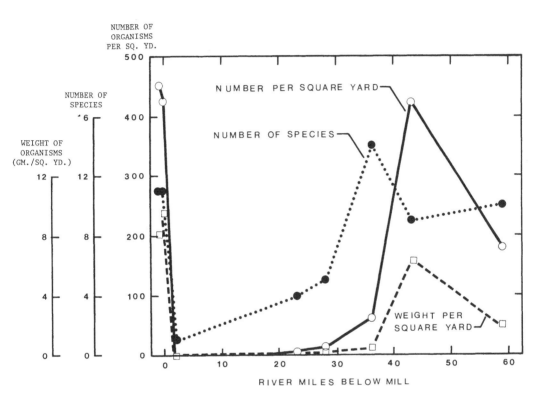

RIVER MILES BELOW MILL

Figure 6.   Number of Species and Density of Bottom Fauna Above and
Below a Uranium-Vanadium Mill on the Animas River,
Colorado, in 1958.

Source: E. C. Tsivoglov, S. D. Schearer, R. M. Shaw, Jr.,
J. D. Jones, J. B. Anderson, C. E. Sponables, and D. A. Clark, Survey
of Interstate Pollution of the Animas River, Colorado-New Mexico
(Cincinnati, Ohio, U.S. Department of Health, Education, and Welfare,
Public Health Service, Robert A. Taft Engineering Center, 1959).

leaching of radioactive tailings and lowered pH (table 11). Sigler and coauthors (1966) also reported that radium was a radioactive material found in water and bottom sediments of the Animas, San Miguel, and Dolores rivers in the Upper Colorado River Basin that limited or extirpated the aquatic biota below uranium mills.

The temperature tolerances, and even optimal temperatures, are fairly well known for salmonids (Giger, 1973; Hooper, 1973), but are not known for the endemic species in the Colorado River. Information on the temperature tolerances for other species, such as carp, channel catfish, and walleyes, is available for other waters in the United States and could be used to predict some responses. Inferences about the responses of the endemic forms to changes in temperature could be drawn from the general behavior of other cyprinids and catostomids.

The water quality requirements of various species of aquatic insects in the Intermountain Region have been studied at the University of Utah since 1946 resulting in considerable data on the effects of pollution on the biota in a number of streams. Since 1946 (e.g., Gaufin, 1973), laboratory studies have been directed at determining the tolerance of twenty species of aquatic insects (Diptera, Ephemeroptera, Plecoptera and Trichoptera) to several adverse environmental factors (high water temperatures, low dissolved oxygen concentrations, and low pH). Additional studies like Gaufin's are needed to predict the responses of aquatic macroinvertebrates to other proposed environmental changes.

Some states are reviewing the water quality of their waters and are establishing plans for maintaining acceptable water quality. Nichols, Skogerboe, and Ward (1972), for example, have outlined decisions concerning water quality management in Colorado.

Table 11. The pH of Animas River Water After the Addition of
Different Quantities of Effluent from Uranium-
Vanadium Mining

Type of effluent	Percentage of river water				
	0	90	96.8	99	100
Main effluent	4.7	6.4	6.7	7.2	7.9
Tailings effluent	<2.0	2.6	4.8	6.5	7.9

Source: Tsivoglov, E. C., S. D. Schearer, R. M. Shaw, Jr.,
J. D. Jones, J. B. Anderson, C. E. Sponagles, and D. A. Clark, Survey
of Interstate Pollution of the Animas River, Colorado-New Mexico
(Cincinnati, Ohio, U.S. Department of Health, Education, and Welfare,
Public Health Service, Robert A. Taft Engineering Center, 1959).

Recommendations for Streamflows

Most methodologies for determining streamflow requirements of fish are intended for the preservation or maintenance of sport fisheries (Stalnaker and Arnette, 1976). However, the streamflow requirements or preferences of other key fish species such as game fish and the threatened and endangered endemic fish are not fully known (Orsborn and Allman, 1967). Streamflow requirements are least known for endemic fish inhabiting the main stem of the Upper Colorado River. In areas where there are great expanses of bare soil, as in the Intermountain West, the runoff would be expected to be high and to result in high peak flows (Branson and Owen, 1970; figure 7). Land use practices such as grazing can greatly change the natural runoff and accompanying sediment load. For example, a ten-year study near Grand Junction, Colorado, demonstrated that ungrazed watersheds had 30 percent less runoff and 45 percent less sediment yield than grazed watersheds (Lusby, 1970). The extent to which peak flows are necessary for flushing action in the main stem streams to produce or maintain backwater areas for the young of endemic species is not known. Peak flows in the higher mountain streams during floods could cause high mortality to young trout such as reported by Seegrist and Gard (1972) and Hoopes (1975). In high mountain streams, reduced streamflows could result in the formation of more anchor and frazil ice that could be detrimental to the bottom organisms and force young fish from their microhabitats, thereby increasing mortality. The microhabitat available in the stream depends to a large extent on the flow. For example, reduction in streamflow could increase the area of the slow-shallow or deep-slow portions of the stream (Hill, 1965; figure 8). Also, a reduction

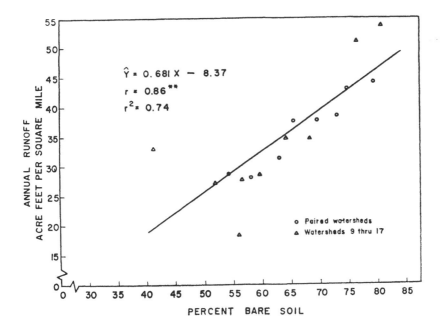

$$\hat{Y} = 0.681 X - 8.37$$
$$r = 0.86^{**}$$
$$r^2 = 0.74$$

o Paired watersheds
▲ Watersheds 9 thru 17

ANNUAL RUNOFF ACRE FEET PER SQUARE MILE

PERCENT BARE SOIL

Figure 7. Relation Between the Percentage of Bare Soil and
Average Annual Runoff for Seventeen Watersheds

Source: F. A. Branson and J. E. Owen, "Plant Cover, Runoff and
Sediment Yield Relationships on Mancos Shale in Western Colorado,"
Water Resources Research vol. 6, no. 3, 1970.

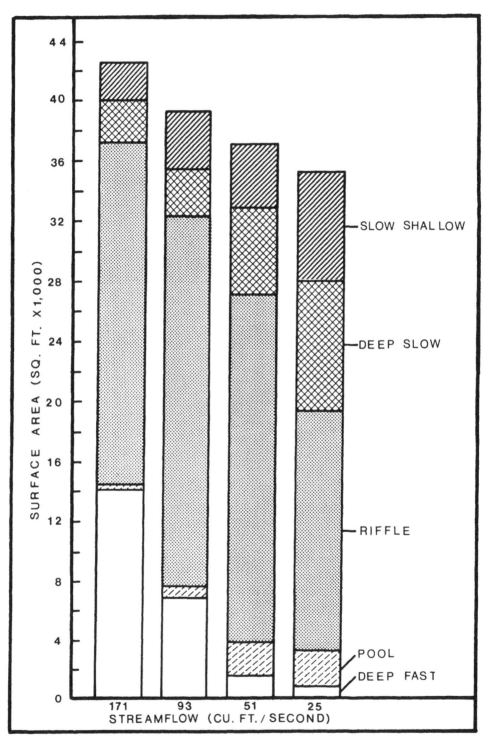

Figure 8. Type of Habitat Under Four Streamflows at One Station in the
North Channel of the White River

Source: R. R. Hill, White River Survey (Denver, Colo., Colorado
Department of Game, Fish, and Parks, Fisheries Research Division, Federal Aid
Report F-26-R-2, Job 2, 1965).

in streamflow could reduce the area of fast water or riffles. However, the physical features of the stream channel and hydraulic gradient determine some of the responses in the type of water under different flows, as shown by Hill (1965) for different reaches in the White River, Colorado.

A quick and easy method to determine flow requirements for the protection of aquatic resources in both warm-water and cold-water streams employs average flows (Tennant, 1976; table 12). This "Montana method," which is based on numerous observations and flow studies, has shown that the condition of the aquatic habitat is remarkably similar in streams of different sizes carrying the same portion of average flow. Using this method, Tennant suggests altering the flow regimen of a stream to fit different hydrologic cycles. Such an alteration could be used to favor game species such as fall-spawning brown trout. Tennant also points out that the results of the application of this method are consistent from stream to stream, and from state to state.

Many recommendations concerning streamflows have been made on rather small rivers but recommendations for larger rivers have been proposed only recently (Banks and coauthors, 1974). From studies of the Green River in Wyoming between Fontenelle Dam and Flaming Gorge Reservoir, with flows of 1,600, 800, 500, and 300 cfs, Banks and coauthors recommended 500 cfs for a winter survival flow (November through March), with an emergency short-term flow of 300 cfs for less than thirty days and a minimum production flow of 800 cfs (April through October). Also, these investigators recommended that a sediment-flushing, or habitat maintenance, flow of 800 to 1,600 cfs be used when necessary.

Table 12. Instream Flow Regimens for Fish, Wildlife, Recreation, and Related
Environmental Resources

Narrative description of flows	Fishery classification	Recommended base flow regimens (percentage of average flows)[a]	
		Oct.-Mar. :	Apr.-Sept.
Flushing or maximum	--	200	
Optimum range[b]	--	60-100	
Outstanding	I	40	60
Excellent	II	30	50
Good	III	20	40
Fair or degrading	IV	10	30
Poor or minimum	--	10	10
Severe degradation	--		

Source: From D. L. Tennant, "Instream Flow Regimens for Fish, Wildlife,
Recreation, and Related Environmental Resources," Fisheries, Bulletin of the
American Fisheries Society vol. 1, no. 4 (1976) pp. 6-10.

[a]Average flow usually fills the active stream channel approximately 1/3
full or to the line of permanent terrestrial vegetation, and 3 times the average
flow often fills the active channel nearly to the point of spilling out on the
first bench of the flood plain.

[b]Optimum is a nebulous term; however, this flow range best covers that definition for all the factors considered by Tennant.

Most requests for streamflows are for maintaining or preserving a
stream fishery, a waterfowl habitat, or a particular species (Enviro Control, Inc., 1976; table 13). In the survey conducted by Enviro Control,
Inc. of the actual effects of various recommended flows on fish and wildlife in the Rocky Mountain region, only about 20 percent resulted in improvement for fish and wildlife; 60 percent were judged to be maintenance,
and thus had no effect; and about 20 percent were detrimental (table 14).
In general, the effects of different streamflows were based on the judgment of biologists who were familiar with the streams, although some effects were determined from rigorous follow-up studies.

The responses of fish and macroinvertebrates to streamflow in the
mainstem Upper Colorado River are little known. Although the past and
probable future variations in streamflow of the Upper Colorado River have
been assessed (Brittan and coauthors, 1961), this assessment did not include fish and wildlife. Research must be directed at understanding the
effects of changes in streamflow on the ecosystem. For prediction, the
use of mathematical models implemented with computers will aid the understanding of the complexity of stream ecosystems. Although ecosystem
modeling is considered to be in its infancy, a literature review by Kadlec
(1971) contains over 600 references. However, none of the models discussed
were capable of simulating a general stream dynamic ecosystem. Sensitivity
analysis must be made to determine the factors that are most responsive to
change in any model. This kind of analysis is especially important in the
formation of models with different resolution that can be used by management agencies. Time and budget constraints (for manpower and equipment)
will dictate what data a management agency is able to collect. Finally,
mathematical models must be calibrated and then validated with empirical

Table 13. Initial and Follow-up Requests for Streamflow in the Rocky Mountain Region, by State

| State | Number of requests | | Biological targets[a] | | | | | |
| | | | Stream fishery or waterfowl habitat | | Population life cycles | | Population class or species | |
	Initial	Follow-up	Initial	Follow-up	Initial	Follow-up	Initial	Follow-up
Region (total)	52	2	88	50	4	0	13	50
Arizona	1	0	100	—	0	—	0	—
Colorado	25	0	96	—	0	—	16	—
Montana	3	0	67	—	33	—	33	—
Utah	9	1	100	100	0	—	0	0
Wyoming	14	1	86	0	7	0	14	100

Source: Enviro Control, Inc., Assessment of Effects of Altered Stream Flow Characteristics on Fish and Wildlife. Task 3: Analysis of Case Study Findings, Identification of Problems, and Recommendations of Remedies. Final Report, U.S. Fish and Wildlife Service Contract No. 14-16-0008-956 (Rockville, Md., 1976).

[a]Data presented as percentage of flow requests.

Table 14. Actual Effects of Various Flows on Fish and Wildlife in the Rocky Mountain Region, by State

State	Total no. of flows	Improvement		Maintenance		Curtailment	
		no. of flows	% of flows	no. of flows	% of flows	no. of flows	% of flows
Region (total)	52	10	19	31	60	11	21
Arizona	1	1	100	0	0	0	0
Colorado	25	0	0	18	72	7	28
Montana	3	2	67	0	0	1	33
Utah	9	3	33	5	56	1	11
Wyoming	14	4	28	8	57	2	15

Source: Enviro Control, Inc., Assessment of Effects of Altered Stream Flow Characteristics on Fish and Wildlife. Task 3: Analysis of Case Study Findings, Identification of Problems, and Recommendations of Remedies. Final Report, U.S. Fish and Wildlife Service Contract No. 14-16-0008-956 (Rockville, Md., 1976).

data if they are to serve as predictors of the responses of fish and macroinvertebrates to changes in streamflow.

Lawrence (1974), who reviewed the political and legal aspects of
streamflows in Colorado, noted that a particular deficiency regarding
minimum streamflows exists in the different laws which apply to federal,
state, and local lands. He also pointed out that federal lands and development projects fall under the Wildlife Coordination Act, and that fish
and wildlife considerations are therefore mandatory, but only on a con-
sultation basis. State and local lands have no such requirements. Others
have also recommended that fish and wildlife values be considered in assessing instream water uses (Binns, 1972; Dewsnup, 1971). These considerations
were also recommended for use of the Upper Colorado River (U.S. Department
of the Interior, Bureau of Reclamation, 1975a; U.S. Water Resources Council,
1971; Neuhold, Herrick, and Patten, 1975). Ellis (1966) reviewed the legal
considerations in regard to the recreational use of water and pointed
out that a 1937 act provided that the Colorado River Water Conservation
District maintain streamflows for fish life. Many western states purchase "conservation pools" in reservoirs to protect the sport fishery,
and some states have begun to establish minimum lake levels (e.g.,
Rhinehart, 1975). Although some effects of water level fluctuations and
minimum conservation pools on fish and other aquatic organisms are known
(Fraser, 1972a), agencies that manage fisheries in the Upper Colorado
River do not have adequate information to make such recommendations. Some
writers have expressed their ideas on the political aspects involving the
multiple use of water (e.g., Dingell, 1972); others have suggested that
standards be proposed to evaluate federal water programs (e.g., Cicchetti
and coauthors, 1973).

Recommendations for Research

The following areas for additional research represent the ideas of a
number of people and groups, such as Lord, Tubbering, and Althen (1975);
McKell (1972a, 1972b); U.S. Department of the Interior, Bureau of Reclamation (1975a); Neuhold, Herrick, and Patten (1975); Stalnaker and Arnette
(1976); Thorne (1973); Utah Water Research Laboratory and Utah Division of
Water Resources (1975); and U.S. Water Resources Council (1968, 1971):

1. Improvement of western water laws to recognize nonconsumptive
uses of water for fish, wildlife, and recreation. This study should include how existing laws and policies affect administrative decisions concerning the management of water resources and the resulting effects on
fish and wildlife, and ultimately, on society.

The U.S. Water Resources Council (1971) recommended that the high quality recreation, fish and wildlife, and the open space of the Upper Colorado
River Basin be recognized as national assets that should be preserved and
given special recognition in land- and water-use planning. Many environmental problems in the Mountain West are truly regional and the resources
are not limited to a specific political region. One recommendation was
that existing institutional arrangements in the region be analyzed to
determine if they adequately address environmental problems (Neuhold,
Herrick, and Patten, 1975). Institutional and legal arrangements may
need to be modified to provide greater flexibility for future uses of the
water and land resources (U.S. Water Resources Council, 1971). Trelease
(1976) pointed out that many current laws can be used to protect instream
uses if society insisted on their implementation by various federal and
state agencies. He also stated that a law is a mechanism for getting

things done for the purposes of society and that the law will protect and
foster the greatest values of water. Therefore, he believes that water
laws can be changed if society so dictates but also warns that institutions
such as law may be resistant to and require time for change. Administrative strategies for legislative improvement in such instances are suggested
by Jensen (1976).

2. Environmental and ecological effects of land management practices.
This research should address ways to reduce the environmental impacts of
watershed management practices on streamflow, water quality, and aesthetics,
and on game and nongame species of fish and on wildlife. The research
should include the effects of logging and grazing on forested watersheds
in the Intermountain West. Grazing and logging result in the release of
sediments to streams. The effects of sediment on the aquatic life (macroinvertebrates and fish) in this region are not known and should be determined.

Thorne (1973) suggested, as a top priority for research, determining
the effects of different levels and types of energy development on the
total environment of the Upper Colorado region, and to critically analyze
alternative strategies. The critical points for decisions on future options in land- and water-use planning must be identified and understood
(Neuhold, Herrick, and Patten, 1975).

3. Effects of salinity increases, water quality degradation, and
sediments on aquatic life. Energy development in the Upper Colorado River
Basin will affect the salinity and water quality of the river system.
Agencies that manage fish and wildlife resources do not have the necessary
information to determine what the effects of these factors will be on game
fish or on threatened and endangered species.

In the past, cost-benefit analyses of water resources projects have not adequately considered all the environmental, or social, costs and benefits. However, more recently, Krutilla and his coworkers have pioneered in this area and now many more of the nonmarket values are being incorporated in these analyses. (See, for example, Krutilla and Fisher, 1975.) The present market system does not properly weigh the short- and long-range costs and benefits that must be applied to the renewable and nonrenewable resources of the Rocky Mountain region (Neuhold, Herrick, and Patten, 1975). Next to research on the environmental impacts of energy development, Thorne (1973) recommended research on agricultural impacts in this region. Thorne emphasized that energy development poses the threat of increasing salinity loads in the Colorado River that will reduce crop production, and also that water quality will enter the political arena through national and international agreements.

4. Minimizing the effects of stream alterations. Streams are altered for various purposes, including highway construction and flood control. Some estimates are that between 30 percent and 70 percent of streams in the intermountain region have been altered by channelization. The benefits of these alterations are often derived over short periods of time, but the impacts on aquatic communities may be felt for long periods. Since energy and other development will result in the channeling of streams, management agencies must be able to predict what the impacts of this will be on fish and wildlife, and what can be done to reduce these impacts.

Streams in the intermountain area are currently being impounded to store runoff for later use. Several reports have recommended that free-flowing streams with unique and/or high qualities be examined for their suitability for inclusion in a national or state system of wild, scenic, or recreational rivers before this option is lost to the American public forever (Neuhold, Herrick, and Patten, 1975; U.S. Water Resources Countil, 1971).

5. Establishing methodologies for determining streamflow requirements for fish and macroinvertebrates. This research should include the impacts of different flows on the entire watershed, including the impacts on water quality, wildlife, and nonconsumptive recreation.

A number of comprehensive studies have recommended that the impact of streamflow on fish be included as a priority for research (Lord, Tubbering, and Althen, 1975; Neuhold, Herrick, and Patten, 1975; Stalnaker and Arnette, 1976). Stalnaker and Arnette recommended a refinement in methodology for streamflow assessment to make it applicable to large rivers. They also recommended that intensive research be conducted to determine the ecological impacts of extreme fluctuations and repetitive short-term fluctuations in streamflows, perhaps using target species of fish or invertebrates. Further, they emphasized the need for long-term research to develop and validate the relationship between streamflow and populations or production of aquatic organisms for reliable, predictive purposes.

6. Determining management strategies for manipulating fish populations (both game and undesirable species) in reservoirs. This problem is considered high on the list of research priorities for the Rocky Mountain states where impoundments and water resources are closely linked. Most of the past management failures can be attributed to a lack of quantitative knowledge of the interrelationships among fish populations, their food chain relationships, and limnology. Selection of sampling gear and the behavior of the fish species are some basic problems in providing quantitative information. Other problems include productivity of the

reservoirs and techniques for reducing or "controlling" the populations of undesirable fish species.

Numerous other recommendations for research based on various goals and objectives have been made by the persons and in the reports listed at the beginning of this section. I believe that the research priorities outlined above for fish and macroinvertebrates are the most critical ones for the Upper Colorado River Basin as they relate to potential energy development there.

Summary

I have attempted to summarize the potential impacts of alterations in streamflow and water quality on fish and macroinvertebrates in the Upper Colorado River, and to offer recommendations for research. The importance of the Colorado River was outlined, and the fish fauna was described with examples of changes that have already resulted in this fauna from the construction of impoundments as well as of problems that may result from the supply and demand for sport fishery in this region. Various examples were used to indicate the relationship between streamflow and water quality and the ecology of fish and macroinvertebrates, with examples used from the Upper Colorado River where possible. Recommendations for priority research were based on a number of comprehensive studies of the region.

Management of the Upper Colorado River Basin must be done on the basis of the goals and objectives to serve society in the entire basin. Development or alterations to this river system upstream can have an effect on fish, macroinvertebrates and people far downstream. Simons

(1976) has concisely stated that "the optimum management of river systems requires utilization of the knowledge and cooperation of all pertinent disciplines."

Development of the upper basin for economic growth and energy production must be based on sound ecological principles if fish and wildlife populations are to be maintained or enhanced.

References

Allen, K. R. 1969. "Distinctive Aspects of the Ecology of Stream Fishes: A Review," Journal of Fisheries Research Board of Canada vol. 26, no. 6, pp. 1429-1438.

Bagley, M. D., C. A. Kroll, and K. Clark. 1973. Aesthetics in Environmental Planning, U.S. Environmental Protection Agency, Socioeconomic Environmental Study Series, EPA-600/5-73-009 (Washington, D.C., EPA).

Baldes, R. J., and R. E. Vincent. 1969. "Physical Parameters of Micro-Habitats Occupied by Brown Trout in an Experimental Flume," Transactions of the American Fisheries Society vol. 98, no. 2, pp. 230-238.

Banks, R. L., J. W. Mullan, R. W. Wiley, and D. J. Dufek. 1974. The Fontenelle Green River Trout Fisheries - Considerations in Its Enhancement and Perpetuation Including Test Flow Studies of 1973 (Salt Lake City, Ut., U.S. Department of the Interior, Fish and Wildlife Service).

Beattie, B. R., E. N. Castle, W. G. Brown, and W. Griffin. 1971. Economic Consequences of Interbasin Water Transfer, Technical Bulletin 116 (Corvallis, Oreg., Oregon State University, Agricultural Experiment Station).

Bessler, M. B. 1975. "Salinity Impacts of Energy Development in Utah," Proceedings, American Water Resources Association, Utah Section, 3rd Annual Conference (Logan, Ut., Utah Water Research Laboratory, Utah State University).

Binns, N. A. 1972. "An Inventory and Evaluation of the Game and Fish Resources of the Upper Green River in Relation to Current and Proposed Water Development Program" (Ph.D dissertation, University of Wyoming, Laramie).

Bishop, A. A. 1971. "Conflicts in Water Management," 42nd Honor Lecture, The Faculty Association (Logan, Ut., Utah State University).

Bishop, A. B., M. D. Chambers, W. O. Mace, and D. W. Mills. 1975. Water as a Factor in Energy Resources Development (Logan, Ut., Utah Water Research Laboratory, Utah State University, PRJER 028-1).

Biswas, A. K., and R. W. Durie. 1971. "Sociological Aspects of Water Development," Water Resources Bulletin vol. 7, no. 6, pp. 1137-1143.

Boccardy, J. A., and W. M. Spaulding. 1968. Effects of Surface Mining on Fish and Wildlife in Appalachia, Bureau of Sport Fisheries and Wildlife, Resource Publication 65 (Washington, D.C., U.S. Department of the Interior).

Bovee, K. D. 1974. The Determination, Assessment and Design of "In-Stream Value" Studies for the Northern Great Plains Region. Final Report, EPA Contract No. 68-01-2413 (Missoula, Mont., University of Montana).

Branson, F. A., and J. E. Owen. 1970. "Plant Cover, Runoff, and Sediment Yield Relationships on Mancos Shale in Western Colorado," Water Resources Research vol. 6, no. 3, pp. 783-790.

Brittan, M. R., L. W. Crow, M. E. Garnsey, P. R. Julian, R. A. Schleusener, and V. M. Yeodjevish. 1961. Past and Probable Future Variations in Stream Flow of the Upper Colorado River (Boulder, Colo., Bureau of Economic Research, University of Colorado).

Chapman, D. W. 1966. "Food and Space as Regulators of Salmonid Populations in Streams," American Naturalist vol. 100, pp. 345-357.

Cicchetti, C. J., R. K. Davis, S. H. Hanke, and R. H. Haveman. 1973. "Evaluating Federal Water Projects: A Critique of Proposed Standards," Science vol. 181, no. 4101, pp. 723-728.

Crandall, D. L. 1974. "Management Objectives in the Colorado River Basin: The Problem of Establishing Priorities and Achieving Cooperation," in A. B. Crawford and D. F. Peterson, eds., Environmental Management of the Colorado River Basin (Logan, Ut., Utah State University Press).

Crawford, A. B., and D. F. Peterson, eds. 1974. Environmental Management of the Colorado River Basin (Logan, Ut., Utah State University Press).

Deacon, J. E., and W. G. Bradley. 1972. "Ecological Distribution of Fishes of Moapa (Muddy) River in Clark County, Nevada," Transactions of the American Fisheries Society vol. 101, no. 3, pp. 408-419.

Dendy, F. E. 1968. "Sedimentation in the Nation's Reservoirs," Soil Water Conservation vol. 23, no. 4, pp. 135-137.

Dewsnup, R. L. 1971. Legal Protection of Instream Water Values (Washington, D.C., U.S. National Water Commission, Report NWC-L-71-023).

Dingell, J. D. 1972. "Political Aspects of Multiple Use," in R. T. Oglesby, C. A. Carlson, and J. A. McCann, eds., River Ecology and Man (New York, Academic Press).

Edmundson, E., F. E. Everest, and D. W. Chapman. 1968. "Permanence of Station in Juvenile Chinook Salmon and Steelhead Trout," Journal of Fisheries Research Board of Canada vol. 25, no. 7, pp. 1453-1464.

Ellis, W. H. 1966. "Watercourses--Recreational Uses for Water Under Prior Appropriation Law," Natural Resources Journal vol. 6, no. 2, pp. 181-185.

Elser, A. A. 1968. "Fish Populations of a Trout Stream in Relation to Major Habitat Zones and Channel Alterations," Transactions of the American Fisheries Society vol. 97, no. 4, pp. 389-397.

Endangered Species Act. 1973. Public Law No. 93-205, 87 U.S. Stat. 884.

Energy Policy Project. 1974. Exploring Energy Choices (Washington, D.C., Energy Policy Project of the Ford Foundation).

Enviro Control, Inc. 1976. Assessment of Effects of Altered Stream Flow Characteristics on Fish and Wildlife. Final Report, U.S. Fish and Wildlife Service Contract No. 14-16-0008-956 (Rockville, Md.).

Etnier, D. A. 1972. "The Effect of Annual Rechanneling on a Stream Fish Population," Transactions of the American Fisheries Society vol. 101, no. 1, pp. 372-375.

Everett, A. G. 1970. "Comprehensive Water Resources Management: The Past and Future," Water Resources Bulletin vol. 7, no. 1, pp. 185-188.

Fitzsimmons, S. J., and O. A. Salama. 1973. A Social Report--Man and Water (Cambridge, Mass., Abt Associates, Inc.).

Fraser, J. C. 1972a. Water Levels, Fluctuations, and Minimum Pools in Reservoirs for Fish and Other Aquatic Resources: An Annotated Bibliography, Food and Agricultural Organization, Fisheries Technical Paper No. 113 (New York, United Nations).

_____. 1972b. Regulated Stream Discharge for Fish and Other Aquatic Resources: An Annotated Bibliography, Food and Agricultural Organization, Fisheries Technical Paper No. 112 (New York, United Nations).

Funk, W. H., and A. R. Gaufin. 1971. "Phytoplankton Productivity in a Wyoming Cooling-Water Reservoir," in G. E. Hall, ed., Reservoir Fisheries and Limnology (Washington, D.C., American Fisheries Society, Special Publication No. 8).

Gaufin, A. R. 1973. Water Quality Requirements of Aquatic Insects (Washington, D.C., U.S. Environmental Protection Agency, Ecological Research Series EPA-660/B-73-004).

Gebhards, S. 1975. "Wild Trout: Not by a Damsite," in W. King, ed., Wild Trout Management (Denver, Colo., Trout Unlimited, Inc.).

Gibbons, D. R., and E. D. Salo. 1973. An Annotated Bibliography of the Effects of Logging on Fish of the Western United States and Canada, Forest Service General Technical Report PNW-10 (Washington, D.C., U.S. Department of Agriculture).

Giger, R. D. 1973. Streamflow Requirements of Salmonids. Final Report, Anadromous Fish Project AFS-62-1 (Portland, Oreg., Oregon Wildlife Commission).

Goldman, C. R. 1973. "Environmental Impact and Water Development," in C. R. Goldman, J. McEvoy, III, and P. J. Richerson, eds., Environmental Quality and Water Development (San Francisco, W. H. Freeman).

Gould, H. R. 1954. _Sedimentology. Lake Mead Comprehensive Survey,_ _1948-49_ (Reston, Va., U.S. Geological Survey) pp. 211-265.

Gunderson, D. R. 1968. "Floodplain Use Related to Stream Morphology and Fish Populations," _Journal of Wildlife Management_ vol. 32,
no. 3, pp. 507-514.

Hall, J. D., and R. L. Lantz. 1969. "Effects of Logging on the Habitat
of Coho Salmon and Cutthroat Trout in Coastal Streams," in
T. G. Northcote, ed., _Symposium on Salmon and Trout in Streams,_
H. R. Macmillan Lectures on Fish (Vancouver, University of British
Columbia).

Hansen, E. A., and G. R. Alexander. 1976. "Effect of Artificially
Increased Sand Bedload in Stream Morphology and Its Implications
on Fish Habitat," _Proceedings of the Third Federal Inter-Agency_
Sedimentation Conference, 1976 (Washington, D.C., U.S. Water Resources
Council, Sedimentation Committee) part 3, pp. 65-76.

Heimstra, N. W., D. K. Kamkot, and N. G. Benson. 1969. _Some Effects of_
Silt Turbidity on Behavior of Juvenile Largemouth Bass and Green
Sunfish, Bureau of Sport Fisheries and Wildlife, Technical Paper
No. 20 (Washington, D.C., U.S. Department of the Interior).

Hickman, G. L. 1975. "Incorporating the Environmental Quality Dimension
in Planning River Management," _Transactions of the North American_
Wildlife Natural Resources Conference vol. 40, pp. 264-272.

Hill, R. R. 1965. _White River Survey_ (Denver, Colo., Colorado Department
of Game, Fish, and Parks, Fisheries Research Division, Federal Aid
Report F-26-R-2, Job 2).

Hoar, W. S., and D. J. Randall, eds. 1969. _Fish Physiology._ Vol. I.
Excretion, Ionic Regulation, and Metabolism (New York, Academic Press).

_____. 1970. _Fish Physiology._ Vol. IV. The Nervous System,
Circulation, and Respiration (New York, Academic Press).

Holden, P. B., and C. B. Stalnaker. 1975. "Distribution and Abundance of
Mainstream Fishes of the Middle and Upper Colorado River Basins,
1967-1973," _Transactions of the American Fisheries Society_ vol. 104,
no. 2, pp. 217-231.

Hooper, D. R. 1973. _Evaluation of the Effects of Flows on Trout Stream_
Ecology (Emeryville, Calif., Pacific Gas and Electric Co., Department
of Engineering Research).

Hoopes, R. L. 1975. "Flooding as a Result of Hurricane Agnes and Its
Effect on a Native Brook Trout Population in an Infertile Headwater
Stream in Central Pennsylvania," _Transactions of the American_
Fisheries Society vol. 104, no. 1, pp. 96-99.

141

Hynes, H. N. 1970. The Ecology of Running Waters (Toronto, University of Toronto Press).

Jensen, D. W. 1976. "Administrative Strategies for Satisfying Instream Flow Needs," in J. F. Orsborn and C. H. Allman, eds., Proceedings, Symposium and Speciality Conference, Instream Flow Needs, Vol. 1 (Washington, D.C., American Fisheries Society).

Joint Committee on Atomic Energy. 1973. Understanding the National Energy Dilemma (Washington, D.C., The Center for Strategic and International Studies, Georgetown University).

Kadlec, J. A. 1971. A Partial Annotated Bibliography of Mathematical Models in Ecology, U.S. International Biological Program, Analysis of Ecosystems (Ann Arbor, Mich., University of Michigan, School of Natural Resources).

Kennedy, H. D. 1967. Seasonal Abundance of Aquatic Invertebrates and Their Utilization by Hatchery-Reared Rainbow Trout, Bureau of Sport Fisheries and Wildlife, Technical Paper 12 (Washington, D.C. U.S. Department of the Interior).

Kindswater, C. E., ed. 1964. Organization and Methodology for River Basin Planning (Atlanta, Ga., Water Resources Center, Georgia Institute of Technology).

Knight, A. W. 1965. "Studies on the Stoneflies (Plecoptera) of the Gunnison River Drainage in Colorado" (Ph.D. dissertation, University of Utah, Salt Lake City).

Krutilla, J. V., and A. Fisher. 1975. The Economics of Natural Environments: Studies in the Valuation of Commodity and Amenity Resources (Baltimore, Johns Hopkins University Press for Resources for the Future).

Lawrence, K. 1974. "The Political and Legal Aspects of Streamflows in Colorado" (Boulder, Colo., Western Interstate Commission for Higher Education). Multilith.

Lewis, S. L. 1969. "Physical Factors Influencing Fish Populations in Pools of a Trout Stream," Transactions of the American Fisheries Society vol. 98, no. 1, pp. 14-19.

Lord, W. B., S. R. Tubbering, and C. Althen. 1975. Fish and Wildlife Implications of Upper Missouri Basin Water Allocation, Program on Technology, Environment, and Man, Monograph No. 22 (Boulder, Colo., Institute of Behavioral Science, University of Colorado).

Luce, C. F. 1973. Water Policies for the Future (Port Washington, N.Y., Water Information Center, Inc.).

Lusby, G. C. 1970. "Hydrologic and Biotic Effects of Grazing vs. Non-Grazing Near Grand Junction, Colorado," Journal of Range Management vol. 23, no. 4, pp. 256-260.

McCloskey, M. 1973. "Alternatives in Water Project Planning: Ecological and Environmental Considerations," in C. R. Goldman, J. McEvoy, III, and P. J. Richerson, eds., Environmental Quality and Water Development (San Francisco, W. H. Freeman).

McKell, C. M. 1972a. Intermountain Universities' Conference on Policy Formulation in the Development of Energy Resources: Proceedings (Logan, Ut., Utah State University, Environment and Man Program).

_____. 1972b. Intermountain Universities' Conference on Policy Formulation in the Development of Energy Resources: Research Needs Statements (Logan, Ut., Utah State University, Environment and Man Program).

McKim, J. M., D. A. Benoit, K. E. Biesinger, W. A. Brungs, and R. E. Siefert. 1975. "Effects of Pollution on Freshwater Fish," Journal of Water Pollution Control Federation vol. 47, no. 6, pp. 1711-1768.

Mihursky, J. A., and L. E. Cronin. 1973. "Balancing Needs of Fisheries and Energy Production," Transactions of the North American Wildlife Natural Resources Conference vol. 38, pp. 459-476.

Miller, R. R. 1959. "Origin and Affinities of the Freshwater Fish Fauna of Western North America," in C. L. Hubbs, ed., Zoogeography (Washington, D.C., American Association for the Advancement of Science, Publication 41).

Minckley, W. L., and J. E. Deacon. 1968. "Southwestern Fishes and the Enigma of 'Endangered Species,'" Science vol. 159, pp. 1424-1432.

Mullan, J. W. 1974. "Impoundments and Their Effects on Aquatic Resources," Proceedings of the Annual Conference of the Western Association of State Game and Fish Commission vol. 56, pp. 367-380. Copies can be obtained from Secretary, Western Association of State Game and Fish Commissioners, 600 South Walnut Street, Box 25, Boise, Idaho 83707.

National Research Council. 1968. Water and Choice in the Colorado River Basin (Washington, D.C., National Academy of Sciences).

Neel, J. K. 1963. "Impact of Reservoirs," in D. G. Frey, ed., Limnology in North America (Madison, Wisc., University of Wisconsin Press).

Neuhold, J. M., D. E. Herrick, and D. I. Patten, eds. 1975. Rocky Mountain Environmental Research--Problems and Research Priorities in the Rocky Mountain Region. Final Report to the National Science Foundation (RANN), U.S. Forest Service, U.S. Environmental Protection Agency, and Ecology Center, Utah State University (Logan Ut., Utah State University).

Nichols, S. R., G. V. Skogerboe, and R. C. Ward. 1972. Water Quality Management Decisions in Colorado (Fort Collins, Colo., Colorado State University, Environmental Resources Center, AER 71-725RN-GVS-RCW8).

Ohmart, R. D., W. O. Deason, and S. J. Freeland. 1975. "Dynamics of Marsh Land Formation and Succession Along the Lower Colorado River and Their Importance and Management Problems as Related to Wildlife in the Arid Southwest," Transactions of the North American Wildlife Natural Resources Conference vol. 40, pp. 240-251.

Orsborn, J. F., and C. H. Allman, eds. 1976. Proceedings of the Symposium and Speciality Conference on Instream Flow Needs (Washington, D.C., American Fisheries Society).

Pacific Northwest River Basins Commission. 1972. Instream Flow Requirement Workshop. A Transcript of Proceedings (Vancouver, Wash.).

Pearson, L. S., K. R. Conover, and R. E. Sams. 1970. "Factors Affecting the Natural Rearing of Juvenile Coho Salmon During the Summer Low Flow Season" (Portland, Oreg., Fish Commission of Oregon). Unpublished manuscript.

Pearson, W. D. 1967. "Distribution of Macro-Invertebrates in the Green River Below Flaming Gorge Dam, 1963-1965" (M.S. thesis, Utah State University, Logan).

Pearson, W. D., and D. R. Franklin. 1968. "Some Factors Affecting Drift Rates of Baetis and Simuliidae in a Large River," Ecology vol. 49, no. 1, pp. 75-81.

Pearson, W. D., R. H. Kramer, and D. R. Franklin. 1968. "Macroinvertebrates in the Green River Below Flaming Gorge Dam, 1964-65 and 1967," Utah Academy of Sciences, Arts and Letters, Proceedings vol. 45, pp. 148-167.

Peters, J. C., and W. Alvord. 1964. "Man-Made Channel Alterations in Thirteen Montana Streams and Rivers," Transactions of the North American Wildlife and Natural Resources Conference vol. 29, pp. 93-102.

Pierce, J. C. 1976. "Participation and Representation in Water Resources Politics," in J. F. Orsborn and C. H. Allman, eds., Proceedings, Symposium and Speciality Conference, Instream Flow Needs, Vol. 1 (Washington, D.C., American Fisheries Society).

Platts, W. S. 1958. "The Natural Reproduction of the Cutthroat Trout, Salmo clarki Richardson, in Strawberry Reservoir, Utah" (M.S. thesis, Utah State University, Logan).

Platts, W. S., and W. F. Megahan. 1975. "Time Trends in Riverbed and Sediment Composition in Salmon and Steelhead Spawning Areas: South Fork Salmon River, Idaho," Transactions of the North American Wildlife Natural Resources Conference vol. 40, pp. 229-289.

Powell, G. C. 1958. "Evaluation of the Effects of a Power Dam Water Release Pattern Upon the Downstream Fishery" (M.S. thesis, Colorado State University, Fort Collins).

Reynolds, P. J., and A. K. Biswas. 1969. "Socio-Economic Simulation for Water Resource System Planning," Congress of the International Association of Hydraulics Research, Proceedings vol. 13, pp. 75-82.

Rhinehart, C. G. 1975. Minimum Stream Flows and Lake Levels in Colorado (Fort Collins, Colo., Colorado State University, Environmental Resources Center, Information Series No. 18).

Richardson, W. M. 1976. "Technical Committee Minutes" (Las Vegas, Nev., Colorado River Wildlife Council) pp. 12-18.

Seegrist, D. W., and R. Gard. 1972. "Effects of Floods on Trout in Sagehen Creek, California," Transactions of the American Fisheries Society vol. 101, no. 3, pp. 478-482.

Shira, D. L. 1976. "Water Resource Planning and Instream Flow Needs--A Reclamation Viewpoint," in J. F. Orsborn and C. H. Allman, eds., Proceedings, Symposium and Speciality Conference, Instream Flow Needs, Vol. 1 (Washington, D.C., American Fisheries Society).

Sigler, W. F., W. T. Helm, J. W. Angelovic, D. W. Linn, and S. S. Martin. 1966. The Effects of Uranium Mill Wastes on Stream Biota, Utah Agricultural Experiment Station Bulletin 462 (Logan, Ut., Utah State University).

Simons, D. B. 1976. "Management of River Systems," in J. F. Orsborn and C. H. Allman, eds., Proceedings, Symposium and Speciality Conference, Instream Flow Needs, Vol. 1 (Washington, D.C., American Fisheries Society).

Smith, A. K. 1973. "Development and Application of Spawning Velocity and Depth Criteria for Oregon Salmonids," Transactions of the American Fisheries Society vol. 102, no. 2, pp. 312-316.

Smith, M. H., H. O. Hillestead, M. V. Manlove, and R. L. Marchinton. 1976. "Use of Population Genetics Data for the Management of Fish and Wildlife Populations," Transactions of the North American Wildlife and Natural Resources Conference vol. 41, pp. 119-133.

Sprules, W. M. 1947. An Ecological Investigation of Stream Insects in Algonquin Park, Ontario (Toronto, University of Toronto, Biology Series No. 56).

Stalnaker, C. B., and J. L. Arnette. 1976. Methodologies for the Determination of Stream Resource Flow Requirements: An Assessment (Washington, D.C., U.S. Department of the Interior, Fish and Wildlife Service, Office of Biological Services).

Surber, E. W. 1951. "Bottom Fauna and Temperature Conditions in Relation to Trout Management in St. Mary's River, Augusta County, Virginia," Virginia Journal of Science vol. 2, pp. 190-202.

Symons, J. M., S. R. Weibel, and G. G. Robeck. 1964. _Influence of Impound-ments on Water Quality: A Review of Literature and Statement of Research Needs_, Public Health Service Publication No. 999-WP-18 (Washington, D.C., U.S. Department of Health, Education, and Welfare).

Tennant, D. L. 1976. "Instream Flow Regimens for Fish, Wildlife, Recreation, and Related Environmental Resources," _Fisheries, Bulletin of the American Fisheries Society_ vol. 1, no. 4, pp. 6-10.

Thorne, D. W. 1973. "The Development and Environmental Protection of the Colorado River Basin," in A. B. Crawford and D. F. Peterson, eds., _Environmental Management of the Colorado River Basin_ (Logan, Ut., Utah State University).

Trelease, F. J. 1976. "The Legal Basis for Instream Flows," in J. F. Orsborn and C. H. Allman, eds., _Proceedings, Symposium and Speciality Conference, Instream Flow Needs_, Vol. 2 (Washington, D.C., American Fisheries Society).

Tsivoglov, E. C., S. D. Schearer, R. M. Shaw, Jr., J. D. Jones, J. B. Anderson, C. E. Sponagles, and D. A. Clark. 1959. _Survey of Interstate Pollution of the Animas River, Colorado-New Mexico_ (Cincinnati, U.S. Department of Health, Education, and Welfare, Public Health Service, Robert A. Taft Engineering Center).

U.S. Department of the Interior, Bureau of Reclamation. 1975a. _Westwide Study Report on Critical Water Problems Facing the Eleven Western States_ (Washington, D.C.).

_____. 1975b. _Environmental Assessment of Proposed Penstock Intake Modifications, Flaming Gorge Dam, Utah_ (Salt Lake City, Ut., Bureau of Reclamation, Upper Colorado Region).

U.S. Department of the Interior, Bureau of Sport Fisheries and Wildlife. 1972. _1970 National Survey of Fishing and Hunting_, Resource Publication 95. (Washington, D.C.).

U.S. Department of the Interior, Fish and Wildlife Service. 1976. "Endangered and Threatened Wildlife and Plants," _Federal Register_ vol. 41, no. 191 (Thursday, 30 September) pp. 43341-43358.

U.S. Department of the Interior, Water for Energy Management Team. 1974. _Water for Energy in the Upper Colorado River Basin_ (Washington, D.C.).

U.S. Environmental Protection Agency. 1971. _The Mineral Quality Problem in the Colorado River_ (Washington, D.C.).

U.S. Water Resources Council. 1968. _The Nation's Water Resources_ (Washington, D.C.).

U.S. Water Resources Council, Upper Colorado Region State-Federal Inter-Agency Group/Pacific Southwest Inter-Agency Committee. 1971. Upper Colorado Region Comprehensive Framework Study, Main Report (Washington, D.C.).

Utah Water Research Laboratory. 1975. Colorado River Regional Assessment Study. Report to the National Committee on Water Quality (Logan, Ut., Utah State University).

Utah Water Research Laboratory and Utah Division of Water Resources. 1975. "Impacts of Energy Development on Utah Water Resources," Proceedings of Third Annual Conference (Logan Ut., American Water Resources Association, Utah Section).

Vanicek, C. D., and R. H. Kramer. 1969. "Life History of the Colorado Squawfish, Ptychocheilus lucius, and the Colorado Chub, Gila robusta, in Dinosaur National Monument, 1964-1966," Transactions of the American Fisheries Society vol. 98, no. 2, pp. 193-208.

Vanicek, C. D., R. H. Kramer, and D. R. Franklin. 1970. "Distribution of Green River Fishes in Utah and Colorado Following Closure of Flaming Gorge Dam," Southwestern Naturalist vol. 14, no. 3, pp. 297-315.

Varley, J. D., A. F. Regenthal, and R. W. Wiley. 1971. "Growth of Rainbow Trout in Flaming Gorge Reservoir During the First Six Years of Impoundment," in G. E. Hall, ed., Reservoir Fisheries and Limnology (Washington, D.C., American Fisheries Society, Special Publication No. 8)

Ward, J. V. 1976. "Effects of Flow Patterns Below Large Dams on Stream Benthos: A Review," in J. F. Orsborn and C. H. Allman, eds., Proceedings, Symposium and Speciality Conference, Instream Flow Needs, Vol. 2 (Washington, D.C., American Fisheries Society).

Waters, T. F. 1969. "Invertebrate Drift-Ecology and Significance to Stream Fishes," in T. G. Northcote, ed., Symposium on Salmon and Trout Streams, H. R. MacMillan Lectures on Fish (Vancouver, University of British Columbia).

Weatherford, G. D., and G. C. Jacoby. 1975. "Impact of Energy Development on the Law of the Colorado River," Natural Resources Journal vol. 15, no. 1, pp. 171-213.

Weber, D. T. 1959. "Effects of Reduced Stream Flows on the Trout Fishery Below Granby Dam, Colorado" (M.S. thesis, Colorado State University, Fort Collins).

Wesche, T. A. 1973. _Parametric Determination of Minimum Stream Flow for Trout_ (Laramie, Wyo., Water Resources Research Institute, University of Wyoming).

Whittlesay, N. K. 1976. "Irrigation: The Competition with Instream Uses of Water," in J. F. Orsborn and C. H. Allman, eds., **Proceedings, Symposium and Speciality Conference, Instream Flow Needs**, Vol. 1 (Washington, D.C., American Fisheries Society).

Wiley, R. W., and J. W. Mullan. 1975. "Philosophy and Management of the Fontenelle Green River Tailwater Trout Fishery," in W. King, ed., _Wild Trout Management_ (Denver, Colo., Trout Unlimited, Inc.).

Williams, J. S. 1975. _The Natural Salinity of the Colorado River_ (Logan, Ut., Utah Water Research Laboratory, Utah State University).

Workman, J. P., and J. E. Keith. 1974. "Economic Aspects of Erosion Control Practices in the Upper Colorado River Basin," **Proceedings, Utah Academy of Sciences, Arts, and Letters** vol. 51, part 1, pp. 102-108.

Wydoski, R. S., K. Gilbert, K. Seethaler, and C. W. McAda. 1976. "Annotated Bibliography for Aquatic Resource Management of the Upper Colorado River Ecosystem" (Logan, Ut., Utah Cooperative Fishery Research Unit, Utah State University). **Multilith**.

POTENTIAL IMPACTS OF ALTERATIONS IN STREAMFLOW
AND WATER QUALITY ON FISH AND MACROINVERTEBRATES
IN THE UPPER COLORADO RIVER BASIN: A DISCUSSION

Donald L. Tennant*

Richard Wydoski has certainly provided an exhaustive review of the
published information on streamflows as they relate to invertebrates,
fishes, and other environmental resources in the Colorado River Basin,
and particularly in the upper basin. Perhaps I could best comment by
saying "Amen" and sitting down. But in the interest of adding some light
to, and maybe even generating some heat from, this subject, I offer the
following observations and comments.

The Western U.S. Water Plan was cited as one of the broad general
streamflow planning efforts (U.S. Department of the Interior, Bureau of
Reclamation, 1975). The Westwide Study Report contained the following
statement: "In the absence of detailed study, a low flow approaching
the historic low flow of record has sometimes been considered as a suitable minimum flow for fishery and related purposes" (U.S. Department of
the Interior, 1973; U.S. Department of the Interior, Bureau of Reclamation, 1975). Other studies, especially those supported by the Federal
Water Pollution Control Administration (FWPCA) and the U.S. Environmental
Protection Agency (EPA), suggest minimum flows equal to the three- or
seven-day low flow of record (Missouri Basin Inter-Agency Committee, 1971a,
1971b). To me these recommendations are all unthinkable and well down in
the range of severe degradation for most elements of our aquatic environments. In my opinion, this is akin to presenting a personal all-time

_____

*U.S. Fish and Wildlife Service, Billings, Montana.

worst health condition as a recommended level for a portion of one's future well-being. This represents one of the real dangers that can result from broad, comprehensive or Level B-type studies.[1]

Richard Wydoski cites many drastic effects that impounding reservoirs and streamflows released through the outlets of dams have on our natural riverine environments. Again and again he stressed the need for definitive data to assess the direct impacts on aquatic life forms of the resulting physical, chemical, and biological changes. I agree with Dr. Wydoski that this is our cardinal need. We must be able to tie the plethora of information we already have, and are continuing to collect, on streamflow quality and quantity directly to the needs of living organisms. This is a very difficult and monumental task. Perhaps computers can assist us here. By storing, sorting, correlating, and retrieving this kind of information, computers may be our biggest aid and turn a near-impossible task into reality. Computers will not solve all our problems, but they may handle our data in an efficient and expedient manner so that the answers we seek may become apparent.

We should not rely solely on computer technology to understand the relationship between streamflows and fish populations. Desirable streamflows for fish can be determined by examining various flow regimens in segments of stream channels for different periods of time, and by monitoring the effects of different streamflows on various aquatic organisms as well as the environment. The data may not tell us how or why the organisms

---

[1]A Level B-type study is defined as a reconnaissance type study of a region or large area that may include many individual water development projects.

died or thrived, but the critters themselves will reveal how well they did, which may be most important. Sometimes we get caught up with the fascination of wanting to know why, and we are often reluctant to accept and use results that do not answer this most difficult of all questions. I suggest, especially for management purposes, that we first assess and calibrate streamflow methodologies whenever and however we can, and then spend the time necessary to answer the specific questions regarding why these relationships and techniques are good or bad for various organisms.

Several problems of sampling large rivers were mentioned by Wydoski. But this situation is improving rapidly. Continued developments in electro-fishing gear, and the recent use of drift nets has added greatly to knowledge about species from trout to sturgeon in Montana's larger streams.

Much attention has been focused on the problems of low flows. But the literature I have read on flood flows also alludes to some very damaging or degrading effects to the aquatic environment. Some of the worst damage I have ever seen in lotic environments was caused by too much water. I recall the cemented substrate of the Delaware River in the vicinity of the Tocks Island Dam after a big hurricane hit the east coast in the mid-1950s; the flushed out stream channels all along the east slope of the Rockies in north-central Montana after the storms of both 1964 and 1975; the gouged up substrate of the upper Yellowstone River after flood waters combined with ice jams had ploughed the channel; and finally the tightly cemented bottom of the Bighorn River below the newly erected Yellowtail Dam after the 1967 flood. It took natural processes five to ten years' time to loosen up, or uncement, the substrates of both the Delaware and Bighorn rivers. I believe these areas were restored to productivity primarily

through the burrowing of benthic invertebrates rather than as a result of
additional flood flows. We need to know for sure to what extent flood or
flushing flows are necessary, because they also require a lot of water.
A one-month flushing flow equal to twice the average annual flow requires
enough water to provide a good base flow equal to 25 percent of the average
flow, for eight months.[2] Many of the chalk streams of England, limestone
creeks of eastern North America, the sandhill-flinthill streams of the
Midwest, and the spring creeks here in the West do not experience annual
floods of several times their average flows. Yet they are some of our most
productive and classic trout streams. Large releases of water from a dam
often result in accelerated erosion due to the ability clear water has for
cutting streambanks and transporting sediment and bedload. Hatchery and
spawning channel managers never run a big flushing flow through the race-
ways to enhance production. It appears that flushing flows are damaging
to aquatic wildlife and we need to learn a lot more about them.

Peaking power operations and other types of regulation at dams that
result in quick, drastic changes in the amount and stage of the streamflow
downstream often cause catastrophic losses to aquatic fauna. In one of
our flow studies on the Missouri River below Holter Dam in Montana, we
found dead and dying organisms ranging from invertebrates to adult carp
in windrows. From this and previous similar experiences, we concluded that
releases from dams should not exceed river stage changes of 6 inches in 6
hours.

---

[2]A one-month flow equal to twice the average annual flow is required
for flushing.

So much attention is being given to the problem of stream channelization that it warrants a brief comment. Currently, we receive about a dozen Section 10 and Section 404 permit requests each week through the U.S. Army Corps of Engineers program for channelization, rip-rap, dikes, levees and other work on the stream channels in Montana and Wyoming. We expect these requests to increase significantly with Phase II of the 404 program now being implemented. The cumulative effects of this type work may drastically change most of our streams as we know them today.

Richard Wydoski cites several references about the damaging effects of heavy metals and salinity. We have also experienced these problems in Montana. On a water development project in the Boulder River watershed, the presence of fines and bedload material containing high concentrations of heavy metals caused substantial concern. These contaminants were finding their way into the main Boulder River from several old mining operations in the basin. Nobody seems to know just if, or how, this material may further degrade the aquatic ecosystem along with the altered streamflows that will occur if and when the project is constructed. The same concerns are associated with irrigation projects, when return flows are likely to have a much higher salinity concentration than the receiving stream.

In summary, a good, practical, systems approach to stream management needs to be developed that will consider all the problems that Richard Wydoski and I have discussed. We especially need to be able to predict the impact of incremental, physical and chemical changes in the riverine environment on the standing crops or annual production of the organisms living in the lotic environment. Given this information, we are in a

better position to make sound specific recommendations regarding changes

to the natural environment by man.

References

Missouri Basin Inter-Agency Committee. 1971a. "Fish and Wildlife Tentative Needs and Problems: Yellowstone River Basin," Comprehensive Framework Study, Missouri River Basin (Washington, D.C., U.S. Department of the Interior).

_____. 1971b. "Fish and Wildlife Tentative Needs and Problems: Upper Missouri River Tributaries Sub-basin," Comprehensive Framework Study, Missouri River Basin (Washington, D.C., U.S. Department of the Interior).

U.S. Department of the Interior. 1973. Western United States Water Plan. Reports of the state study teams for Montana and Wyoming (Washington, D.C.).

U.S. Department of the Interior, Bureau of Reclamation. 1975. Westwide Study Report on Critical Water Problems Facing the Eleven Western States (Washington, D.C.).

POTENTIAL IMPACT OF ALTERATIONS IN STREAMFLOW
AND WATER QUALITY ON FISH AND MACROINVERTEBRATES
IN THE UPPER COLORADO RIVER BASIN: A DISCUSSION

Carl B. Schreck*

Richard Wydoski has broadly outlined the potential impacts of alterations of streamflows and water quality on fish and macroinvertebrates in
the Upper Colorado River Basin. He stated the basic management problem
as one of balancing social and economic needs with the maintenance of a
"quality" environment.

A variety of factors that affect consideration of fish, macroinvertebrates, and the fisheries of the upper basin were mentioned. These
include: (1) the existence of rare and endangered species; (2) direct and
indirect competition of exotic (introduced) fish with the endemic (native)
forms; (3) creation of reservoirs which alter thermal regimes, create new
habitat, alter streamflows, and change siltation and turbidity patterns;
(4) channelization; (5) shifts in food chains; (6) alterations in water
quality; and (7) creation of different types of fisheries, and the impact
on traditional fisheries in the upper basin.

The habitat of both fish and macroinvertebrates can be altered
through habitat destruction and changes in water quality. Wydoski discussed potential effects on fish and macroinvertebrates of sedimentation,
water loss, increased concentrations of heavy metals, decreased pH, release of radioactive materials, increased or decreased temperature and

*Oregon Cooperative Fishery Research Unit, Department of Fisheries
and Wildlife, Oregon State University, Corvallis, Oregon.

its subsequent impacts on diurnal rhythms, reduced dissolved oxygen levels, and finally, elevated salinity levels which he considered to be of paramount importance. The paper further pointed out that habitat factors can be modified. In addition, the author notes that instream flows are most often managed for preservation or maintenance, but that minimum flow requirements for endemic fishes have not yet been established. The impacts on fish and macroinvertebrates of the natural flushing action of peak flows, and more generally, the hydraulics of river systems, are also often not considered in water resources planning. Comprehensive land use planning is necessary if society's economic and social goals, on the one hand, are to be balanced with a "quality" environment, on the other.

Several recommendations for research were made at the end of the paper. These included changes in, and improvement of, laws concerning western water, and determination of how these laws affect decision making and social outcomes. Models need to be developed that can predict the impacts of channelization, altered flows, and minimum instream flows on fish and macroinvertebrates. The effects of increased salinity on endemic fishes need to be determined. Current land management practices should be evaluated, and their effects on the aquatic system--primarily via sedimentation--elucidated. Means for manipulating fish populations for fisheries management purposes need to be established.

If the paper has a shortcoming, it would be in the lack of analytical structure for organizing the information and demonstration of how this information can be used to better manage the fish and wildlife resources of the upper basin. In my opinion, one of the prime problems facing fish and wildlife managers is the maintenance of habitat and water quality so that fitness, in the broad sense, of endemic fish populations will not be reduced either directly or indirectly through shifts in macroinvertebrate production. Prevention of this problem entails the establishment of reasonable and attainable goals and objectives.

To help give both the needs of society and the biota their proper consideration and priorities, tactical planning (Phenicie and Lyons, 1973) should be used to aid in the formulation of management and research strategies. Such planning should involve development of step-down plans to clearly establish (1) the goals and objectives for which the Colorado River system should be managed, and (2) the information needed for management decisions on alternatives directed toward attainment of these goals and objectives. It is meaningless to outline specific areas of research before overall management goals and objectives concerning the fish and fisheries of the Colorado River system are established. Implied in this is the establishment of standards for many attributes of the system such as water quality and flow, and human use. I am not criticizing Wydoski's paper, for his is an overview of what is known already. But I wonder if this information is known in the context of meeting the management objectives. That is, specifically how is this information going to help the river basin "manager"? What is the value to the river basin "manager" of the information presented by the studies that Wydoski reviewed?

For directing thinking toward the ordering of research objectives the "Step-down Plan on Stream Alteration--Channelization," developed in 1974 by the Division of Population Ecology Research and others in the Division of Ecological Services of the U.S. Fish and Wildlife Service, coordinated by Dr. J. F. Watson, could be of value. The goal of this plan is to "develop methods and/or models to predict the impact of a proposed stream alteration project upon a watershed ecosystem and its associated fish and wildlife resources."

Once the research objectives have been established, with their priorities, sensitivity analyses need to be undertaken. These analyses should not only be used to increase the predictive power of models, but also to assist in developing sound experimental design. Sensitivity analyses should ensure that data collection leads to more accurate information for management purposes, within necessary confidence limits, so that sound decisions can be made.

Richard Wydoski has addressed several key issues. Elaboration on a few of these may be of value to the reader. First, it would continue to be a useful exercise to try to develop a common denominator for the value to be attached to different habitats and species; to the energy produced; to the water stored; and to recreation. To assess these different uses of the same resource, management needs some sort of workable accounting system. Such a system would, as a minimum, keep open communication among resource scientists, engineers, and social scientists.

Consideration of water quality should not neglect potential synergisms between various environmental factors. The relationship between hardness and the toxicity of a variety of compounds has been well established, but the role of temperature, pH, and dissolved gases needs further evaluation for their part in affecting the toxicity of other toxicants. Pesticides also should not be neglected as a potential problem. The quality of groundwater needs to be considered as well as that of surface waters. The lowering of the water table has obvious impacts, but contamination of subterranean water with toxicants such as pesticides, radioactive materials, and heavy metals could have far-reaching consequences.

Preservation of endemic genotypes is one of the more important issues addressed by Wydoski. Impacts on gene pools can be more subtle than extinction of a species. It might be of interest to note that many stream-inhabiting species are migratory or have both mobile and sedentary components to their populations (Funk, 1955; Gunning and Shoop, 1964). Creation of dams, impoundments, unnatural thermal conditions, and siltation could disrupt this natural phenomenon and perhaps serve as a selective agent for either mobile or sedentary individuals.

Maintenance of genotypes implies maintenance of the heterozygosity (variability) of a species. Altered thermal regimes could be a selective factor in species with variable forms of the same protein-type (polymorphic biochemical genotypes). The desert sucker (Catostomus clarki) in the lower Colorado River system has various forms of serum esterase protein. The frequency and activity of one allele (one form of a gene) necessary to produce one type of esterase, increases in southern populations with warmer water. The other form of the gene for esterase is more frequent in northern populations and has elevated activity in colder waters (Koehn and Rasmussen, 1967; Koehn, 1969). Changes in temperature could perhaps alter the gene frequencies of a species, reduce heterozygosity, and by definition reduce fitness.

Maintenance of a diversity of fish and macroinvertebrate habitats would provide more stability to the system. One must know in a general sense how a river is trophically structured before one can predict how the biota may respond to an alteration; for instance, is the river primarily autotrophic or heterotrophic? General thermal conditions, changes in diurnal temperature regimes, and sedimentation are all extremely important to trophic ecology. The stability of the substrate is also extremely important.

For management purposes, it is important to view the Colorado River system as a continuum in the sense of Vannote (1977); that is, alterations may not only directly affect the impacted area, but may also affect trophic relationships many miles downstream. Fluvial geomorphology, terrestrial ecology, and aquatic trophic structure all need to be taken into account when studying aquatic ecosystems and particularly when drawing comparisons between systems. The management of rivers should take into account the interactions between the river continuum and the natural or perturbed terrestrial ecosystems. Studies should include all orders of streams from small heterotrophic streams where production (P) is less than respiration (R) to larger autotrophic streams where the P/R ratio is greater than one, to even larger streams where the P/R ratio may be less than one again. The ecology of various running waters can be vastly different, and stability and energy dynamics of these lotic ecosystems can vary with stream order and geomorphology (Cummins, 1975). A predictive model of a stream under consideration must be based on data from that stream or other streams similar in bottom-type, gradient, and energy inputs.

Fish and wildlife resources should be given full consideration in decision making, and data should be made available to allow proposal of reasonable mitigation. Because of the relatively short time periods under which decision makers usually must operate, it might be worthwhile to determine if, and how well, a biologist could predict the impact of a project given the data currently available. In other words, without undertaking a thorough ten- to twenty-year study of a proposed project, how well can a biologist predict the outcome of the project with respect to impact on fish and the fishery given the current state-of-the-art of stream biology.

In conclusion, Wydoski has outlined in general the potential impacts of alterations of streamflows and water quality on fish and macroinvertebrates in the Upper Colorado River Basin. He also reviewed a number of studies that consider the various factors that could affect the biota and fisheries of the upper basin. I believe the establishment of well-founded goals and objectives for management of the river system by resource custodians is a prerequisite to focusing on the kinds of information outlined by Wydoski for assisting in management decisions. Once these goals and objectives have been established, the availability of information necessary for making these management decisions can be established, and new studies can be undertaken to provide the lacking information. In this way, only the information that is essential for making management decisions concerning the Colorado River Basin will be obtained.

References

Cummins, K. W. 1975. "The Ecology of Running Waters; Theory and Practice,"
    in D. B. Baker, W. B. Jackson, and B. L. Prater, eds., Proceedings of
    the Sandusky River Symposium (Tiffin, Ohio, Heidelberg College).

Funk, J. L. 1955. "Movement of Stream Fishes in Missouri," Transactions
    of the American Fisheries Society vol. 85, pp. 39-57.

Gunning, G. E., and C. R. Shoop. 1964. "Stability in a Headwater Stream
    Population of the Sharpfin Chubsucker," Progressive Fish-Culturist
    vol. 26, pp. 76-79.

Koehn, R. K. 1969. "Esterase Heterogeneity: Dynamics of a Polymorphism,"
    Science vol. 163, pp. 943-944.

Koehn, R. K., and D. I. Rasmussen. 1967. "Polymorphic and Monomorphic
    Serum Esterase Heterogeneity in Catostomid Fish Populations,"
    Biochemical Genetics vol. 1, pp. 131-144.

Phenicie, C. K., and J. R. Lyons. 1973. Tactical Planning in Fish and
    Wildlife Management and Research, Bureau of Sport Fisheries and
    Wildlife, Research Publication 123 (Washington, D.C., U.S. Department
    of the Interior).

Vannote, R. L. 1977. "The River Continuum: A Theoretical Construct for
    Analysis of River Ecosystems," BioScience. In press.

Chapter 9

THE IMPACTS OF HABITAT ALTERATIONS
AND INTRODUCED SPECIES ON THE NATIVE FISHES
OF THE UPPER COLORADO RIVER BASIN

Manuel Molles*

Introduction

The Colorado River drainage contains the largest percentage (74 percent) of endemic fish species of any major drainage in North America
(Miller, 1959). This high degree of endemicity is testimony to the long
history of isolation of the basin. As a consequence of this isolation,
the native fishes form a fundamentally insular fauna. This insular nature
is proposed as a major reason underlying the precarious position of major
components of the original fish community in the face of habitat alterations and species introductions. This is the thesis I wish to develop
in the belief that current theories of insular biogeography and community
structure are keys to understanding the history, present status, and
future prospects for the native fishes of the Upper Colorado River Basin.

Native Fish Fauna

Miller (1959) listed thirty-five native fish species for the Colorado River drainage (table 1). The majority of these species (twenty-
three) belong to only two families--Catostomidae and Cyprinidae. Adaptation within these families has produced species which perform
a full range of ecological roles from omniverous bottom feeders to large

*Assistant Professor and Curator of Fishes, Department of Biology,
University of New Mexico, Albuquerque, New Mexico.

Table 1. Native Fishes of the Colorado River

Family	Number of genera	Number of species
Cyprinidae	9	14
Catostomidae	3	9
Cyprinodonidae	2	2
Poecilidae	1	1
Elopidae	1	1
Salmonidae	1	2
Coregonidae	1	1
Mugilidae	1	1
Cottidae	1	2
Eleotridae	1	1
Gobiidae	1	1

Source: R. R. Miller, "Origin and Affinities of the Freshwater Fish Fauna of Western North America," in C. L. Hubbs, ed., Zoogeography (Washington, D.C., American Association for the Advancement of Science, 1959).

streamlined carnivores. The latter species (the chubs and the Colorado squawfish) seem to be evolutionarily convergent with the Pike and Salmon families. However, these predators lack jaw teeth and are likely at a competitive disadvantage relative to the more heavily armed introduced species.

Holden and Stalnaker (1975) collected only ten species of native fishes from the Upper Colorado River Basin (table 2). However, among these are some of the most spectacular and interesting of the native fishes. Two of these species, the Colorado squawfish, _Ptychocheilus_ _lucius_, and the humpback chub, _Gila_ _cypha_, have been listed as endangered (U.S. Department of the Interior, 1977; Miller, 1972).

Fish Community of the Lower Basin

It seems reasonable that the effects of habitat alteration and introduced species on the native fishes of the Upper Colorado River Basin can be predicted, using the history of the lower basin as a model. The Lower Colorado has undergone much more habitat manipulation than the upper basin and has an earlier history of species introductions. In addition, most fishes native to the upper basin are also native to the lower basin.

If the lower basin is used as a model, the future of the native fishes of the Upper Colorado seems grim indeed. The history of the Lower Colorado has been one of extirpation of major components of the fish fauna in every major drainage (Minckley and Deacon, 1968). In most cases, species native to the lower basin have rapidly gone from abundance to rarity or extinction following habitat alteration and introduction of exotic species. This pattern has been repeated in some sections of the Upper Colorado.

Table 2. Native Mainstream Fishes of the Upper Colorado River Basin

Species	Common name	Number of sites collected	Relative abundance[a]
Salmonidae			
Prosopium williamsoni	Mountain whitefish	4/10[b]	0-A
Cyprinidae			
Gila cypha	Humpback chub	3/10	R
Gila elegans	Bonytail chub	4/10	R
Gila robusta	Roundtail chub	7/10	C-A
Ptychocheilus lucius	Colorado squawfish	9/10	R
Rinichthys osculus	Speckled dace	9/10	R-A
Catostomidae			
Catostomus discobolus	Bluehead sucker	10/10	R-A
Catostomus latipinnis	Flannel mouth sucker	10/10	C-A
Xyrauchen texanus	Humpback sucker	6/10	R
Cottidae			
Cottus bairdi	Mottled sculpin	2/10	R

Source: P. B. Holden and C. B. Stalnaker, "Distribution and Abundance of Mainstream Fishes of the Middle and Upper Colorado River Basin, 1967-1973," Transactions of the American Fisheries Society vol. 104, pp. 217-231, 1975.

[a] R = rare, 0 = occasional, C = common, A = abundant.

[b] Indicates number of the indicated species collected from number of sites tested.

One of the best documented case studies on the Lower Colorado is that
of the Salt River. The Salt River, which flows through south-central
Arizona, is a major tributary of the Gila River. The Gila formerly joined
the mainstem of the Colorado at Yuma, Arizona. Now, however, both the
Gila and Salt rivers have been dammed and are intermittent streams at
their lower ends.

Minckley and Deacon (1968) reviewed the changes in fish species composition in the Salt River at Tempe, Arizona, which took place between
1890 and 1967. Fourteen species were listed as occurring in the river
in 1890. Seven of these species had disappeared from the area by the
late 1920s. The introduction of exotic species began about this same
time. Shortly thereafter, five more native species disappeared. By 1967
only two native species remained. During this period the number of fish
species had grown from fourteen to twenty-two as the net outcome of twelve
extinctions and twenty introductions. One of the most significant points
concerning these extinctions is the rapidity with which they occurred.

Drastic changes in the flow of the Salt River at Tempe coincided with
the above period of faunal change. Roosevelt Dam was built on the Salt
River in 1910. The result was a decrease in discharge at Tempe. Most of
the remaining flow came from the Verde River. However, even discharge
from the Verde River was eventually eliminated by the construction of
dams. As a consequence, the Salt River channel at Tempe was almost completely dry by the late 1950s.

The extirpation of fishes native to the Salt River has probably been
due to a combination of habitat alteration and competition with introduced species. Decreases in discharge have been so extreme in the Tempe

area that changes in the native fish fauna might be entirely attributed to them. However, decimation of native species has occurred in the Salt River above Roosevelt Lake, in spite of continued availability of apparently suitable habitat (Minckley and Deacon, 1968). Competition with introduced species must be implicated in this case.

The Fishes of Champaign County, Illinois

The changes in fish communities observed on the Lower Colorado is in marked contrast to the patterns observed in Champaign County, Illinois. The fishes of Champaign County were surveyed at intervals for sixty years (Forbes and Richardson, 1908; Thompson and Hunt, 1930; Larimore and Smith, 1963). During this period changes in land and water use resulted in marked habitat alterations. Despite these changes, the fish communities of Champaign County have shown a surprising constancy of structure (Larimore and Smith, 1963). Most of the fishes of the county showed no changes in abundance or distribution over the period. Those species that did disappear seemed to be replaced by others in an almost one-to-one fashion. Approximately the same number of species were recorded in the last two surveys of the county (Thompson and Hunt, 1930; Larimore and Smith, 1963).

The question that immediately comes to mind is what makes the central Illinois fish communities so different from those of the Colorado River? It might be pointed out that habitat changes on the Colorado have been much greater in magnitude. However, it was noted previously that extinctions in the Colorado are suspected even where alterations of habitat have been minimal. These eliminations can probably be attributed to competition with introduced species. Some species introductions have

169

been documented in Champaign County. However, the number of successful invaders has been rather small. The most obvious difference between the two areas is a difference in the diversity of fishes in the respective regions. Miller (1959) listed twenty-three primary freshwater species for the Colorado. These belonged to two families. In contrast, 260 primary freshwater species belonging to thirteen families were listed for the Mississippi drainage. The watercourses of Champaign County are part of the Mississippi system. Only seventy-four of these species were collected from Champaign County in the separate censuses of Thompson and Hunt (1930) and Larimore and Smith (1963). Hence, there is tremendous potential for invasion of the Champaign County fish fauna that is apparently not realized.

The greater diversity of fishes of the Mississippi drainage at both the familial and species levels offers a large variety of phylogenetic material from which to construct fish communities. I postulate that eco-logical roles within such communities (e.g., those of Champaign County, Illinois) would be performed in a more efficient fashion than within communities like those originally found in the Colorado. As a consequence, these communities would probably be less vulnerable to invasion than the simpler fish communities of the Colorado River. If the Colorado fauna can be considered as basically insular, I suggest that the fishes of Champaign County are part of a "continental" fauna.

The Colorado River as an Insular System

The view of aquatic systems as islands is not novel. Several authors have taken this view (e.g., MacArthur and Wilson, 1967; Maguire, 1971; Barbour and Brown, 1974; etc.). As one colleague put it, aquatic systems are "aquatic islands in a terrestrial sea" (Dr. Elisabeth A. Stull, University of Arizona). This seems to be an especially apt analogy to the semiarid Southwest. I think this view can lead to productive insights. This is due primarily to the amount of attention that the theory of island biogeography has received in recent years (see Simberloff, 1974).

There are several features of the Colorado that would lead one to view it as an insular system: (1) high levels of endemism in the fish fauna are evidence of a long history of isolation; (2) the vulnerability of the fish community to outside invasion is reminiscent of island communities (Elton, 1958); and (3) the fragmentation of the Colorado by dams has increased the resemblance of local fish communities to bounded insular systems. With these observations offered as support, the effects of habitat alteration and species introductions on the native fishes of the Colorado are discussed below, relative to pertinent aspects of insular biogeography and community ecology.

Competition with Introduced Fishes

Holden and Stalnaker (1975) conducted an extensive survey of the fishes of the Middle and Upper Colorado River Basins. Ten of their twelve collecting sites were located in the upper basin above Lake Powell. Only these collections will be discussed here. As was noted above, Holden and Stalnaker collected ten native species from the upper basin during their study (table 2). In addition, they collected nineteen introduced species (table 3). Species introductions have, therefore, resulted in a tripling of species numbers. Six of the ten native species collected were listed as rare. These low population densities probably resulted from

Table 3. Introduced Mainstream Fishes of the Upper Colorado River Basin

Species	Common name	Number of sites collected	Relative abundance[a]
Salmonidae			
Salmo clarki	Cutthroat trout	2/10[b]	O
Salmo gairdneri	Rainbow trout	4/10	O
Salmo trutta	Brown trout	4/10	O
Cyprinidae			
Cyprinus carpio	Carp	9/10	C-A
Gila atraria	Utah chub	2/10	R
Notropis lutrensis	Red shiner	8/10	R-A
Notropis stramineus	Sand shiner	2/10	C
Pimephales promelas	Fathead minnow	10/10	C
Richardsonius balteatus	Redside shiner	3/10	C-A
Semotilus atromaculatus	Creek chub	2/10	C
Catostomidae			
Catostomus catostomus	Longnose sucker	1/10	R
Catostomus commersoni	White sucker	4/10	R-A
Ictaluridae			
Ictalurus melas	Black bullhead	5/10	R-C
Ictalurus punctatus	Channel catfish	9/10	R-A
Cyprinodontidae			
Fundulus zebrinus	Rio Grande killifish	3/10	R
Centrarchidae			
Lepomis cyanellus	Green sunfish	8/10	R-C
Lepomis macrochirus	Bluegill	1/10	O
Micropterus salmoides	Largemouth bass	5/10	O-C
Percidae			
Sitzostedion vitreum	Walleye	2/10	O

Source: P. B. Holden and C. B. Stalnaker, "Distribution and Abundance of Mainstream Fishes of the Middle and Upper Colorado River Basin, 1967-1973," Transactions of the American Fisheries Society vol. 104, 1975.

[a]R = rare, O = occasional, C = common, A = abundant.

[b]Indicates number of the indicated species collected from number of sites tested.

competition with exotics since areas significantly affected by Flaming Gorge Reservoir were not included in Holden and Stalnaker's study.

Competition with introduced species will probably eventually result in the replacement of most, if not all, of the native fishes of the Upper Colorado River Basin. That is the pattern that has been consistently observed in the lower basin as was noted above. However, there is justification for such a prediction even without these historical data. The insular characteristics of the native fauna plus the variety and intensity of introductions (past and future) makes competitive replacement highly probable.

The characteristics of many introduced species make them good competitors. Introductions are not, for the most part, drawn randomly from outside species pools, even though many of them are unplanned. The unplanned variety is typified by the leftovers of a fisherman's bait bucket. This type of introduction seems to contrast sharply with those carefully planned by government agencies. However, I think the species introduced in these vastly different fashions have something in common. The probability of collecting a fish species for use as bait is surely proportional to its numerical dominance in the collecting area. This may be taken as a rough measure of a species' competitive abilities. The chance that such a species will survive to be dumped into some body of water by a fisherman is dependent on the range of its physiological tolerances. Planned introductions for the purpose of providing sport fisheries usually involve those species that have consistently produced adequate returns for the effort. Such consistent performance is the result of

characteristics that make species good invaders and competitors--
high productivity, generalized life cycles, and wide physiological
tolerance.

All of the above is not to say that any and all species could invade
communities of native Colorado River fishes. This is surely not the case.
The dramatic and seemingly deterministic replacement of these native fish
communities is partially due to our observing only successful invasions.
Most species which have been placed in the Colorado via bait buckets or as
stowaways in hatchery trucks have probably been unsuccessful in establishing
populations.

Predation

Fishes native to the Colorado are probably especially vulnerable to
predation by exotics, since they have only recently established contact with
these species.   Predation by introduced species has been responsible for
local extinctions of native fishes in at least two cases (Zaret and Paine,
1973). Largemouth bass, _Micropterous salmoides_, introduced into Lake Atitlán,
Guatemala, eliminated populations of _Poecilia sphenops_ and _Cichlasoma nigro-
fasciatum_ in less than fifteen years. These native fishes were formerly
abundant enough to support a thriving fishery. The introduction of the peacock
bass, _Cichla ocellaris_, has eliminated several species of native fishes from
sections of Gatun Lake, Panama. The dramatic decline of native fishes in
the Colorado following the introduction of exotics probably results, at
least partially, from predation by some of these species.

Habitat Alteration

The main forms of habitat alteration on the Upper Colorado will probably
be the construction of dams and consequent regulation of discharge. These

alterations have had a number of direct and indirect effects. The construction of dams restricts the passage of fishes along the length of a
river, especially in an upstream direction. The result is a series of
fragmented populations in the place of formerly contiguous ones. Stream
fishes in a variety of systems are known to undertake extensive migrations
for the purpose of spawning and feeding (Hynes, 1970). The Colorado squawfish probably made such migrations. A member of the same genus, _Ptychocheilus_
grandis, the Sacramento squawfish, is known to engage in extensive movements
(Taft and Murphy, 1950). A congregation of male Colorado squawfish in
spawning condition observed by Holden and Stalnaker (1975) suggests that the
Colorado species also engages in similar migrations. The blockage of such
movements can be shown to theoretically increase probabilities of extinction
(see section on extinction theory below).

With the stabilization of flows below dams (e.g., the Green River below
Flaming Gorge Dam (Vanicek and coauthors, 1970)), selection for changes in
reproductive strategies and foraging efficiencies which would increase
competitive abilities near equilibrium would be expected to follow (MacArthur
and Wilson, 1967; Pianka, 1970, 1972). The number of fish species in the
community would be expected to increase over time and to consist of relatively
more specialized species. However, since invasion of these communities has
coincided with flow stabilization, the native fishes of the Colorado will not
have sufficient time to make such an evolutionary response. The development
of a diverse community in a stabilized Colorado River will take place over
ecological time, not evolutionary time. As a consequence, these communities
can be expected to be composed mainly of exotics of superior competitive
ability in an altered Colorado River.

The direct effect of cold-water discharge below reservoirs is a rather straightforward proposition. Such waters are unsuitable for reproduction by most native fishes, and these fishes will be eliminated from reaches where water temperatures are significantly reduced below normal levels. This response by native fishes to cold-water releases below a reservoir has already been recorded in the Green River of the upper basin below Flaming Gorge Dam. Reduced water temperatures below the dam have eliminated reproduction by native fishes for 104 kilometers downstream, and native species are either rare or absent throughout this stretch (Vanicek and coauthors, 1970). Cold-water discharges may also have important indirect effects. The elimination of native fishes from reaches below reservoirs reduces total population size and further restricts native fishes to relatively small, local populations. This would predictably lead to increased probabilities of extinction (see section on extinction theory below).

The standing water habitats of reservoirs appear unsuitable for native fishes for several reasons. Flowing water is probably necessary for reproduction for a majority of the native species. Minckley and Deacon (1968) present several pieces of evidence which indicate that humpback suckers, Xyrauchen texanus, fail to reproduce in reservoirs. The rarity of Colorado squawfish in the large reservoirs of the Colorado, such as Lake Powell,[1] is probably at least partially due to reproductive failure. Their rarity may also be due to avoidance of standing water areas. Adaptation to standing water conditions by native species might be expected over time. However, even where a natural lake population of the Sacramento squawfish, Ptychocheilus grandis, occurred, it was eventually replaced by exotic competitors better adapted to lentic conditions (Taft and Murphy, 1950).

---

[1]Personal communication with Steve Gloss, University of New Mexico, Albuquerque, New Mexico.

The amount of river habitat made unsuitable by an impoundment in the narrow canyons of the Colorado River can be considerable.   For example, the reservoir above Flaming Gorge Dam extends for 144 kilometers upstream.

Reservoirs might also affect native fish populations in very unexpected ways.   Even species which were at a competitive disadvantage in the original river environment could exert significant competitive pressures on native fishes with the addition of reservoirs. Primarily, lentic species could use reservoirs as refugia. Excess reproduction by these populations could serve as a source of colonists to adjacent river areas, maintaining populations of introduced species in the river habitat despite competitive inferiority relative to native fishes. The presence of these immigrants would predictably depress native fish populations and further increase probabilities of extinction.

Extinction Theory

The theory of extinction developed by MacArthur and Wilson (1967), and later refined by MacArthur (1972), is relevant to the problem being considered. Their model expresses the expected time to extinction of a species on an island as a function of the birth rate (λ) and death rate (μ) of the species and the carrying capacity (k) of the island for the species. The expected time to extinction of a single propagule colonizing an island (isolated stretch of the Colorado River) is given as:

$$T_1 \sim \frac{1}{k\lambda} \left(\frac{\lambda}{\mu}\right)^k$$

The expected time to extinction of a population of the same species at carrying capacity is a multiple of the above expression:

$$T_k \sim \frac{\lambda}{\lambda-\mu} \left[\frac{1}{k\lambda} \left(\frac{\lambda}{\mu}\right)^k\right]$$

These models provide a more precise way of analyzing the effects of the
habitat alterations and species introductions discussed above.

The dependence of time to extinction in the above model on birth rates,
death rates, and carrying capacity is intuitively pleasing. However, the
form of population growth assumed in this model is clearly unrealistic.
A population is allowed to grow exponentially until carrying capacity is
reached and then assumed to fall instantaneously to zero. Notice, however,
that this assumption maximizes the ability of a population to recover from
reductions and, therefore, maximizes the probability of survival. The result is that time to extinction for natural populations should always be
less than that predicted by the model.

As MacArthur and Wilson pointed out, we are in no position to estimate
the parameters λ, μ, or k for most species. This is particularly true for
a poorly-known fauna like the fishes of the Colorado. The value of the
theory lies in the qualitative predictions it makes. The most important
prediction is that for all relative values of λ and μ, there is some value of
k below which the expected time to extinction decreases from a very long
period to a very short one. For example, the reduction of a native species
carrying capacity to one-third some former level would reduce the expected
time to extinction by the cube root of its previous value. In the face of
such a reduction, a species that would have been expected to persist for
one million years could be expected to go extinct in only 100 years! The
native species of the Upper Colorado now considered either rare or endangered have probably suffered even greater population reductions than
presented in the above example.

Priorities for the Upper Colorado River Basin

The first priority for the native fishes of the Upper Colorado River Basin must be preservation. This area is the last stronghold of some of the unique species of the Colorado system. However, the word stronghold must be taken in a very relative sense. Even in the Upper Colorado where some of the more vulnerable native species are most common, they are rare at best. The extinction of many of these species may not be far off. Development of the upper basin will probably accelerate their demise.

Maintenance of populations of most native fishes is probably possible only in semi-natural environments in the absence of competitors. What is needed is a preserve where a portion of the Upper Colorado is allowed to flow freely and from which introduced species are excluded. What I am suggesting is not preservation of individual native species but preservation of the entire community. This may seem far more ambitious an undertaking than the maintenance of hatchery populations of these species with periodic introductions to sections of the Colorado. However, once such a preserve was established, less management input would be required to maintain the populations of native fishes. As stated above, the time to extinction of a species introduced to a habitat already saturated with competitor species is less than that of a population near its carrying capacity. If (and this may be a big if) exotic species could be excluded from such a preserve, most native species could be expected to persist indefinitely.

The subject of nature preserves has received much attention in the ecological literature in recent years (Terborgh, 1974; Diamond, 1975; Wilson and Willis, 1975). Most of this literature views the design of nature preserves relative to island biogeographical theory in recognition of the increasing fragmentation of natural systems. Because of this, most of this literature is highly relevant to the establishment of a preserve for Colorado River fishes.

One of the best documented features of island biogeography is the dependence of island diversity on island size (MacArthur and Wilson, 1967; Simberloff, 1974, 1976). The relationship is: the larger the island (or nature preserve) the more species it will support. What size preserve would be required to support all the native fishes of the Upper Colorado? The data required to answer this question do not exist. In any case, the cost of the enterprise would probably set the size limit of such a preserve. Hopefully, a preserve of adequate size will be economically feasible.

The data available are insufficient to answer the questions raised in this paper concerning the fate of the native fishes of the Upper Colorado River. The very lack of this critical information provides justification for establishing a preserve for these fishes. The prerequisite to under- standing natural communities is the availability of numerous systems of contrasting structure and evolutionary history. The probability of ever reaching such an understanding is diminished with the destruction of any natural community. The elimination of a unique system like the native fish community of the Upper Colorado River would diminish that prospect im- measurably.

References

Barbour, C. D., and J. H. Brown. 1974. "Fish Species Diversity in Lakes," *American Naturalist* vol. 108, pp. 473-489.

Diamond, J. M. 1975. "The Island Dilemma: Lessons of Modern Biogeographic Studies for the Design of Natural Reserves," *Biological Conservation* vol. 7, pp. 129-146.

Elton, C. S. 1958. *The Ecology of Invasions by Animals and Plants* (London, (Mathuen and Co. Ltd.).

Forbes, S. A., and R. E. Richardson. 1908. *The Fishes of Illinois* (Urbana, Ill., Illinois State Laboratory of Natural History).

Holden, P. B., and C. B. Stalnaker. 1975. "Distribution and Abundance of Mainstream Fishes of the Middle and Upper Colorado River Basin, 1967-1973," *Transactions of the American Fisheries Society* vol. 104, no. 2, pp. 217-231.

Hynes, H. B. N. 1970. *The Ecology of Running Waters* (Toronto, University of Toronto Press).

Larimore, R. W., and P. W. Smith. 1963. "The Fishes of Champaign County, Illinois, as Affected by 60 Years of Stream Changes," *Illinois Natural History Survey Bulletin* vol. 28, pp. 299-382.

MacArthur, R. H. 1972. *Geographical Ecology* (New York, Harper and Row).

MacArthur, R. H., and E. O. Wilson. 1967. *Theory of Island Biogeography* (Princeton, N.J., Princeton University Press).

Maguire, B., Jr. 1971. "Phytotelmata: Biota and Community Structure Determination in Plant-Held Waters," *Annual Review of Ecological Systems* vol. 2, pp. 439-464.

Miller, R. R. 1959. "Origin and Affinities of the Freshwater Fish Fauna of Western North America," in C. L. Hubbs, ed., *Zoogeography* (Washington, D.C., American Association for the Advancement of Science No. 41).

_____. 1972. "Threatened Freshwater Fishes of the United States," *Transactions of the American Fisheries Society* vol. 101, pp. 239-252.

Minckley, W. L., and J. E. Deacon. 1968. "Southwestern Fishes and the Enigma of 'Endangered Species,'" *Science* vol. 159, pp. 1424-1432.

Pianka, E. R. 1970. "On r- and K- Selection," *American Naturalist* vol. 104, pp. 592-597.

_____. 1972. "r- and K- Selection or b and d Selection?" *American Naturalist* vol. 106, pp. 581-588.

Simberloff, D. S. 1974. "Equilibrium Theory of Island Biogeography and Ecology," Annual Review of Ecological Systems vol. 5, pp. 161-182.

_____. 1976. "Experimental Zoogeography of Islands: Effects of Island Size," Ecology vol. 57, pp. 629-648.

Taft, A. C., and G. I. Murphy. 1950. "The Life History of the Sacramento Squawfish (Ptychocheilus grandis)," California Fish and Game vol. 36, pp. 147-164.

Terborgh, J. W. 1974. "Preservation of Natural Diversity: The Problem of Extinction Prone Species," BioScience vol. 24, pp. 715-722.

Thompson, D. H., and F. D. Hunt. 1930. "The Fishes of Champaign County: A Study of the Distribution and Abundance of Fishes in Small Streams," Illinois Natural History Survey Bulletin vol. 19, pp. 1-101.

U.S. Department of the Interior, Fish and Wildlife Service. 1977. Federal Register vol. 42, no. 135 (Thursday, 14 July).

Vanicek, C. D., R. H. Kramer, and D. R. Franklin. 1970. "Distribution of Green River Fishes in Utah and Colorado Following Closure of Flaming Gorge Dam," Southwestern Naturalist vol. 14, pp. 297-315.

Wilson, E. O., and E. O. Willis. 1975. "Applied Biogeography," in M. L. Cody and J. M. Diamond, eds., Ecology and Evolution of Communities (Cambridge, Mass., Harvard University Press).

Zaret, T. M., and R. T. Paine. 1973. "Species Introduction in a Tropical Lake," Science vol. 182, pp. 449-455.

THE IMPACTS OF HABITAT ALTERATIONS AND
INTRODUCED SPECIES ON THE NATIVE FISHES
OF THE UPPER COLORADO RIVER BASIN: A DISCUSSION

John P. Hubbard*

Theories are the skeletons that give reason to facts, allowing a
complex body of data to be fleshed into some definitive and usable form.
Recent interest in facts about the biotas of islands has given rise to
much theory and has stimulated a great deal of thought. At times, the
facts involving insular biotas are merely aggregations of numbers, or at
best, rude categorizations of some aspects of niches; thus, theories derived from such "facts" may be open to question. In view of the high
stakes--i.e., survival or extinction--involving the ichthyofauna of the
Colorado River Basin, a careful review of facts as well as theories is
clearly in order.

The first question is, what _is_ the ichthyofauna of the Colorado
River Basin? Dr. Molles cites Miller (1959) as the source for such data,
to the effect that some eleven families, twenty-two genera, and thirty-
five species comprise that fauna. Of the species, 74 percent are deemed
to be endemic to the system, the highest level of any major drainage in
North America. I have reexamined data on the native fishes of the Colorado Basin, and in the light of more modern treatment and after eliminating marine or brackish water forms, I find that the fauna consists of
thirty species, eighteen genera, and six families of native, freshwater

*New Mexico Department of Game and Fish, Santa Fe, New Mexico.

fishes (table 1). Endemism is still high, at 73 percent of the species and 39 percent of the genera. The genus _Agosia_ is a near-endemic, as is the species _Poeciliopsis occidentalis_; if these are considered as _autochthonous_, then the levels of endemism rise to 80 percent of the species and 44 percent of the genera. Also endemic to the basin is the minnow subfamily, Plagoterini, composed of six species in three genera.

Such a high degree of endemism carries with it several evolution-ary implications, two being that the basin has been (1) well isolated from other drainages, and (2) isolated for a considerable period of time. These conclusions are supported by the facts that even the non-endemics tend to be represented by distinct subspecies (e.g., in _Salmo clarki_, _Rhinichthys_, _Cottus bairdi_) or may be the result of rather re-cent stream captures (e.g., _Catostomus platyrhynchus_), or both.

The implications of strict isolation for a long period of time may con-vey overtones of the existence of a certain evolutionary senility or vulner-ability in a fauna. A fauna evolving in such a situation, especially from a limited stable of ancestors, might logically be expected to be less fit to compete than a fauna derived from a more open situation or a more diverse ancestry, or both. Indeed, the Colorado Basin must have been rather an evolutionary vacuum, and the stable of ancestors comprised only six surviving families--two , the Cyprinidae (minnows) and the Catostomidae (suckers), encompassing 67 percent of all the species and genera in the fauna. Given the above scenario, the tentative conclusion arises that the fauna might be highly vulnerable to such things as the

Table 1. Native, Nonmarine Fishes of the Colorado River Basin

	Subbasins of the Colorado Drainage						
	Green River-Colorado River junction northward	San Juan River to junction of Green River-Colorado River	Lake Mead to junction of San Juan River	Little Colorado River	Lake Mead to White River, Nevada	Lake Mead to Yuma, Arizona	Gila River, Arizona and New Mexico
Salmonidae							
Salmo clarki	x	?					
Salmo *gilae					x		x
Salmo *apache				x			x
Prosopium williamsoni	x	?					
Catostomidae							
*Xyrauchen texanus	x	x	x	?	x	x	x
Pantosteus *discobolus	x	x	x	x			
Pantosteus platyrhynchus	x						
Pantosteus *clarki					x		x
Catostomus *insignis							x
Catostomus *latipinnis	x	x	x	x	x	x	x
Cyprinidae							
Rhinichthys osculus	x	x	x	x	x	x	x
*Tiaroga cobitis							x
Agosia chrysogaster					x	x	x
*Moapa coriacea					x		
Ptychocheilus *lucius	x	x	x	?	x	x	x
Gila *robusta	x	x	x	x	x	x	x
Gila *elegans	x	x	x	?	x	x	x
Gila *cypha	x	x	x	?	x	?	

Table 1. (continued)

	Subbasins of the Colorado drainage						
	Green River-Colorado River junction northward	San Juan River to junction of Green River-Colorado River	Lake Mead to junction of San Juan River	Little Colorado River	Lake Mead to White River, Nevada	Lake Mead to Yuma, Arizona	Gila River, Arizona and New Mexico
Cyprinidae (continued)							
*Lepidomeda vittata				x			
*Lepidomeda mollispinis			x		x		
*Lepidomeda albivallis					x		
*Lepidomeda altivelis					x		
*Plagopterus argentissimus			x	?	x	x	x
*Meda fulgida							x
Cyprinodontidae							
*Crenichthys baileyi					x		
*Crenichthys nevadae					x		
Cyprinodon *macularius							x
Poeciliidae							
Poeciliopsis occidentalis							x
Cottidae							
Cottus bairdi	x	x					
Cottus *annae	x						
Species totals (30 possible)	13	9(11?)	10	6(10?)	16	3(9?)	16
Genera totals (18 possible)	9	7(9?)	8	6(10?)	11	7	13

* indicates genus and species or species is endemic to the area(s) indicated.
x indicates species has occurred.
? indicates status questionable.

introduction of exotics, particularly as the latter confront native species that might be regarded as overly specialized, such as a minnow
(Ptychocheilus) that is a large predator.

On the face of it, a scenario of isolation-depauperateness-vulner-
ability seems made to order for the Colorado Basin ichthyofauna. Furthermore, severe declines in the populations of some of the species, plus the
extinction of at least one (Lepidomeda altivelis), seem to show that the
scenario is, in fact, at work. A good correlation exists between the
declines and the increasing number and diversity of exotics in the Colorado ecosystem. However, as mentioned at the outset, mere numbers may
not be a sufficient or satisfactory basis for theorizing in a case such
as this; consequently, let us look beyond.

One thing that treating a fauna as mere numbers does do is to convey
a connotation of that assemblage as a monolith. The fact is that the
fauna of the Colorado Basin are not a monolith, e.g., only a few species
attain even relatively wide distributions in the system; these are mainly
fishes that can successfully confront "big water" situations of the main
streams. Only 20 percent of the fauna achieve this status, these being
Xyrauchen, Catostomus latipinnis, Rhinichthys, Ptychocheilus, and Gila spp.
Even when one includes the moderately widespread species and those that
are geographic replacement members of species complexes, at most half of
the fauna can be regarded as wide-ranging.

Conversely, many of the Coloradan species have narrow ranges in the
system, and in any given subbasin an average of only 11 species (37 percent)
and 8.5 genera (47 percent) occur. The maxima are in the Lake Mead to
White River (includes the Moapa and Virgin drainages) and the Gila

subbasins, both of which also feature several examples of very narrow endemism (table 1). Each subbasin hosts 53 percent of the species: the former also hosts 61 percent of the genera, plus five endemic species and two genera, while the Gila hosts 72 percent of the genera, plus four endemic species and two genera. Other subbasins harboring narrow endemics are the Little Colorado (one species) and the Upper Colorado (one species); all other endemics (twelve species, three genera) are more widespread. In total, taxa that are confined to a single subbasin comprise 37 percent of the endemic species and 22 percent of the endemic genera.

What do the distributional data tell us about the fishes of the Colorado Basin and their evolution? For one thing, the facts that many species are not widespread and that high degrees of endemism exist in single subbasins suggest that, even in what has until rather recent time been one continuous ecosystem, factors may be at work that cause a marked partitioning in ranges of fishes in the system. Among these factors may be barriers to dispersal, such as the "big waters" of main streams; however, it is inconceivable that at some time during recent millenia such barriers would not have moderated to the point of allowing dispersal to occur.

It is my opinion that the marked partitioning in the ranges of fishes in the Colorado Basin reflects something more complicated than the presence of physical barriers to dispersal. In fact, it is even likely that dispersal opportunities have probably been frequent, at least through early historic time. Given that dispersal, then, is not a major problem, the question arises as to what other factors are at work to produce the observed partitioning. I believe that a major factor may be in the ecological limitations in the system itself, i.e., a dearth of niches. It may be that, even following successful dispersal, niches are simply not abundant enough to allow distribution or diversity, or both, to increase beyond a moderate level. This is perhaps to be expected, as the Colorado Basin seems to have severe limitations on fish habitat, with its dearth of aquatic vegetation, highly variable flows, and seemingly low productivity.

If the above hypothesis is accurate, the Colorado Basin may be viewed as a depauperate environment for fishes, with the result that even with limited faunal diversity, the native fauna can obtain neither a universally wide range in the basin nor maximal diversity in any one area. Of course, one would not expect that all species could become widespread nor that any single site could host all species; however, when both aspects remain so far from fulfillment, and assuming that barriers to dispersal are less than fully effectual, one may well conclude that a reduced carrying capacity is involved. If this is true, then far from being regarded as senescent and vulnerable, the ichthyofauna of the Colorado drainage is more likely a well-adapted assemblage of species functioning well until recent time in a less-than-outstanding environment for most fishes. Further, it is likely that this relationship would have persisted indefinitely had conditions remained the same or shifted only slowly, as the fauna may well have had a sufficient degree of internal diversity and adaptability to exploit most of the available array of niches. Unfortunately for the fauna, conditions have neither remained the same nor shifted slowly; rather, they have been altered rapidly and drastically, thereby stressing the fauna and bringing it to the point of decline.

In my opinion, the major detrimental shift in conditions for the native ichthyofauna of the Colorado Basin has been the drastic physical alteration of the environment. In particular, the construction of reservoirs would appear a (or the) major problem, producing an important habitat type that was lacking before. Reservoirs call for adaptations to a lacustrine life, and while some native species have been able to make the transition, many others have not. The creation of reservoirs has been closely followed by large scale, successful introductions of exotic fishes, the majority of which are lacustrine-adapted. In many (most?) cases in reservoirs, the exotics appear to out-compete native species, and consequently the former group has shown a marked tendency to expand and increase--frequently at the expense of natives. As exotics have come to dominate reservoirs, they are also able to expand into adjacent, more primeval habitats that harbor native species. In time, the populations of exotics build up to levels that severely impact on native species, aided and abetted by, but not confined to, reservoirs.

Reservoirs also negatively impact on native species in other ways, especially through cold-water discharge downstream, and prevention or curtailment of migration past the damsites. These additional impacts couple with competition from exotics to stress native species even further. Reproductive success is a particular area of sensitivity for natives, which may fail to spawn in the cold waters in and below reservoirs, or which may have their young preyed upon by exotics. Exotics have the advantage, too, of being augmented by hatchery plantings, so that artificially high numbers are attained, further compounding the negative impacts on native species.

It is my contention that drastic habitat change, especially reservoir construction, has been the prelude to the decline of the native ichthyofauna of the Colorado Basin. The mechanisms by which reservoirs negatively impact on the native fishes are both biological (e.g., introduction of exotics) and physical (e.g., lower water temperatures). However, without

such drastic habitat changes, it seems unlikely that the native fauna would
be so severely impacted. Certainly, it is unlikely that exotics would be
the threat they are at present, for most do not appear to be superior competitors outside areas influenced by environmental alteration. This is not
to say that some exotics might not out-compete natives in some more or less
natural habitats, e.g., Gambusia affinis vs. Poeciliopsis occidentalis,
and Salmo trutta vs. S. gilae/apache/clarki; however, even these cases may
have their roots in a preluding habitat change.

In summation, a more detailed examination of the distribution of the
native fishes of the Colorado drainage provides some insight into the valid-
ity of applying concepts of insular biogeography to this system and its
declining ichthyofauna. One cannot quarrel with the probability that
exotic fishes have been a major element in the decline of native species,
but the success of the former has probably been rooted in drastic habitat
changes, rather than in real or imagined shortcomings in the native fauna--
especially those resulting from its insularity. With the persistence of
prehistoric habitats, the native fauna could probably cope with most exotics.
Even in the face of gradual changes in habitats, a sufficient adaptability
and diversity may exist in the native fauna to allow it to exploit new
niches and compete successfully with exotics.

I believe that it is likely that, until recent times, the diversity of
fishes in the Colorado Basin was geared to habitat availability. While
there were relatively fewer species per unit area than in, say, Champaign
County, Illinois, the former nevertheless may well have been sufficiently
diverse to achieve an ecological saturation of the Colorado drainage.
How efficiently Coloradan species exploited their habitats, compared to
Illinoisan species, is quite unknown, but there is no reason to assume

that it was less efficient in either case. A comparison that may reflect on this question is with regard to the native fish fauna of the Rio Grande drainage of New Mexico, a system that bears more than a passing ecological similarity to the Colorado drainage. In the Colorado drainage, the minnow and sucker families, respectively, constitute 26 percent and 21 percent of the genera, and 7.5 percent and 10 percent of the species that occur in the United States. In the Rio Grande drainage, the respective figures are 26 percent and 43 percent for genera, and 9.5 percent and 10 percent for species—percentages that show the two drainages to be rather similar in the diversity of the two families. This is interesting because the Rio Grande drainage is far closer and more accessible to the very rich ichthyofaunas of the southeastern United States, and a far greater diversity would seem possible. Could it be that in the upper Rio Grande, niche saturation has also been reached—and at a similar level to that of the Colorado (and notably below its seeming potential)?

In closing, let me say that any differences in my interpretations and those of Dr. Molles are mainly ones of philosophy rather than substance. My main concern is that we not slander the native fishes of the Colorado system by downgrading them evolutionarily or ecologically, for there are those who would seize on the opportunity to give up on them. The root of the problems for these fishes is not their shortcomings, but rather it is those of mankind. Realistically, we cannot undo all of the damage that we have done to the native fishes of the Colorado system, but a good start would, indeed, be in the form of preserves for surviving segments of that ichthyofauna. Let us begin now to select such sites, with an eye especially for those in which primeval-like habitats prevail, such as the Yampa in the upper basin and the Gila and San Francisco in the lower basin.

References

Miller, R. R. 1959. "Origin and Affinities of the Freshwater Fish Fauna
    of Western North America," in C. L. Hubbs, ed., Zoogeography (Washington,
    D.C., American Association for the Advancement of Science No. 41).

THE IMPACTS OF HABITAT ALTERATIONS AND INTRODUCED SPECIES ON THE NATIVE FISHES OF THE UPPER COLORADO RIVER BASIN: A DISCUSSION

Robert F. Raleigh*

The Colorado River presents a varied habitat to support the fish fauna located within its basin. The river begins as a series of cold, clear streams in the Rocky Mountains. It drains an area of about a quarter of a million square miles as it travels across portions of seven states and a piece of Mexico to reach the Gulf of California. As the river flows across this arid to semiarid land with its hot climate and highly erodible soils, the water temperature rises and the silt load increases. Historically, annual and seasonal flow regimes appear to have been quite variable. This river, with its highly varied and some-what harsh habitat conditions, existed in a virtually isolated condition until the early 1900s. Under these conditions, the Colorado River eventually supported the largest assemblage of endemic fishes of any known river system in North America (Miller, 1959; Holden and Stalnaker, 1975). Dr. Molles very appropriately treats the Colorado River system as an island. I am in basic agreement with him. This insular concept reasonably explains the highly endemic nature of the fish fauna of the Colorado River.

Water has probably been the most consistently sought after natural resource in the entire Colorado River Basin. Fortunately, the region has

---

*U.S. Fish and Wildlife Service, Fort Collins, Colorado.

been only sparsely settled, but the demand for water, primarily for irri

gation, has been great. The extensive use of Colorado River water began

with development in the early 1900s and has continued to grow until, for

many years now, no water from the Colorado has spilled into the Gulf of

California (Crandall, 1974). Even so, the upper basin states still have

legal rights for additional consumptive water use developments.

The Upper Colorado River Basin is presently attracting considerable

attention due to the conflict between water resource development and in-

stream flow needs to preserve the remaining segments of the river's uniq

native fish fauna. As Dr. Molles pointed out, Holden and Stalnaker (197

list ten species of fishes that are native to the upper mainstream Colo-

rado River. A second list of native fishes published by the Utah Water

Research Laboratory (1975) and recent collections that include some tribi

tary fishes yield an estimate of at least thirteen native and thirty-one

introduced fish species for the upper Colorado River drainage system

(table 1).[1]

Oil shale, coal mining, powerplants, hydroelectric and irrigation

developments in the Upper Colorado River Basin presently consume about

3,700,000 acre-feet of water per year. The Bureau of Reclamation has

estimated that an additional 870,000 acre-feet will be needed annually--

primarily for energy development--by the year 2000 (U.S. Department of

the Interior, 1974). According to my best estimate, there are about

twenty-two dams and water diversion structures presently in existence

on the Upper Colorado River, two additional dams are under construction,

---

[1]Personal communication with Robert J. Behnke, Department of Fisher
and Wildlife Biology, Colorado State University, Fort Collins, Colorado.

Table 1.   Native and Introduced Fishes of the Upper Colorado River
           and Tributaries

Common name	Scientific name
Native fishes	
Cutthroat trout[a]	Salmo clarki pleuriticus
Whitefish	Prosopium williamsoni
Roundtail chub	Gila robusta
Bonytail chub[a]	Gila elegans
Humpback chub[a]	Gila cypha
Colorado squawfish[a]	Ptychocheilus lucius
Speckled dace	Rhynichthys osculus
Kendall Warm Springs dace[a]	Rhynichthys osculus thermalis
Flannelmouth sucker	Catostomus latipinnis
Humpback sucker[a]	Xyrauchen texanus
Bluehead sucker	Catostomus discobolus
Mountain sucker[b]	Catostomus platyrhynchus
Mottled sculpin	Cottus bairdi
Piute sculpin[c]	Cottus beldingi
Introduced fishes	
Rainbow trout	Salmo gairdneri
Brown trout	Salmo trutta
Golden trout	Salmo aquabonita
Ochrid trout	Salmo letnica
Brook trout	Salvelinus fontinalis
Lake trout	Salvelinus namaycush
Kokanee	Onchorynchus nerka
Coho	Onchorynchus kisutch
Grayling	Thymallus arcticus
Threadfin shad	Dorosoma petenense
Northern pike	Esox lucius
Carp	Cyprinus carpio
Redside shiner	Notropis lutrensis
Sand shiner	Notropis stramineus

Table 1. (continued)

Common name	Scientific name
Introduced fishes (continued)	
Utah chub	Gila atraria
Creek chub	Semotilus atromaculatus
Fathead minnow	Pimephales promelas
Brassy minnow	Hybognathus hankinsoni
White sucker	Catostomus commersoni
Longnose sucker	Catostomus catostomus
Mountain sucker[b]	Catostomus platyrhinchus
Rio Grande killifish	Fundulus zebrinus
Plains killifish	Fundulus sciadicus
Channel catfish	Ictalurus punctatus
Black bullhead	Ictalurus melas
Yellow bullhead	Ictalurus natalis
Smallmouth bass	Micropterus dolomieui
Largemouth bass	Micropterus salmoides
White crappie	Pomoxis annularis
Black crappie	Pomoxis nigromaculatus
Green sunfish	Lepomis cyanellus
Bluegill	Lepomis macrochirus
Walleye	Stizostedion vitreum
Yellow perch	Perca flavescens
Mosquito fish	Gambusia affinis
Striped bass	Morone saxatilis

For footnotes see following page.

Table 1. (continued)

[a]Fishes generally considered by fisheries authorities to be
endangered or threatened.

[b]The position of the mountain sucker as a native or introduced fish of the Colorado River is not fully resolved.

[c]The Eagle Sculpin (Cottus annae) of Jordan and Starks as
reported in Reeve Bailey and Carl Bond, Four New Species of
Sculpins from Western North America, Occasional Paper No. 639
(Ann Arbor, Mich., Museum of Zoology, University of Michigan,
1963) p. 27.

and about twelve more are in some stage of serious planning. Some of these projects divert water entirely out of the Colorado River Basin. All of the projects bring about subtle or dramatic changes to the river ecosystem. Changes in water chemistry, stream productivity, transparency, flow and temperature regimes, and river bottom materials commonly accompany water use developments. All of the above changes affect the well-being and survival of the fish community.

Dr. Molles cites several investigators who have documented changes in the fish fauna that accompanied water development projects and the introduction of exotic fish species in the Lower Colorado River Basin. These same changes are currently taking place in the Upper Colorado Basin; at the present time, nearly half of the native fishes are considered by authorities to be endangered or threatened (table 1). Five of the six species listed as endangered or threatened in table 1 have probably achieved that status due to recent man-made changes in the aquatic ecosystem either through habitat alterations or through the introduction of exotic fish species.

Some years ago, Carlander (1955) observed that the total biomass of fishes was greatest when fish species diversity was greatest, but the biomass of a particular species was greatest when species diversity in the system was least. In essence, this means that no single species can possibly fully utilize a diverse aquatic system but that individual species thrive best under conditions of little or no interspecific competition. Hence, as the number of species increases, the more efficiently the ecosystem is utilized and the greater the total fish biomass becomes.

But, as species diversity increases, competition among species also increases and the abundance and realized niche of individual species is reduced.

Adaptive changes in the gene pool of a species takes place slowly. As indicated by Dr. Molles, over half of the native fish species of the Colorado River Basin belong to two families, Catostomidae and Cyprinidae. Apparently the native fish fauna of the basin began with few kinds, and over long time spans speciation took place by adaptive radiation. Western United States aquatic systems typically have less diverse fish species communities than eastern or central states aquatic systems. A less diverse species complex would generally mean a less competitive situation, a more expanded and diverse realized niche, and, consequently, the possibility of a more diverse gene pool within a given species. Such a system would permit further speciation and would also provide relatively easy access to introduced species. As additional species occur, however, species diversity would increase, competition would increase, the available niches in the system would be more fully utilized, and a more highly specialized, competitive species complex would result. Thus, as shown by Dr. Molles, species introductions in central states aquatic systems apparently met with great resistance. In the few cases where introduced species were successful, the native species were not merely displaced, as apparently happened earlier in the less species diverse Colorado system, but they were eventually competitively replaced.

The total species numbers of the central states aquatic systems remained relatively stable over the years. In the case of Colorado River fishes, however, the total species numbers increased as exotics were introduced. At first the native fishes of the Colorado system were simply displaced by introduced fishes. More recently, however, as available niches were taken up and man-made habitat changes further reduced suitable habitat, native species have tended to either disappear or become endangered or threatened.

Mortalities induced on a fish population either by competition or by environmental stress usually have a selective effect on the gene pool of that population. Let me illustrate. Salmon hatcheries normally receive their egg stocks from spawning runs of salmon that home to individual hatcheries. In most years, while some hatcheries receive more eggs than they can accommodate, other hatcheries may not receive an adequate egg supply. This situation has been rectified by transferring eggs from one hatchery to another. Two observations were made by hatchery managers: (1) salmon embryos originating from a foreign hatchery stock often suffered higher hatching and adult return mortality than native hatchery stocks; and (2) in instances where there is survival of transplanted stocks, survival success increased rapidly from generation to generation.[2]

In my opinion, this adjustment took place because the less adaptive genotypes in the gene pool of the transferred stocks were virtually eliminated within the first few generations. And further, the more harsh the stress upon a population, the larger the segment of the gene pool that will be lost. Man-made habitat changes and fish introductions along the Upper Colorado River are beginning to severely stress the native fish stocks. The reduction in native Colorado River fishes that we are currently observing is more than just a quantitative reduction in numbers of

---

[2]Personal communication with Fred C. Cleaver, Project Coordinator, Columbia River Fisheries Council.

fish. The really serious loss is a qualitative reduction of genetic variability in the fish populations, and hence, a reduction in genetic fitness, the ability of these populations to adapt and survive. These stresses are increasing. Dr. Molles has suggested the establishment of a preserve or series of preserves for native Colorado fishes in the upper basin. I would like to support Dr. Molles' suggestion and add the following suggestions of my own:

1. Set critical habitat preserves in key habitat areas for selected Upper Colorado native fishes as suggested by Dr. Molles.

2. Obtain a clear delineation of water development policy for the Upper Colorado River by all concerned agencies. That policy should include the priority position of the fisheries resource.

3. Conduct research studies designed to assess the habitat requirements of selected native fishes. We really know very little about their habitat requirements at present.

4. Initiate an immediate active participation of fisheries biologists early in the planning process for water development projects.

The fisheries biologist typically does not enter the decision-making system until far too late to be really effective. Some of the presently planned projects probably should not be built; the biological and economic costs may be too great. Others should and will be built, but environmental damage can often be minimized and a better project obtained by careful, cooperative planning among biologists, economists, and engineers. In some cases, we may be able to enhance previously impaired areas of the river ecosystem by specifying design features in proposed dams that will yield water temperatures, flow regimes, and maintenance of adequate in-stream flows to meet the requirements of important fishes.

References

Carlander, K. D. 1955. "The Standing Crop of Fish in Lakes," Journal of Fisheries, Research Board of Canada vol. 12, no. 4.

Crandall, D. L. 1974. "Management Objectives in the Colorado River Basin: The Problem of Establishing Priorities and Achieving Coordination," in A. B. Crawford and D. F. Peterson, eds., Environmental Management in the Colorado River Basin (Logan, Ut., Utah State University Press) pp. 12-23.

Holden, P. B., and C. B. Stalnaker. 1975. "Distribution and Abundance of Mainstream Fishes of the Middle and Upper Colorado River Basins, 1967-1973," Transactions of the American Fisheries Society vol. 104, no. 2, pp. 217-231.

Miller, R. R. 1959. "Origin and Affinities of the Freshwater Fish Fauna of Western North America," in C. L. Hubbs, ed., Zoogeography (Washington, D.C., American Association for the Advancement of Science, Publication 41).

U.S. Department of the Interior, Water for Energy Management Team. 1974. Water for Energy in the Upper Colorado River Basin (Washington, D.C.).

Utah Water Research Laboratory. 1975. Colorado River Regional Assessment Study: Part 3, National Commission on Water Quality, WQ5A0054 (Logan, Ut., Utah State University).

Chapter 10

THE IMPACTS OF HABITAT ALTERATIONS ON THE ENDANGERED
AND THREATENED FISHES OF THE UPPER COLORADO RIVER BASIN

Robert J. Behnke*

Introduction

To understand the basic reasons for the endangered and threatened

status of some of the fishes native to the Colorado River Basin, it is

helpful to have some concept of the geologic history of the basin and the

long evolutionary history of the species specializing and adapting to an

environment which essentially no longer exists. That is, the evolutionary

programming resulting in such unusual species as the squawfish, the razor-

back sucker, and the bonytail and humpback chubs has dictated life histo-

ries and ecologies discordant with present conditions.

The Environment and the Fishes

The Colorado River Basin extends approximately 1,700 miles from the

Gulf of California to the headwaters of the Green River, Wyoming. The

present drainage was established when two separate river systems forged

a connection by cutting through the present Grand Canyon several million

years ago in Pliocene times (McKee and coauthors, 1967). Except for main-

stream species, there has always been a sharp faunistic separation between

upper and lower basin fishes (above and below the Grand Canyon). The

Colorado River Basin probably lacked direct connections with any other

_____

*Department of Fishery and Wildlife Biology, Colorado State University, Fort Collins, Colorado.

major drainage for millions of years. This resulted in long isolation of
the fish fauna. Except for species inhabiting headwater streams such as
trout, sculpins, speckled dace, and mountain suckers, which can be transferred between drainage basins by stream capture, the majority of the native species of the Colorado River Basin are endemic, that is, they have
been so long isolated from their nearest relatives they have evolved into
species now restricted to the Colorado Basin and found nowhere else. The
great antiquity of the native fauna is revealed by chub and squawfish
fossils of mid-Pliocene age found in Arizona (Miller, 1959). The Colorado
Basin fish fauna exhibit the highest degree of endemism of any major drainage in North America. The minnow and sucker families (Cyprinidae and
Catostomidae) comprise about 70 percent of the freshwater fish species
native to the Colorado Basin. Miller (1959) claimed 87 percent of the
twenty-three species of minnows and suckers known to be native to the basin
at that time are endemic to the basin.

Of the thirty-five-plus species of freshwater fishes native to the
Colorado River Basin (table 1, Manuel Molles' paper, this volume), thirteen
are native to the upper basin (table 2, Manuel Molles' paper, this volume;
and table 1, Robert Raleigh's discussion, this volume). Some of these
species are now extremely rare. The squawfish and humpback chub are listed
as endangered under the 1973 Endangered Species Act. The state of Colorado
also considers the razorback sucker (<u>Xyrauchen texanus</u>)[1] and bonytail chub

---

[1]The species <u>Xyrauchen texanus</u> is more commonly known as the humpback
sucker. The author prefers to use the name razorback sucker because it is
more descriptive of the sharp-edged structure forming the "hump" of the
fish, and because it avoids confusion with the humpback chub.

as endangered species and the Colorado River cutthroat trout as a threatened species. (See also table 3, Bishop's and Porcella's paper, this volume.) Four of these fishes will be discussed below. It should be pointed out that a comprehensive, basic ichthyological survey of the upper and lower basins has yet to be made. Long ago, Miller (1946) stressed the need for such a survey. For example, both the flannelmouth sucker and the blue-head sucker are represented by two morphologically distinct forms. It is not known if these "ecotypes" are reproductively isolated species or local environmental modifications of a common genotype.

The distribution of some of the native upper basin species is disjunct and sporadic. For example, the mountain sucker, Catostomus (Pantosteus) platyrynchus, a typical inhabitant of small tributary streams, is common in such habitat in the Green River drainage of Wyoming, but in Colorado is known only from Piceance Creek (tributary to the White River) and Trout Creek (tributary to the Yampa River). The Piute sculpin, Cottus beldingi, (formerly the Eagle sculpin, Cottus annae) is known only from a few localities in Colorado, despite an abundance of small, tributary habitat apparently ideal for this species' requirements in the basin.

More detailed information on species distribution and faunal associations would be most useful information for interpreting and predicting effects of environmental change, and to recognize specific areas where significant numbers of native species yet persist. Holden and Stalnaker (1975) reviewed collections made in the upper basin from 1967 to 1973. Of twenty-nine species found, nineteen were nonnative fishes.

Prior to the civilizing impact of humans, the Colorado River system was characterized by tremendous fluctuations in flow and turbidity.

Miller (1961) cites flows recorded in the Colorado River at Yuma, Arizona, ranging from 18 cubic feet per second (cfs) in 1934 to 250,000 cfs in 1916. The drainage basin lacked large natural lakes so the native fishes lacked evolutionary specializations for lacustrine environments. There were no barriers to prevent free movement along the main channels and into major tributaries, and it is likely migratory movements were a regular part of the life history of the mainstream species.

For millions of years the unique environment of the Colorado River, with its great diversity and torrential flows through canyon areas, directed the evolutionary pathways followed by the native fishes and molded the bizarre morphologies of the razorback sucker and the humpback and bonytail chubs, and produced the largest of all North American minnows, the giant squawfish.

The major tributaries draining the mountains and foothills formed meandering streams with quiet backwater areas which were likely important reproductive and nursery areas for the native fishes.

The effects of mainstream dams on the large, mainstream species are well known because the dramatic environmental changes can be characterized as sudden and catastrophic in relation to the survival of these species both in the new reservoirs and in the cold, clear tailwaters below the dams. Other gradual, cumulative impacts relating to land-use practices have been occurring for more than 100 years in the basin. These more subtle environmental changes are difficult to quantify, and the assumption of their negative influence on the presently endangered and threatened species is largely theoretical and circumstantial. According to predictable channel development dictated by the principles of fluvial geomorphology, major tributaries such as the Yampa, White, Gunnison, and San Juan rivers would continually cut new channels and create oxbows and side channels of quiet backwater habitat separated from the new main channels. It is logical to assume that the utilization of such predictably-occurring backwater environments as nursery areas for the young of the main channel fishes (analogous to the importance of estuarine areas for early life history stages of many marine fishes) would be incorporated as an intrinsic part of their life history.

Because of fertility and ease of irrigation, the valley bottom lands were the first lands to pass into private ownership. To protect their investments from the natural encroachment of the rivers and to prevent flooding and facilitate irrigation, landowners began to channelize the rivers and riprap the banks to better confine and control the rivers' flow. The advent of the bulldozer has greatly accelerated this process.

There is no documentation, to my knowledge, estimating the loss of backwater habitat in the upper basin. Also, there is no documentation in reference to the significance of such habitat in the life history of any of the endangered or threatened species.

The circumstantial evidence of the significance of quiet, backwater environments is based on the frequent field observations of squawfish and, particularly, razorback suckers in such areas. Because these backwater habitats are so rapidly vanishing, all existing areas possibly associated with any of the endangered and threatened species should be identified and, if possible, protected--at least until we have a better understanding of the significance of such environments.

The appearance of squawfish and razorback suckers in the backwater pond (an artificially created backwater pond resulting from gravel excavation

and subsequent inundation by the Colorado River) in the Walter Walker Wildlife Area near Grant Junction, Colorado, in 1975 indicates the importance
of offstream backwater habitat in the life history of these species. It
also points to the obvious potential of the creation of artificial backwater habitats in an effort to enhance the survival and abundance of endangered species. The basic characteristics of "ideal" backwater habitat are
yet to be discovered, however, in relation to optimum utilization for all
of the critical life history stages of the endangered and threatened species.

The construction of mainstream dams formed large lakes which regulated
flow regimes, precipitated out the silt load, and released cold, clear
water, creating new environments for which the native mainstream fishes
were ill-adapted. The four mainstream specialized species (razorback sucker,
squawfish, bonytail chub, and humpback chub) have suffered substantial declines from their former abundance.

The Razorback Sucker

The razorback sucker (_Xyrauchen texanus_) has maintained limited populations sporadically throughout its former range, but is particularly rare
in the upper basin. The incidence of hybridization between the razorback
and flannelmouth suckers is increasing in a changing environment. Such
hybrids have long been known (Hubbs and Miller, 1953). Recently, forty
of ninety-three specimens of razorback sucker specimens collected from 1967 to
1973 in the upper basin were found to be hybrids (Holden and Stalnaker,
1975). The razorback sucker is currently proposed for "threatened" status
under the Endangered Species Act (1973).

The Squawfish

The squawfish (<u>Ptychocheilus</u> <u>lucius</u>) has not been found in the lower
basin since 1968, and has been continually declining in the upper basin.
The decline of the squawfish was apparently well underway in the upper basin
prior to the construction of Flaming Gorge Reservoir (1962) and Lake Powell
(1963). Squawfish were rarely encountered in the pre-impoundment surveys of
1959 and 1960. This earlier decline can be attributed to the gradual, cumulative effects of a changing environment, pollution, and a changing fish
fauna, becoming increasingly dominated by nonnative fishes. A nonnative species I would single out as particularly inimical to the squawfish and other
native fishes is the redside shiner, <u>Richardonius</u> <u>balteatus</u>. The redside
shiner was first found in the upper Green River, Wyoming, in 1938. By 1959
it was the dominant species in the upper Green River drainage as determined
by the Flaming Gorge pre-impoundment surveys (Smith, 1960; Bosley, 1960).

The redside shiner was not found in the Yampa River drainage during a
1952 survey (Bailey and Alberti, 1952). This species was first recorded from
the lower Yampa in 1961 (Banks, 1963), was judged as moderately abundant in
1966 samples,[2] and was the dominant species in collections made in 1975-76
(Prewitt and coauthors, 1976). No indication of successful squawfish reproduction (finding young) has been observed in the Yampa River since 1969
(Holden, 1973; Prewitt and coauthors, 1976), and adult specimens continue
to decline in fish collections of various surveys made from 1967 to 1976.

---

[2]Personal communication with Dr. Kent Andrews, U.S. Fish and Wildlife
Service, Denver, Colorado (February 1977).

The Yampa River has not been altered by large, mainstream dams or any other modification which may be characterized as resulting in sudden and significant change. The decline here is associated with gradual, cumulative changes favoring the introduced species. The squawfish is currently listed as an endangered species under the Endangered Species Act (1973).

The Bonytail Chub

The bonytail chub (Gila elegans) was at one time widely distributed and abundant in all of the mainstream of the Colorado and Green rivers and their major tributaries. Its virtual demise came about rapidly after the construction of large, mainstream dams and is now one of the rarest species in the basin, holding on in small numbers in Lake Mojave in the lower basin and perhaps in the Green River in Desolation Canyon, Utah. I know of no authenticated records of Gila elegans from other areas in recent years. The problem of documenting data on the bonytail chub is complicated by the fact that this species has long been confused with the common roundtail chub, Gila robusta, and also by the increase in the incidence of hybridization among the Gila chubs (bonytail, humpback and roundtail) since the completion of Lake Powell and Flaming Gorge Reservoir. The bonytail chub has been proposed for endangered status under the Endangered Species Act (1973).

The Humpback Chub

The humpback chub (Gila cypha) was probably always relatively rare and of local occurrence in deepwater canyon areas of the Colorado and Green rivers. Most of the former prime humpback chub habitat was inundated by the numerous mainstream reservoirs, and the few known present populations in the upper basin appear to be affected by hybridization with Gila elegans and Gila *robusta.* As with the razorback sucker, a changing environment with a preponderance of nonnative species has stimulated the breakdown of reproductive isolation and threatens further the existence of the original species. The humpback chub is recognized as an endangered species under the Endangered Species Act (1973).

Cutthroat Trout

The native cutthroat trout of the upper basin should be considered as threatened. This fish is virtually extinct as pure populations. Although also suffering from habitat loss, the major factor in the decline of native trout has been the introduction of nonnative trouts which have replaced or hybridized with the native subspecies.

Energy Development, Coexistence, and Realities

Considering the history of extinction of other animal species in relation to the prospects for continued survival of the four endangered and threatened species, it should be recognized that if well-documented curves of historical abundance were available for these species, they would likely be beyond the inflection point where the process of extinction proceeds rapidly toward zero abundance.

Another doleful aspect for the future of the endangered and threatened species is that natural resource- and conservation-oriented groups and agencies are not likely to make a unified stand to champion the cause of endangered and threatened Colorado River fishes because these species are of the minnow and sucker families and of little direct economic significance. Although the environmental changes, unwise planning, mismanagement of Colorado River water, and the fate of the native fishes make a tragic story, the fact remains that the new reservoirs support multi-million dollar recreational fisheries--based entirely on nonnative fishes, a fishery that was not possible with the native fishes in their original environmental setting.

The Endangered Species Act provides some protection of the habitat of the species listed, particularly section 7 of the act which prohibits the activities of any federal agency from jeopardizing the continued existence of these species. However, a mere "holding the line" on preserving what is left of the former environment may not be sufficient to prevent extinction, and certainly will not serve to increase the distribution and abundance of these species to a point where they are no longer considered endangered or threatened; these species can be restored only by creating conditions which are conducive to successful reproduction and survival of early life history stages in areas where such conditions no longer exist.

Because of the above considerations, I have come to believe that the most hopeful option for avoiding extinction of species such as the squawfish is for future development projects to build in endangered species mitigation and enhancement plans from the earliest planning stages, analogous to salmon restoration projects on Pacific Coast rivers. For example, squawfish and razorback suckers have been artificially propagated but the problem of where to stock the young for any survival is yet to be solved. Excavations for dam construction could be made to create artificial backwater areas and to serve as spawning and nursery grounds. Control structures could be designed to exclude nonnative competitors and predators. Flow regimes from a dam could be controlled to produce a downstream minimum temperature of 68° F by mid-June and maintain a predetermined minimum flow for the reproductive season.

Unfortunately, so little is known about the life histories of the endangered and threatened species that there is no assurance that any enhancement project will work at all, or what the best project design would be. But time is running out and to avoid initiating some attempt to perpetuate endangered species, while awaiting the results of years of basic research to learn some subtle aspects of life history and ecology, is a delaying tactic. We may find that, although the new information is useful, it is too late.

References

Bailey, C., and R. Alberti. 1952. "Fish Population Inventory," in Colorado Game and Fish Department, Lower Yampa River and Tributaries Study, Federal Aid Project F-3-R-1 (Denver, Colo.).

Banks, J. L. 1963. "Fish Species Distribution in Dinosaur National Monument During 1961-1962" (M.S. thesis, Colorado State University, Fort Collins).

Bosley, C. E. 1960. Pre-Impoundment Study of the Flaming Gorge Reservoir, Fish Technical Report No. 9 (Cheyenne, Wyo., Wyoming Game Fish Commission).

Endangered Species Act. 1973. Public Law No. 93-205, 87 U.S. Statute 884.

Holden, P. B. 1973. "Distribution, Abundance and Life History of the Fishes of the Upper Colorado River Basin" (Ph.D. dissertation, Utah State University, Logan).

Holden, P. B., and C. B. Stalnaker. 1975. "Distribution and Abundance of Mainstream Fishes of the Middle and Upper Colorado River Basins, 1967-1973," Transactions of the American Fisheries Society vol. 104, no. 2, pp. 217-231.

Hubbs, C. L., and R. R. Miller. 1953. Hybridization in Nature Between the Genera Catostomus and Xyrauchen (Ann Arbor, Mich., Michigan Academy of Science, Arts, and Letters, Paper 38) pp. 207-233.

McKee, E. D., R. F. Wilson, W. J. Breed, and C. S. Breed, eds. 1967. "Evolution of the Colorado River in Arizona," Bulletin of the Museum of Northern Arizona vol. 44.

Miller, R. R. 1946. "The Need for Ichthyological Surveys for the Major Rivers of Western North America," Science vol. 104, no. 2710, pp. 517-519.

_____. 1959. "Origin and Affinities of the Freshwater Fish Fauna of Western North America," in C. L. Hubbs, ed., Zoogeography (Washington, D.C., American Association for the Advancement of Science, Publication 41).

_____. 1961. Man and the Changing Fish Fauna of the American Southwest (Ann Arbor, Mich., Michigan Academy of Science, Arts, and Letters, Paper 46) pp. 365-404.

Prewitt, C. G., D. E. Snyder, E. J. Wick, C. A. Carlson, L. Ames, and W. D. Fronk. 1976. Baseline Survey of Aquatic Macroinvertebrates and Fishes of the Yampa and White Rivers, Colorado Annual Report to the Bureau of Land Management (Denver, Colo., Colorado Department of Game, Fish, and Parks).

Smith, G. R. 1960. "Annotated List of Fishes of the Flaming Gorge Reservior Basin 1959," in C. E. Dibble, ed., <u>Ecological Studies of the Flora and Fauna of Flaming Gorge Reservoir Basin,</u> Utah and Wyoming (Salt Lake City, Ut., University of Utah Anthropological Papers 48) pp. 163-168.

THE IMPACTS OF HABITAT ALTERATIONS ON THE ENDANGERED
AND THREATENED FISHES OF THE UPPER COLORADO RIVER BASIN:
A DISCUSSION

Paul B. Holden*

It is often difficult for most of us to critically judge the position
of another scientist in an area where we know very little. We often look
for faults in their reasoning, but seldom know enough about the data underlying their position to adequately critique it. Occasionally, we hear or
read a paper in our field and can evaluate not only the logic, but also the
underlying data. For many of us in specialized fields, where information
is constantly being added and is poorly circulated, it is common to find a
position that logically sounds good, but is predicated on erroneous information. This situation often establishes an opinion as fact rather than as
a testable hypothesis.

It is rather important that this type of situation be avoided in the
Upper Colorado River Basin. The endemic fish fauna has suffered great declines and is in danger of extinction. Considerable energy development is
planned for the upper basin and, if initiated on incorrect information,
could cause the Colorado River Basin ecosystem to be lost. Therefore,
decision makers must have in hand all the facts necessary to develop energy
resources without losing the rare fishes. My comments are directed at
Robert Behnke's basic data and are an attempt to provide the reader with
an evaluation of the available data.

Dr. Behnke's analysis of the evolution of the river and recent alterations and their effects on native fishes is accurate. The larger tributaries

*Aquatic Section Manager, BIO/WEST, Incorporated, Logan, Utah.

and main channels of the Colorado River system are geologically old. The
San Juan, Yampa, Gunnison, and White rivers were old, meandering rivers before the uplift of the Colorado Plateau. This geologically recent uplift
increased their gradient substantially, yet the rapid cutting action of the
rivers maintained their meandering configuration. These rivers are by and
large entrenched in canyons and do not have the opportunity to cut extensive
new channels or to create oxbows of quiet backwater habitat. These habitats
occurred naturally in limited portions of the basin. One such area in the
upper basin is the Grand Valley near Grand Junction. In this area, the
Colorado River may have been channelized to allow agricultural development
of the bottomland. However, recent aerial photographs indicate little such
activity. Several other relatively flat areas exist, such as the Green
River in the Uinta Basin. I have personally floated these areas many times
and flown over them and have yet to see an oxbow. The available evidence
rather conclusively shows that natural, large backwaters, or oxbows, were not
common in the upper basin.

Natural backwaters in the river system covered relatively small areas,
usually mouths of washes flooded during high flows, and deeper, high-water
channels during low flows. They lasted only short periods of time and constantly changed locations. These areas still exist in much of the upper
basin, especially along the Green River, and are used extensively by Colorado
squawfish and razorback suckers as feeding areas.

Dr. Behnke's inference that large, artificial backwaters near Grand
Junction are good for rare fishes is erroneous. He infers that these species spawn there by mentioning "the appearance of young." Kidd (1977) has

worked extensively in this section of river and mentions nothing about young-of-the-year, or juvenile rare fish, in these areas; nor does he mention anything about spawning in these areas. McAda and Seethaler (1975) noted razorback suckers spawning in the Walker Wildlife Area, but could find no young after repeated sampling efforts. They also noted razorbacks spawning on cobble riffles in the Yampa and Colorado rivers. Several investigators have noted razorback suckers spawning near the shores of lower basin reservoirs (Minckley, 1973). Young have not been found in these areas either. It appears that razorbacks, trapped in ponded areas, spawn near shores where wave action causes some current, but success is very poor. Jordan (1891) and others in early reports suggested that razorbacks and squawfish migrated upstream to spawn. No reference in any of the literature to date suggests anything other than spawning in running water. Areas where Vanicek (1967) and Holden (1973) found young-of-the-year squawfish did not contain large backwaters.

The Walker Wildlife Area and the adjacent Colorado River have extremely abundant green sunfish and largemouth bass populations (Kidd, 1977). This is the only area in an upper basin main channel where these exotic piscivores are abundant. The large, permanent backwaters that warm considerably in the summer are ideal spawning and rearing areas for these fish. The natural backwaters of the river system are too ephemeral for these species, hence they do poorly in most natural areas. It appears very likely that the large numbers of exotic fishes play a major role in the reduced population recruitment of rare native fishes.

220

Carl Seethaler and Chuck McAda[1] have both mentioned the high incidence
of parasitism in the Walker Wildlife Area. They believe, as I do, that the
high temperatures and stagnant water are centers for disease and parasites
that the native fishes are not adapted to, but which the introduced species
probably evolved with.

All the evidence suggests that large backwaters such as the Walker Wildlife Area are not used as spawning or rearing areas for rare fish. They are
used by adult or subadult squawfish and razorback suckers as feeding areas.
Some of these fish apparently become trapped and end up trying to spawn
there. Parasitism, disease, and predation are high in these backwaters and
they therefore exert a strong negative influence on the native fish populations of the adjacent river.

Dr. Behnke has singled out the redside shiner as a serious problem for
squawfish reproduction in the Green and Yampa rivers. The redside was found
in the Little Snake River, a tributary of the Yampa River, in the late 1950s
(Wyoming Game and Fish Department, 1958). It was abundant in the Green
River of Dinosaur National Monument in the mid-1960s when Vanicek (1967)
found abundant reproduction of Colorado squawfish and bonytail chubs. Holden
and Stalnaker (1975) considered the redside shiner abundant in the Yampa
River during studies in 1971. Just when this species became abundant in the
Yampa is questionable. Differences in collecting techniques may bias individual collector's assessments of true abundance. It appears that the redside has been in the Yampa drainage for over twenty years and, even when
abundant, it has not affected the success of squawfish reproduction in the

---

[1] Utah Cooperative Fishery Unit, Utah State University, Logan (personal
communication, 1976).

Green River. Colorado squawfish were reproducing in the Green River of Dinosaur National Monument below the mouth of the Yampa River until 1968 (Holden
and Stalnaker, 1975b). Tailwaters of Flaming Gorge Dam became colder in about
1967 when the reservoir became filled to near capacity. This colder water
apparently caused the loss of squawfish reproduction in the Green River below
the mouth of the Yampa River in 1968, and no young squawfish have been found
there since.

The problem of cold water and its effects on native fishes is best shown
by Vanicek, Kramer, and Franklin (1970). The problem I am addressing here is
the delayed effect of the tailwaters getting colder and colder as the reservoir fills and the intakes for the tailwaters become deeper and deeper. I
discussed this as a hypothesis in earlier research (Holden, 1973). Since
that time, this problem has become well documented and I will be initiating
a project to evaluate the effect of penstock modifications of Flaming Gorge
Dam (which will warm up the tailwaters) on native fishes and invertebrates in
the Green River.

Based on the data available to us at present, much additional information
is needed in order to protect the rare fishes. Fortunately, there is sufficient time to collect and analyze the necessary data if we start now and
do not delay any longer. There is no need to become a doomsday advocate yet,
except perhaps for the bonytail chub. More intense searches for bonytails
may yet find reproducing populations.

Hopefully, my discussion has provided a look at the other side of the
coin. A considerable amount of information is known about the upper basin
fishes and the reasons for their declines. Unfortunately, only a very few

people are familiar with this information. Public policy makers must be aware of all the data available and must have all interpretations of the data before deciding on a course of action.

Several instances have already occurred where plans were made concerning fisheries in the Upper Colorado Basin based upon inaccurate information. It costs time and money to rectify these situations--situations that need not have occurred. Considering the status of the fishes in the upper basin, it is imperative that future decisions regarding development be based upon accurate interpretations of the available biological information. Information currently available strongly suggests that instream flows, and their effects on habitat, may be the single most important factor in the survival of native fishes in the upper basin. I recommend that we stop wasting money on large backwaters and water chemistry-bioassay studies. We should place the major effort into habitat changes occurring because of reduced flows. In addition, studies on sites and factors necessary for adequate spawning are needed. These are the areas that most future developments will be impacting and, therefore, need to be studied now. Finally, competent researchers, familiar with the Colorado basin and its fishes, should do the work or much time and money will be wasted.

My major difference with Dr. Behnke is that I believe we can save the remaining native fish populations in the upper basin and actually increase their size and range in the future. There is no need to talk of having these species only in hatcheries or large man-made backwaters.

References

Holden, P. B. 1973. "Distribution, Abundance and Life History of the Fishes of the Upper Colorado River Basin" (Ph.D. dissertation, Utah State University, Logan).

Holden, P. B., and C. B. Stalnaker. 1975. "Distribution of Fishes in the Dolores and Yampa River Systems of the Upper Colorado Basin," Southwestern Naturalist vol. 19, no. 4, pp. 403-412.

_____. 1975b. "Distribution and Abundance of Mainstream Fishes of Middle and Upper Colorado River Basins, 1967-1973," Transactions of the American Fisheries Society vol. 104, no. 2, pp. 217-231.

Jordan, D. S. 1891. "Report of Explorations in Colorado and Utah During the Summer of 1889 with an Account of the Fishes Found in Each of the River Basins Examined," Bulletin of the U.S. Fisheries Commission vol. 9, pp. 1-40.

Kidd, G. 1977. An Investigation of Endangered and Threatened Fish Species in the Upper Colorado River as Related to Bureau of Reclamation Projects. Final Report to the U.S. Bureau of Reclamation, Grand Junction, Colo.

McAda, C., and K. Seethaler. 1975. Investigations of the Movements and Ecological Requirements of the Colorado Squawfish and Humpback Sucker in the Yampa and Green Rivers. Progress Report to the Fish and Wildlife Service, Salt Lake City, Utah (Logan, Ut., Utah Cooperative Fishery Unit, Utah State University).

Minckley, W. L. 1973. Fishes of Arizona (Phoenix, Az., Arizona Game and Fish Department).

Vanicek, C. D. 1967. "Ecological Studies of Native Green River Fishes Below Flaming Gorge Dam, 1964-1966" (Ph.D. dissertation, Utah State University, Logan).

Vanicek, C. D., R. H. Kramer, and D. R. Franklin. 1970. "Distribution of Green River Fishes in Utah and Colorado Following Closure of Flaming Gorge Dam," Southwestern Naturalist vol. 14, pp. 297-315.

Wyoming Game and Fish Department. 1958. A Fisheries Survey of Streams in the Little Snake River Drainage. Fisheries Technical Report No. 6 (Cheyenne, Wyo.).

THE IMPACTS OF HABITAT ALTERATIONS ON THE ENDANGERED
AND THREATENED FISHES OF THE UPPER COLORADO RIVER BASIN:
A DISCUSSION

James E. Johnson*

At present, a diverse assemblage of groups and agencies is evaluating
the waters of the Upper Colorado River Basin for a variety of uses. This
variety is possible because, to date, water developers in the upper basin
have selected only the most valuable, available and easily exploited resources.
This has resulted in only scattered irrigation, energy development, and flood
control projects. But now the best damsites and most easily irrigated fields
are in use, and demand for the remaining water is rapidly increasing. Planners are looking to the Upper Colorado River Basin for the water needed to
develop future energy supplies of coal, shale oil, and hydroelectric power.
A growing population needs water to drink and water to grow food; reservoir
recreation is becoming very popular; and native, endangered fish need habitat.
Potential uses for the remaining upper basin waters are highly competitive
because of these expanding needs, and the demand is rapidly approaching and
perhaps even exceeding the supply. All uses alter the resource, some of
course more than others, and all changes place limitations on future options.
When water is used for one purpose, its potential use is limited for others.
The use itself changes the nature of the resource.

Recognizing that the demand for upper basin waters is increasing,
potential users of the Colorado River attempt to justify their proposals by
asserting the many secondary benefits that will be derived from the develop-

*Endangered Species Biologist, U.S. Fish and Wildlife Service, Albu-
querque, New Mexico.

ment. This game of multiple use is interesting but unrealistic. The waters
of the upper basin of the Colorado River are a valuable resource and they are
becoming more valuable daily; they should be put to their most beneficial uses,
assuming that they can be determined, and these uses alone should justify the
alteration of the resource.

Thus, we come to the first problem I would like to address--how should
priorities for the uses of the upper basin waters be determined? Is it possible to assess impartially a proposed water development project while at the
same time considering the value of the resource as habitat for endangered
species like the Colorado squawfish (Ptychocheilus lucius) and the humpback
chub (Gila cypha)? It is difficult not to begin comparing renewable and nonrenewable resources directly at this point, but the totally different outlook
on their utilization appears to make this difficult, if not impossible. Nonrenewable resources are, by definition, finite in quantity and their utilization apparently is regulated by a combination of supply, demand, and conservation towards an endpoint of total exploitation. Renewable resources
have a nearly infinite potential if managed properly. However, without
management and under the control of supply and demand, renewable resources
can also be rapidly exploited to the endpoint of extinction, eliminating all
of their future potentials, the majority of which are probably yet to be
determined. It therefore seems more realistic not to compare these two types
of resources directly, but to keep in mind the advantages of both the potential of the renewable resource and the immediate satisfaction of the nonrenewable resource.

Using Colorado River water as a finite resource, two scenarios are
likely in assessing the benefits of an economic water development project

and conservation of endangered fish species. If the policy decision is made to protect the resource as a natural habitat for endangered species, the results are just that the area is not developed, the nonrenewable resources are exploited at a slower rate than they would have been otherwise, and society is forced to either find alternative sources, lower their demands and perhaps their standards of living too, or even to do without. In addition, several resources, including the endangered species, are conserved for the future. If the decision goes the other way and the area is developed, immediate material gains are realized, the demand for more of the same resource increases, and the use alters the resource. The habitat is destroyed for the endangered species and the potential of the renewable resource eliminated, not directly by exploitation, but incidentally as the habitat is changed. Note that a third scenario is not suggested, one that allows both the development of the water resource and continuation of the endangered fish species. Of course, this is dependent upon the type and extent of the development, but past experience in the lower basin of the Colorado River indicates a mutual exclusion of major water developments and endangered species of fish.

I suggest the question of the priority use of Colorado River water is not as simple as some make it out to be. One of the first questions usually asked in determining priorities is, "Which is more important, people or fish?" Of course, this is not the real question. It would better be phrased, "Which is more important, a better way of life for me or the very survival of several unique and little understood species of fish and perhaps even an entire ecosystem?" I cannot argue that most people would like an increased lifestyle, but a realistic estimation of the costs of that increase must be carefully considered. Native fish are a renewable resource and, with proper management, the supply will last forever. What may be looked upon as having little value to us today may be extremely important 100 or 1,000 years from now. In my opinion we cannot afford to fortuitously destroy any renewable resource, no matter how unimportant it may seem in light of our present needs, and eliminate it from future options.

I believe that even the most ardent conservationists agree that we must continue to exploit our nonrenewable resources, even though there may be impacts on our renewable resources as well. But only the most essential developments should be permitted to cause the extinction of a renewable resource, such as fishes, and then only with forewarned knowledge of the outcome and the best possible estimate of benefits and losses, and to whom they accrue.

The most acceptable method of developing any resource is to minimize the impact on the surrounding resources. In part, this is what the Endangered Species Act (1973) hopes to accomplish. For example, a recommendation went into Washington early in 1976 nominating 623 miles of the Green, Yampa, Colorado, and Gunnison rivers in Colorado and Utah as critical habitat for the Colorado squawfish (figure 1 and table 1). It is the belief of most biologists working with the remnant Colorado River fish fauna that if these remaining river reaches are protected from the massive alterations that destroyed the lower basin habitats, the native fish may survive. If these reaches are not protected, the species will surely disappear. If the Secretary of the Interior accepts the recommendations of critical habitat for the Colorado squawfish, it does not mean that those 623 miles of rivers will be closed to development, or even fully protected against major alterations.

But what it does mean is that any project that is funded, authorized,

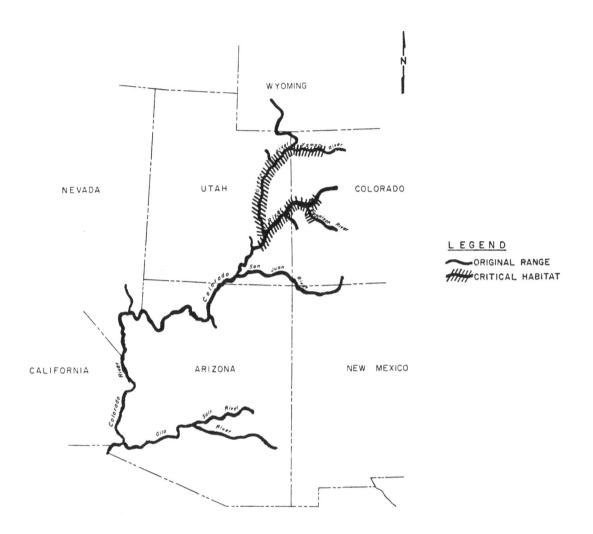

Figure 1. Map of the Colorado River Basin Showing the Original Range
of the Colorado Squawfish (Ptychocheilus lucius) and the
Reaches of the Four Streams Recommended for Critical Habitat

Table 1. Stream Reaches in the Colorado River Basin Recommended as
          Critical Habitat for the Colorado Squawfish (Ptychocheilus lucius)

Stream	Location	Stream miles
Colorado River	From its junction with Plateau Creek in Colorado to the high water level of Lake Mead, Utah	225
Green River	From its junction with the Yampa River in Dinosaur National Monument, Colorado, to its confluence with the Colorado River in Utah	294
Yampa River	From its junction with Milk Creek in Colorado to its confluence with the Green River in Dinosaur National Monument, Colorado	92
Gunnison River	From the State Highway 141 bridge crossing in Colorado to its confluence with the Colorado River at Grand Junction, Colorado	12
Total		623

or carried out by a federal agency will not be allowed to jeopardize the squawfish or destroy its critical habitat. According to the act, any group, private or federal, is free to develop the resources of a region as long as the project does not negatively impact the endangered species or their critical habitat. For example, the Bureau of Land Management (BLM) in Colorado is initiating coal leasing in the Yampa River Basin. In the stipulations to private firms bidding on these leases, BLM must carefully state that mining, cleaning, and shipping of the coal must not negatively impact the critical habitat portion of the Yampa River. The private firms must then take these stipulations into account when estimating their bids for the development of the resource.

This brings up the next topic I would like to discuss, the economics of preserving parts of the Upper Colorado River Basin as critical habitat for endangered species of fish. The economic system in the United States seems built upon the necessity for growth, development, and an unlimited supply of natural, nonrenewable resources. In the past, we have been able to ignore sustained yield and concentrate on exploitation because of the vast quantities of resources available relative to the demand. But now some nonrenewable resources are dwindling, and new sources are becoming more difficult to locate. As supplies continue to decrease, prices will rise, making it profitable to dig deeper holes and build larger dams, further exploiting the nonrenewable resources and increasing the chances of destroying the nearby renewable resources.

In the past, we have been willing to set aside small tracts of public land as wilderness, primitive areas, and national parks as long as economic progress was not impeded. Nearly all of these areas, by definition, were undeveloped before they were reserved, indicating that at the time it was unprofitable to develop them. This is not the case with western water, and especially Colorado River water. This is a resource everyone agrees we cannot afford to waste. Economic growth is straining this resource. Thus, we are starting to face another problem--what are the limits of this finite resource? There are already reports that Colorado River water is overallocated; in essence, the fish there are already living in someone else's water. We must address the fact that Colorado River water, like any other finite resource, has limits. How many people, projects, developments, acres, etc. can this finite resource support? Developers refer to the abundance of barren land in the drainage, the number of open damsites, the coal deposits, and the unirrigated fields, and point out that they are not being utilized; that there are still resources available and there is still room for development. The limiting factor in any development in the Upper Colorado River Basin will not be land, minerals, manpower, energy, or money--it will be water. Water has always been the limiting factor in this basin, and there are many indications that those limits have already been reached. If we continue to develop water-using projects in this area, the long-term losses will begin to outweigh the near-term benefits. At present, I believe that there is still enough natural habitat left in the upper basin to support remnant populations of the endemic fauna, using the water allocated to lower basin states. If development in the upper basin continues uncontrolled, the native fauna of the entire drainage will be destroyed and the projects will be appropriating water from downstream users. A more satisfactory agreement among the uses of the Colorado River waters must be reached, one that takes into account both the finite limits of the water and the infinite potentials of the native fish.

The legal mandate to provide for representation of the native fishes has already been created in the Endangered Species Act. It is now the law of the land that we protect those species faced with possible extinction, and preserve habitat critical to their survival. In this case, the least altered reaches of the Upper Colorado River Basin will be protected by reducing the impacts of some federal projects and by the total ban of other federal projects with histories of significant negative impacts.

I have two specific criticisms of the proposal Dr. Behnke presents in his paper. These concerns lead me to a totally different conclusion from the one Dr. Behnke reaches. However, I do believe his ideas are new and creative, and should be presented to the scientific community for review and comment.

1. I question the statement that the Colorado River and its major tributaries "according to predictable channel development dictated by the principles of fluvial geomorphology...would continually cut new channels and create oxbows of quiet backwater habitat separated from the new main channels." The Colorado River and its major tributaries are generally described as classic examples of youthful watercourses, expending energy by downcutting and vertical erosion rather than lateral erosion and deposition that cause low gradient river systems to form oxbows. The canyons of this region are legendary (e.g., Grand Canyon, Desolation Canyon, Black Canyon of the Gunnison), and backwater areas are usually limited to a few stream mouths and connecting arroyos. It seems unlikely that oxbows and extensive backwaters were vital parts of a habitat in which the endemic Colorado River fishes evolved.

2. I agree that young squawfish and chubs do not normally inhabit mainstream channels and torrential runs, but rather seek quieter waters

behind spits and bars, and even backwaters when available. However, I cannot
believe that a lack of these areas is now causing the decline of Colorado
squawfish, humpback chubs, and bonytail chubs (Gila elegans). The endemic
Colorado River fishes evolved under a low competitive pressure, partially
because the environment was so extreme as to drastically limit the available
niches, and because the extremes quickly eliminated rather than propagated
most genetic diversity. The few species that did evolve in the Colorado River
Basin were uniquely adapted to its harshness.

Today most of the Colorado River and its tributaries have been changed.
In the lower basin, many backwaters (called mainstream reservoirs) have been
formed. But the squawfish and chubs have not profited from these backwaters;
in fact, they have been totally and unquestionably eliminated from the entire
lower basin. Only in the upper basin, where backwater areas and exotic species are still scarce, do the native species remain. I therefore find it
difficult to understand the suggestion that backwater areas planned in conjunction with dam building in the upper basin might be the only salvation for
the native Colorado River fish species. Instead, I would suggest backwater
areas in the upper basin would be far more suitable to the many exotic fish
species that now totally prevail in the lower basin and are slowly encroach-
ing upon the upper basin. I further suggest that giving these exotic fish
species a haven in areas now more suitable to native species might result
in just the opposite effect than the one desired, by increasing competition.
Perhaps the 623 miles of Green, Colorado, Yampa, and Gunnison rivers are not
ideally suited for the endemic fish species, but they are still less suitable
for the exotic species.

The conclusion I draw as to the future of the endangered Colorado River

fish species is in opposition to Dr. Behnke's. Healthy reproducing populations of Colorado squawfish still exist in the Green River between the Yampa
River and the Colorado River, and perhaps in the Colorado River itself below
the Green River. The same is true for humpback chubs in the Colorado River
in at least three separate localities. Adult squawfish are found throughout
the areas recommended for their critical habitat, as are razorback suckers
(Xyrauchen texanus). The Endangered Species Act (1973) and associated newly-
awakening state interests in these native species can protect this habitat.
Given this protection and the natural tendency of this drainage to exclude
poorly-adapted species, I believe the native fish species can survive and
even expand their present range and numbers. But if we give up on them now,
and fail to give existing legislation and natural selection a chance, we
doom the native fish species to extinction.

References

Endangered Species Act. 1973. Public Law No. 93-205, 87 U.S. Statute 884.

236

Chapter 11

THE ECOLOGY OF COLORADO RIVER
RESERVOIR SHORELINES

Loren D. Potter*

Introduction

Except for the headwaters of the Colorado River drainage system, the

main part of the Upper Colorado River Basin is within the Colorado Plateau.

The most recent and extensive reservoir of this region is Lake Powell formed

behind Glen Canyon dam. Centrally located in the Colorado Plateau, its geo-

logy, physiography, climate, and vegetation are typical. Because of its

recency and because of the research efforts to study its impact on a variety

of shoreline features, Lake Powell is being used here as an example to illus-

trate some of the important researchable programs, results, and interactions

of the physical and biological environment, and highlights of significant

concerns relating to fluctuating water levels as they affect the aquatic-

*Professor of Biology, Department of Biology, University of New Mexico,
Albuquerque, **New Mexico.**

Research for the shoreline investigations was supported by grants
NSF GI-34831 and NSF-AEN 72-03462 A03 to the Shoreline Ecology Subproject
of the Lake Powell Research Project at the University of New Mexico, under
the Research Applied to National Needs Program of the National Science
Foundation. This is Contribution No. 43 of the Lake Powell Research Project
Contribution Series.

Appreciation is expressed to Natalie Pattison, who has been associated
with the project from its inception, and to graduate students Ellen Louder-
bough, Don Standiford, and Ed Kelley for their excellent cooperation and
productivity in the field and laboratory.

terrestrial interface--the shoreline. The impact of man and the effect of the dynamic physical and biological processes occurring along the shoreline are discussed as exemplary of other reservoirs in the upper basin. The contrast of shorelines of the river above a reservoir, and mention of the transition from river to reservoir, and the distinct effects of the reservoir on the nature of the river below the dam are alluded to.

It is not the intent here to present an academic, total, comprehensive outline of a complete ecosystem analysis, the completion of which would exceed any reasonable funding or time available. Instead, an example of an ecological approach is presented, based on field perception of the many physical and biological aspects which are most critical and specific to the interactions of the shoreline. These are deemed to be significant to the management of the reservoir and are related to man's use. It is important to remember that although reservoirs may have been constructed for water flow regulation, irrigation, and power production, Lake Powell is in the Glen Canyon National Recreation Area and the use of the lake itself, and its shoreline, is for recreational purposes. Fishing involves people and people use the shoreline. This paper addresses itself to that relationship. The need is stressed for some shoreline management in order to preserve some desirable, limited sandy shores for recreational use, rather than allowing their development into impenetrable thickets of exotics with noxious insect populations and with submerged off-shore masses of flooded vegetation.

River and Reservoir Setting

Geology and Physiography of the Lake Powell Area

In the arid Southwest the sparseness of vegetational cover and the harshness of the environment result in a close and direct correlation of the geologic strata and physiography to the development of vegetation and to the physical features of reservoir shorelines. Because of important interactions, a brief discussion of the geology, climate, and physiography of the Colorado Plateau is essential as a base for understanding the current shoreline ecology. By mid-Miocene the general course of the Colorado River had developed through the faulting, folding, and uplift of the Tertiary. The early Miocene channel is today represented by the broad upper gorge with its own steep cliffs in the upper sediments.

Into this channel a second cutting cycle in the Pliocene and Pleistocene carved an inner gorge throughout much of the region into the Glen Canyon group--Navajo, Kayenta, and Wingate sandstones. The overlying, marine-deposited carmel sandstone was left to form a broad irregular platform between the inner and outer gorges. Alluvial mantles of cobble and gravel of the Pleistocene stream activity were left on several levels of terraces. Above rise the massive walls of mesas, buttes, and monuments of Entrada, Summerville, and Morrison formations. The location and extent of the reservoir and some of the major physiographic features are indicated in figure 1.

Cataract Canyon is the area of the Colorado River from the junction of the Green River to the mouth of the Dirty Devil River. Glen Canyon is that portion of the Colorado River from the Dirty Devil River to **Lees Ferry.** While Cataract Canyon lies across major anticlinal features and has a

239

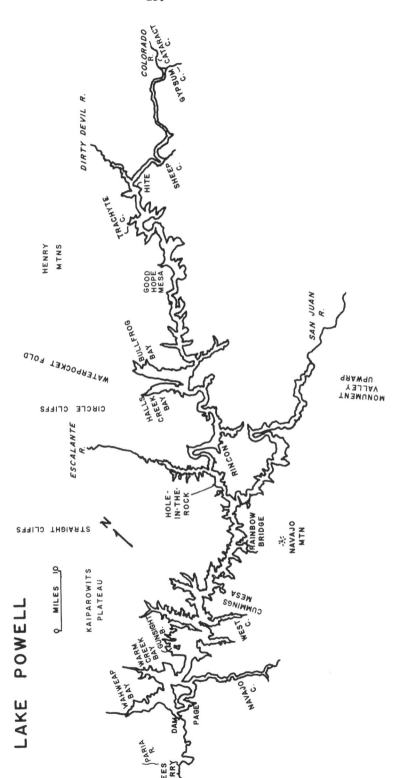

Figure 1. Map of Lake Powell Showing Principal Features

gradient of about 8 feet per mile (with rapids), Glen Canyon has a low average gradient of the Colorado River bed of 2 feet per mile. Because of major structural features of uplift and dip, strata appear and disappear as seen along the shoreline. Glen Canyon was the longest continuous canyon along the Colorado River. Because of an overall dip of strata to the west, the river bed which started on Permian Cedar Mesa sandstone at the Dirty Devil ended at the present damsite on the younger Jurassic Navajo sandstone.

In the Lake Powell area expanses are vast, desolate, and relatively uninhabited. Plateaus are deeply incised by tortuous, vertical-walled gorges, or saw-cut canyons, especially distinctive in such rocks as Navajo and Wingate sandstones. Erosional remnants of plateaus remain as isolated mesas or buttes. Upthrust areas, or laccolithic structures, appear as steeply rising mountains, e.g., the Henry Mountains and Navajo Mountain. The sparseness of vegetation, the erodability of the sandstone, and the force of intense summer thunderstorms result in a rapid runoff, the scouring and polishing of surfaces to form "slickrock," and the accumulation of eroded sand into shallow basins to form sandpockets which become the "garden spots" of the area. As the runoff water accumulates on the plateaus it develops a shallow drainage pattern moving down-slope across the mesas, pouring across the steep surface of the cliff face and swirling out **erosional pits in soft spots. These pits vary from shallow, temporally** wet depressions to deep, extensive ponds with marsh vegetation. Some occur at the base of waterfalls several hundred feet high which may flow only occasionally after a severe local rainstorm. The physiography of the Glen Canyon area and the previous geological studies have been summarized by Cooley (1958, 1959a, 1959b).

Climate

Because of the great differences in elevation within the Lake Powell region there are varying climatic conditions. The Henry Mountains to the northwest of the upper end of Lake Powell have peaks rising to elevations of 11,615 feet (Mt. Ellen). The lower portion of the lake is affected by the extensive high plateaus, e.g., the Kaiparowits to the north, and Navajo Mountain rising to 10,388 feet to the southwest.

Most of the Lake Powell area is in the zone classified as Upper Sonoran, at elevations of 3,000 to 5,000 feet. Late summer thunderstorms from air masses of the Gulf of Mexico account for the highest monthly precipitation in August and September. The annual average precipitation is 4.78 inches. The erratic aspect of the rainfall is illustrated by that of October 1972 when nearly 4.0 inches, almost the yearly average, fell that month. Winter rains from Pacific air masses are usually slow drizzles accompanied by snow flurries.

The average maximum temperatures occur in July at 97° F; the average minimum is in January at 24° F. The comfort factor in the summer when hot days often exceed 100° F is somewhat ameliorated by the cooling effect of the dry air when in the shade but the effect of the brilliant sunlight in the clear air can be extremely enervating. Comfort at night is largely affected by the physiographic situation as it controls air movement.

Strong continuous winds are most common from February to May, while summer winds are strong and gusty with the possibility of accompanying thunderstorms in the afternoon.

Regional Vegetation

Surrounding mountains have vegetation ranging from subalpine grassland meadows, spruce-fir forest, western yellow pine/Douglas-fir forest,
to pinyon-juniper woodland. The upland plateau areas are principally
pinyon-juniper woodland plus a complex of types often classified as northern desert shrub.

Because moisture is the prime limiting factor for plant growth and
many species are close to their moisture tolerance limits, any features
of topography, soils, or slope exposure which influence the moisture
regime of the microclimate become dramatically important in affecting the
nature of the vegetation. For example, where cavation results in overhanging alcoves coincident with a seepage line where water may perenni-
ally or seasonally flow to the surface, a mesic, calciphilous vegetation
dominated by maidenhair fern (Adiantum capillus-veneris) or rock mat
(Petrophytum caespitosum) develops. Sandy soils, once stabilized, provide
for deep moisture infiltration, excellent root growth, and high biomass
production. Stabilization of alluvial soils along tributary canyons has
led to their development into the only areas with a true soil profile,
supporting stands of oak woods with typical forest litter and understory
vegetation.

The most mesic vegetation occurs at the heads of tributaries, large
alcoves where water is supplied from springs or seeps, or in temporary
pools supplied by runoff water. Marshy vegetation may develop, surrounded
by a deciduous woody grove--the counterpart of a moist eastern deciduous
forest stand.

From a River to a Lake

Many of the aspects of the Colorado River as it was before the formation of Lake Powell can be seen today in areas of Cataract Canyon and above
it. The principal source of flow is from the spring melt of snow from the
Rocky Mountains, with about 70 percent of the annual runoff during April to
July. A secondary time of inflow occurs during summer thunderstorms when
there is rapid runoff from the sedimentary rock strata with minimal absorptive soils. The many aspects of Glen Canyon prior to flooding have been
excellently summarized by Jennings (1966).

A dense woody vegetation is found on old, stranded river terraces,
some 25 feet above the river bed. Below is a lower floodplain terrace
occasionally covered during flood stage. The scouring effect of the silt-
laden, rushing water and tumbling rocks effectively destroys most of the
vegetation which has developed during low-water stages. With the retreat
of the floodwaters, an ideal, newly silted, moist seedbed is available for
the typical plant species of sandbar and river floodplain succession. In
addition to willow (Salix spp.), tamarisk or salt cedar (Tamarix pentandra)
is now a common dominant pioneer where, after three months, seedlings of
this exotic may form a solid green mat on the lower terrace.

Above the level of Lake Powell the river remains the same as it has
been. Where the river flows into the lake the changes are variable. Previously established shoreline vegetation becomes permanently submerged and a
new seasonal cycle of flooding occurs, with lake level fluctuations devoid
of the impact of rushing floodwaters. In some areas there is silt deposition and throughout the lake there is the occasional impact of wave action.
Water levels of the lake now are the result of controlled releases at the

dam superimposed upon natural inflow rates and evaporational losses. On entering the lake, the force of the current diminishes; the silt load gradually settles; and turbid, cold, saline water tends to sink toward the bottom of the lake.

The transition from river characteristics to those of a lake (involving such features as flow, salinity, temperature, and biological regimes) is reported in bulletins on physical and biological limnology (Reynolds and Johnson, 1974; Hansmann, Kidd, and Gilbert, 1974; Kidd, 1976). In spite of the great length and volume of the lake, it exhibits lateral transport and mixing features typical of a river.

Glen Canyon dam, a structure to a height of 583 feet above the former river bed, was designed to form a reservoir which, at a 3,700-foot elevation, will comprise 28,040,000 acre-feet of water, an area of 164,000 acres, a length of 186 miles, and a total shoreline of about 1,800 miles. There is an abrupt change from lake back to river immediately below the dam. From the dam and some 15 miles through lower Glen Canyon to Lees Ferry, which is the dividing line between the Lower and Upper Colorado River Basins, the river has taken on new characteristics and will continue to do so because the released water is cold, clear, and controlled. Some aspects of change in the river and shoreline have been summarized by Dolan, Howard, and Gallenson (1974).

Shoreline Surface Materials

Since 1964 the reservoir has been in the process of filling--with patterns of yearly net gains and normal seasonal fluctuations. While increased levels in one tributary may result in a shift to extensive

shallow bays with great increases in surface area, this is usually balanced by other areas where the shoreline becomes steeper with little increase in area.

The nature of the surface materials of the shoreline, the physiography, texture, rock composition, and erosion obviously affect the importance of the shoreline for aesthetics, boat mooring, camping, shoreline
recreation, and vegetational development above the water line and of the
seasonal drawdown zone. There is also a significant effect, both direct
and indirect, on the physical and biological features of the reservoir
itself as there are changes in sedimentation, turbidity, wave action,
aquatic vegetation, nutrient addition, and fisheries.

Shoreline Mapping

As part of the Lake Powell Research Project sponsored by the National
Science Foundation, the shoreline surface of Lake Powell was surveyed and
mapped. For purposes of mapping, the shoreline surface types were divided
into seven basic categories: cliff face, domed terrace, shelfy terrace,
talus, alluvium, sand, and rockslide. Each was indicated by a type symbol
on the map. A variation was added if a thin mantle of talus, alluvium, or
sand was found over a basic type. The entire shoreline of about 1,800 miles
was mapped in detail on 10-foot contour maps at a scale of 1:4,000, or 13.6
inches per mile. Type differences as small as 200 feet, represented on the
map as 0.5 inch, were mapped.

Description

The cliff face is the shoreline type for which the Canyon lands and Glen Canyon area, particularly, are most famous. The classification was used for both sheer vertical cliffs and for those too steep on which to easily land a boat and get ashore.

Domed terraces refer to areas of smooth undulating contours or domes with sloping sides and gently carved-out depressions between. These are favorite rock campsites.

Shelfy terraces of Upper Navajo sandstone, a shoreline type of limited extent, are most common to the lower end of the reservoir. They are finely cross-bedded with alternate hard and soft layers which are frequently tilted and erode into projecting shelves. Beaching and mooring a boat here requires selection of very local sites.

The talus classification was used for the weathered rubble of varying size which had eroded and slid downslope but did not include the large, blocky landslides classified as rockslides.

Dune sand was used for areas with sand to several feet or more in depth. These shoreline areas are the most popular for many vacationers because of the lack of rocks, the ease of mooring a boat and pitching a tent, the generally level topography, the sandy playground for children, and the fulfillment of the "traditional" concept of a sandy beach at the lake. To others, the disadvantage of windblown sand mixed with bedroll and food makes these sites least desirable.

Throughout the extent of Lake Powell there are areas geologically designated as Pleistocene alluvium, often as a mantle of coarse gravel and boulders over old river terraces.

Rockslides are of minor extent and are most common in areas of exposure of the soft, crumbling, Chinle formation underlying the blocky
Wingate sandstone.

Extent of Shoreline Types

The mileage and percentage of each shoreline type are given in table 1.
The 74 percent of the shoreline comprised of cliff, talus, and rockslide
(54, 19, and 1 percent, respectively) provides much of the scenic aspect
but is of little significance in recreational shoreline use. A portion of
the 21 percent of domed terrace, and especially the 3 percent that is sand
and alluvium, receive the major recreational impact. These are also the
areas most readily invaded by tamarisk when exposed as drawdown zone.

Shoreline Succession

Biomass Studies

The entire area of Lake Powell is within the region designated by
Kuchler (1964) as type no. 59, blackbrush shrub (<u>Coleogyne</u> <u>ramosissima</u>).
A variety of associations occur in differing physiographic situations
dependent principally upon the depth of weathered soil, the soil texture,
the microclimatic conditions, and the availability of water. Because of
the major influence of physiography and surface materials on vegetation,
the studies of biomass were conducted under the same classification as
the types used in mapping the surface materials around the lake. These
vegetational studies provide a basis for evaluating the effects of flooding and the process of land succession into the seasonal drawdown zone.
Specific results are published elsewhere (Potter and Pattison, 1976).

Table 1. Percent Distance of Shoreline Surface Types and Total Mileage at Four Contour Levels of Lake Powell

Total shoreline surface types

Percent distance

Contour (feet)	Cliff	Domed terrace mantle of				Shelfy terrace	Talus	Sand	Alluv-ium	Rock slide	Total mileage
		Barren	Talus	Sand	Alluvium						
3,620	52.58	9.99	0.02	0.75	10.94	2.45	17.93	2.99	1.43	0.95	1,472
3,660	52.69	9.04	0.01	0.85	11.48	2.38	17.22	2.63	2.87	0.85	1,638
3,680	53.72	8.89	0.04	0.81	11.37	2.31	18.87	2.31	0.98	0.69	1,733
3,700	54.42	8.67	0.04	0.71	11.14	2.25	18.98	2.19	0.93	0.66	1,823

The richest vegetation, both in biomass and in number of species represented, of the shoreline types sampled was that of the sandy shore composed
of dune sand. The biomass weight and foliage cover were highest of all
types. In these areas there is deep infiltration of precipitation and almost no runoff. Although blackbrush, because of its woody perennial nature,
provides over one-third of the biomass, the greatest coverage is provided
by Indian ricegrass (Oryzopsis hymenoides).

The principal shoreline type which is invaded by vegetation upon exposure during the seasonal drawdown is the sandy shore. In the successional
zone, the total foliage cover in one year was nearly the same, 4.5 percent
versus 5.0 percent, and the yearly production of vegetal biomass amounted
to 13.3 grams per square meter in comparison with 86.5 for the standing
accumulated biomass of the unflooded sandy sites. The notable addition to
the flora was that of salt cedar, or tamarisk. The seasonal lowering of
water level in 1971 exposed a drawdown zone to one-year succession of which
tamarisk comprised 66 percent of the biomass and 56 percent of the foliage
cover. Secondly, there was a great increase of pale primrose (Oenothera
pallida). The third most important invader was sand dropseed (Sporobolus
cryptandrus), which is well known for its high rate of seed production and
dissemination into sandy areas. In total, about eighteen of the fifty taxa
from the dune sand area were represented in the one-year successional drawdown vegetation below.

It is this biomass which, upon reflooding and decomposition in the
spring, will lead to nutrient enrichment of the lake and which meanwhile
serves as a growth substrate for a rich periphyton flora and fauna. It is
also a major factor in recreational use of the favorite sandy shoreline

sites. The one-year growth, dominated by tamarisk, may be up to 6 feet tall. If not flooded for two years, it can become 10 to 12 feet tall. When flooded, a submerged woody vegetation will be left offshore, resulting in serious objections by boaters and swimmers.

Riparian Succession of the Colorado River

Where the Colorado River originally flowed through relatively wide valleys in the Glen Canyon area, the densest and most extensive development of vegetation occurred. On the upper terraces, forest stands were dominated by Fremont cottonwood, netleaf hackberry, and Gambel oak with a rather dense cover of understory shrubs. Along the lower river terraces and the stream banks was a vegetation dominated by thickets of willow and typical of sandy shorelines of the Southwest. Tamarisk, the ubiquitous introduced exotic, was then and is today a dominant shrub and tree, forming nearly impenetrable thickets and finally forest stands along river terraces both above and below the area of Lake Powell.[1] Woodbury and coauthors (1959) listed some fifteen species of grasses and eighty-nine species of forbs common to river terraces in Glen Canyon.

---

[1]Tamarisk is native to the Mediterranean area of Africa, Europe, and Asia. According to Horton (1964) it was not reported in the Southwest in the early 1800s but was for sale in New York City nurseries in 1823. It was reported along the Gila River in 1898 and along the Salt River in 1901. Since 1900, the spread of tamarisk throughout the Southwest has been phenomenal.

Lakeshore Succession of Lake Powell

With the flooding of Glen Canyon, many of the above types of stream-
side vegetation have been almost entirely eliminated. Along the main channel there are no comparable physiographic sites, even though the rising
water of the reservoir is now flooding ancient river terraces and meanders.
These have been exposed and eroded for thousands of years with the resulting desiccation and eluviation having removed the former vegetation and the
soils which had been produced there. Most tributary streams or canyons
have a steep descent into Glen Canyon and are without the typical river-
bank vegetation. Undoubtedly then, the formerly riparian vegetational
types most decreased and lacking today as lakeshore types are those which
occupied sandbars, floodplains, and lower river terraces. The vegetational types of plateaus, dissected canyons, seeps, alcoves, talus slopes,
dunes, and ancient terraces remain much as before.

Effect of Fluctuating Water Levels on Shoreline Types. The impact of
the rising water level into an arid desert shrub vegetation is very slight
over the vast area of the shoreline represented by cliff wall, domed terrace, and shelfy terrace. Here the changes due to flooding and recession
of water level are principally chemical and physical. With the lowering
of the reservoir level there is left a whitish "bathtub ring" on cliff
walls. The surface layer consists of a coating of friable, granular de-
posit rich in calcium carbonate but is usually mixed with a large percentage of organic material resulting from the growth of diatoms and blue-green
and green algae.

Outstanding are the tremendously large talus slopes which have developed from weathering of canyon walls and from sand blown over the edge from

the plateau above. With water level fluctuations, some talus slopes have slumped downward as much as 20 feet; others have completely disappeared into the bottom sediments.

Along shorelines previously covered with Pleistocene alluvium submergence also results in a change in texture. The area of the yearly drawdown zone has been subject to wave action and removal by suspension of fine-textured material of sand and silt which has left a coarse, boulder field as a shoreline--a much less desirable place for recreational use.

Shoreline areas consisting of several feet of dune sand may slump, be wave-cut into terraces, or be completely washed away.

The maximal period of shoreline use for camping occurs from May to September. It should be noted that in the normal yearly cycle of fluctuating water levels, the times when levels are at the upper or maximal heights correspond to this period of maximal use. Although visitor use for fishing remains relatively stable from April to October, recreational shoreline use is more summer-oriented. During the reservoir's filling stages, the yearly peak of water level usually exceeds that of the year before. The most-used shorelines are sandy beaches which, during the summer seasons, have had new water levels rising into native vegetation. During this period there is no evidence of either previously submerged, dead, partially decomposed vegetation, or the result of successional invaders into the sandy areas of the drawdown zone. From the viewpoint of maintaining a natural appearance of the summer aspect of the reservoir's shoreline and of avoiding the apparent intrusion of exotic successional invaders, this period of reservoir life is the "honeymoon" period.

Growth Effects of Peak Water Levels. The vegetation of those areas submerged by rising waters is soon killed. Most plants succumb upon complete submergence--it is biologically unreasonable to expect a desert-adapted species to quickly be successful as a submerged aquatic. As the water level rises into the native vegetation there is an increase of available soil moisture along the new shoreline. The distance and rate of lateral movement of moisture is greatest and fastest in dune sand. Here a zone of growth stimulation occurs for a distance of 10 to 20 feet upslope. The dominant species, such as Indian ricegrass, Mormon tea, goldenweed, and sand sagebrush (Artemisia filifolia) show an increase in vegetative growth and flowering.

Vegetational Succession of Drawdown Zone. As the water level retreats in sandy areas, there is left a moist, sandy seedbed ideal for germination of seeds and the development of roots of seedlings of species the migrules of which are either blown by wind or are water-dispersed and left along the shore.

The dominant plant of these stands is tamarisk. This introduced Eurasian exotic has been successful in spreading throughout the Southwest, being found along the channels of most rivers, and to the headwaters of most tributaries. Its light, airborne seeds, also capable of water transport, have allowed it to develop a discontinuous distribution. Its establishment is favored by a moist, sandy seedbed. It is tolerant of alkaline conditions common in the Southwest. It has all the attributes common to an invader of sandbars and sand dunes. It flowers and matures seed over a long period, its seeds are quick to germinate after dispersal, and the seedling root system consists of a rapidly growing taproot which allows it

to penetrate deeply and to keep ahead of the surface drying of sandy soils.
During seedling development there is a favorable root:shoot ratio which
provides for maximal water absorption and reduced loss by transpiration.
Seedlings with shoots 1 to 2 centimeters high may have taproots 30 to 38
centimeters long. As lateral roots develop, shoot growth increases elonga-
tion of stems, basal buds produce a bushy growth form, absorptive power of
roots is increased, and the roots become capable of adding to vegetative
proliferation by producing root sprouts, thus avoiding the more susceptible
seedling stage. After germination in July to September, the seedlings grow
slowly during the winter and do not become deciduous. They reach a height
of 1 to 2 meters by May and start to flower by May or June. If allowed to
grow through the summer growing season, they mature and continue to flower
the first year and reach heights of over 2 meters by September. Thus, the
species has all the attributes needed for competitive invasion and rapid,
dense establishment, and it succeeds in reproducing itself both vegeta-
tively by root sprouts and by dispersing seeds within the first year of
establishment. During the winter months when the water table is lowered
because of reservoir drawdown, tamarisk has the advantage after the first
year of being deciduous and losing its transpiring leaves.

As the water level is lowered after its peak in early July, those
areas having a sandy shoreline are receptive seed beds for seeds which
are dispersed by water and are washed up on the shore. Germination occurs
and the seedlings are repeatedly moistened for several weeks by occasional
waves washing ashore, even though water levels are lowered. In a few weeks,
rows of tamarisk seedlings appear on contour lines. Meanwhile, those at
the maximal water levels have been growing just below the wave-cut terrace

representing the peak water level. The net result of one fall season of
invasion and growth in the drawdown zone is a series of contour rows of
seedlings and a gradient in height.

If the vegetation of the drawdown zone is not completely submerged by
the reservoir levels the next year, as in 1972 when the peak was 2 feet
lower than the 1971 maximum, a belt of tamarisk remains along the shoreline, is already mature, and provides a local source of seeds to reinvade
the drawdown zone. With the abundance of nearby seed supply, the density
of seedling establishment increases. And where there is a gently sloping
sandy shoreline, vast meadows of tamarisk seedlings may develop. The lower
part of the drawdown zone exposed in late fall and early winter is either
barren of vegetation or is invaded by bugseed (Corispermum nitidum)--a
common weed naturalized from Europe, or by Russian thistle (Salsola kali)--
a common invader of the West and naturalized from Eurasia. So, the drawdown zone becomes dominated by three exotic pioneer species--quite in contrast to the native vegetation.

In figure 2 is a summary of the number and distribution of tamarisk
seedlings as they relate to shoreline elevations and dates of water elevations. In the zone between 3,622 and 3,620 feet was found the tamarisk belt
established in the previous year which was not inundated in 1972 and which
was a principal source of seeds. However, the number of seedlings within
this belt of tamarisk grown to 1 or 2 meters in height was very small.
Here, most of the reproduction was from root sprouts. After reaching the
maximum level of 3,620 feet in late June (1972), the water level receded.
During July there was an increasing number of seedlings developed, reaching
a maximum density at the water level attained about 25 July. Irregular

256

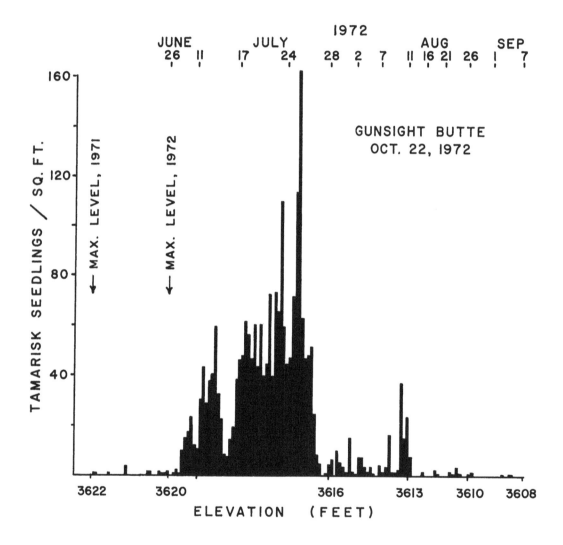

Figure 2.   The Number of Tamarisk Seedlings of the
Drawdown Zone Correlated to Time of Beach
Exposure and Contour Level

germination occurred at decreasing water levels until the second week in August after which only occasional seedlings developed. Beyond the first week in September practically no seedlings were found. Thus, the upper vertical 7 feet of the drawdown zone exposed from the end of June until the second week in August contained nearly all the seedlings of tamarisk.

The above pattern of reproduction during July and August suggests a possible method of control with proper water level manipulation. If a narrow zone was exposed during this time of seedling establishment and then the water level was raised sufficiently to completely submerge the seedlings and held for an adequate period to cause their death, the water level could then be lowered and the exposed zone not be suspect to tamarisk invasions. Plans were made to experimentally determine the limitations and susceptibility of tamarisk seedlings to this method of control.

Submergence Effects

Breakdown and Decomposition of Flooded Vegetation

The rising water of the reservoir which floods vegetation, organic litter, and in some areas soil with organic matter has effects which are both detrimental and beneficial. The decomposition of all organic matter provides a source of increased nutrients and organic acids into the reservoir which are added to the continued influx of nutrients washed in from the tributary canyons and the Colorado River. The rate of physical breakdown and biological decomposition of different species varies with their anatomical structure and with the chemical composition of tissues, e.g., the degree of lignification. Some underwater studies at intervals of time after submergence have provided an insight into the relative rates of breakdown.

Among the trees common to the area, the more resistant include single-leaf ash, Gambel oak, and netleaf hackberry. Less resistant to decay are western redbud (Cercis occidentalis), squawberry (Rhus trilobata), Fremont cottonwood, and species of willow.

Among the common shrubs, the most prevalent, blackbrush, is also the most resistant to decay. It is followed in resistance by species of Mormon tea whose photosynthetic stems, which are heavily cutinized and lignified, remain intact. The deep, woody taproots of several species resist removal by wave action and the surface erosion by undertow during periods of stormy weather. Species of cacti rapidly decompose after submergence due to their high percentage of soft, succulent tissue. Unique is the breakdown of the leaves of narrow-leaved species of Spanish bayonet (Yucca spp.). The stiff leaves are resistant to mechanical breakage by the waves. However, after several weeks of submergence, microbial action and retting leave only the long strands of thick-walled, lignified cells of the fibro-vascular bundles.

Most broad-leaved herbs (forbs) are quickly decomposed after submergence and none remain photosynthetically active for more than a few days or weeks. Grass leaves quickly become limp strands of fibrous tissue. Some of the fine leaf and stem material form floating mats of straw-like chaff which are areas of active microbial activity and mercury accumulation (Potter, Kidd, and Standiford, 1975).

The chemical and biological decomposition of the submerged vegetation is important in adding to the enrichment of nitrates and phosphates of the reservoir. Other breakdown products such as tannic acids and a variety of soluble organic carbon compounds may have complex interacting effects on such things as salt precipitation rates, an interaction being investigated

by Dr. Robert Reynolds of Dartmouth College. Finally, the submerged vegetation provides a substrate for the growth of periphyton and invertebrate
organisms. These organisms include macro and microinvertebrates, diatoms,
green and blue-green algae, and bacteria. During the spring rise of water
level, submerged vegetation may quickly develop a coating of periphyton
within several days, appearing as a misty shroud of greenish, gelatinous,
and filamentous growth. With the accumulation of the algal coating, the
population of invertebrates increases. These are important as a basic food
supply to the fish populations and are thus an important aspect of fishery
management. Simultaneous with the development of periphyton is the development of benthic (bottom) organisms on the sand, silt, and rock substrate of
the bottom. Silt and organisms combine into a highly flocculated, granular
complex, olive-green in color.

When the water level is lowered, the flocculated silt, clay, and algae
are left to dry on cliff walls and shores of rock and sand. On the cliffs
a thin layer of whitish, crusty material forms part of the whitening referred
to as the "bathtub ring." On sloping shorelines the benthic material,
which may wash into depressions, dries into a light-weight, porous, foam-
like material, greenish-gray in color.

This benthic organic mat, surprisingly dense in coverage and high in
organic matter, is a major source of organic nutrients. Where pockets of
silt and organic matter occur under shallow water of only several feet in
depth and even aerated by wave action, the bottom may become anaerobic and
the chemical process of reduction may occur. This is evidenced by a black
layer of reduction just below the surface which is further evidenced by
the odor of hydrogen sulphide.

Impact of Submergence on Tamarisk Seedlings

Because most of the invading tamarisk seedlings become established from the time of peak water level in July until the first part of September, an experiment was designed to test the possible control of tamarisk seedlings one, two, and three months old by submergence in late September.

After the number of seedlings in each container was counted, they were maintained off a sandy shoreline in Lake Powell under about 50 centimeters of water. At appropriate intervals, containers of plants of each age were removed and partially buried in the sandy shore just above the waterline to allow them to receive some moisture from below and for the soil to become drained to field capacity. Several days after removal from the water the live plants were counted. Figure 3 is based upon the data obtained from these measurements. It is obvious that a few days of submergence is not an effective control measure for any age of seedlings, elimination being 50 percent after three days for one-month-old seedlings, a level not reached until about six days for two-month-old seedlings, and not until more than twenty-four days for three-month-old seedlings. About 95 percent mortality was attained for one-month-old seedlings at twelve days and 100 percent elimination for both one- and two-month-old seedlings after twenty-four days. At that time, however, only 40 percent of the three-month-old seedlings had succumbed.

Although the plants in the experimental pots were subject to some stresses such as wave action and coating with silt and algal growth, natural seedlings would probably be exposed to greater physical deterioration caused by the surge of sand back and forth across the bottom, and to undercutting and burial.

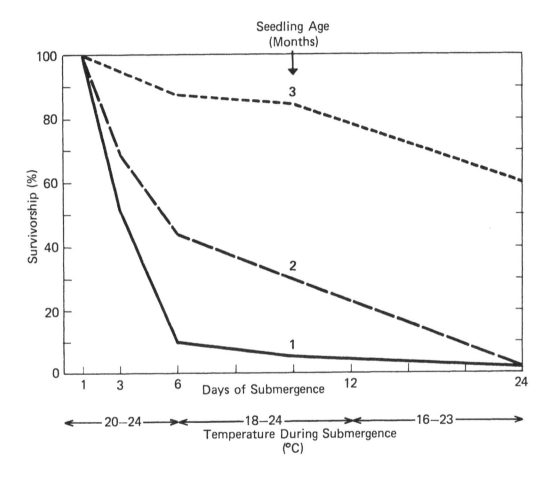

Note: The number of 1-, 2-, and 3-month-old seedlings at the beginning of each submergence trial averaged 1,099, 263, and 33, respectively.

Figure 3.   The Effect on Tamarisk Seedlings of Submergence Under Natural Conditions in Lake Powell, 26 September-20 October 1975

This preliminary experiment showed a definite increased resistance to flooding with increase in age of the seedlings from one to three months. The submergence study was done in the fall of the year at the time when it would be necessary to use water level regulation as a possible control of tamarisk invasion of the drawdown zone. This management practice would require lowering the water after the peak level to expose a narrow zone for tamarisk seedling establishment, then in mid-September raising the water sufficiently to submerge the newly established seedlings. From the data presented above, this procedure would be quite successful to control one- to two-month-old seedlings if they were submerged for a period of about three weeks.

A stand of tamarisk on a sandy shore in Wahweap Bay became established in July to September 1974, and plants were 1 to 2 meters tall by the time the bases were flooded on 26 June 1975. The water level in 1975 completely covered the tops of 2-meter-tall tamarisk plants for a period of thirty-four days. During this period of submergence many twigs and some leaves stayed alive; adventitious roots were developed from the submerged stems and functioned as aquatic absorptive roots. Upon emergence with the drawdown from 17 August to 9 October 1975, many shoots showed increased chlorophyll in surviving leaves and others produced new shoots. Within a month's time after exposure of the upper twigs, some plants were in flower and even fruit. As the water retreated to the ground level, the adventitious stem roots (now aerial) became nonfunctional, and survival was dependent on the activity of the original root system growing in the sand base. During submergence this sandy substratum had developed a layer 7 centimeters thick just below the surface which was jet black and emitted hydrogen sulphide as a result of chemical reduction. Roots in this layer had sloughed off. Below the zone the roots remained alive and, upon soil drainage with lowered water level, continued growth.

The tremendous adaptability of this riparian species to flooding once plants reach several months of age is obvious and is further evidenced by its mechanisms for rapid recovery on exposure. These mechanisms also include the habit of proliferation by producing basal shoots and new stems by root sprouts from lateral roots. Simultaneously, most of the herbs which had invaded the area before flooding were killed by the submergence, removing them from interspecific competition. The moist sand left by the retreating water also provides an ideal seed bed for the seeds of tamarisk produced by the recovered plants.

Shoreline Predictions and Water Level Management

At the time of this writing (in the fall of 1976) the predictions are that with normal runoff and water use, the water level of Lake Powell will reach 3,700 feet in a period of two to three years. There is also a generally accepted opinion that after several years, with increased water demand of a variety of projects, the operating level will be lowered to somewhere around 3,660 feet. A normal seasonal drawdown of 15 to 20 feet can be anticipated. Based on these predictions, it is felt advisable to indicate the likely sequence of events which will occur after a reexposure of 40 vertical feet along the shoreline.

The shoreline least affected will be the cliff walls which occupy about 54 percent of the shoreline. The principal change will be superficial but very visual. Even one year of submergence and then exposure with lowering of water level results in a distinct white "bathtub ring." Because many of the vertical cliff walls are protected from weathering, after an initial sloughing off of encrusted material, the "whitewash" appearance can be expected to remain for several decades. On cliff faces subject to streaking by patina formation, the horizontal bands of whitening will become streaked.

Also slightly affected, but in a different way, will be the 21 percent of domed terrace shoreline. About one-third of this area had a thin mantle of talus, sand, or alluvium covering the rounded surface of sandstone. Some of this mantle will have been removed by submergence and wave action. Some will have washed into deeper sediments; some will be redeposited in hollows within the domed terrace. Upon exposure and aeration, these pockets will be reinvaded by vegetation and will slowly tend to develop toward the rich isolated flora which was there previously. Many, however, may be occupied by tamarisk during the period of lowering the water level. The depth of sediment and its moisture capacity may then be a selective factor in the survival of tamarisk. The longer it survives, the longer it will take, if ever, for a succession to take place back to a flora composed principally of native species.

Areas of shelfy terrace are minor, about 2 percent. The sharp, jagged contours will be somewhat softened and will be more favorable for beaching a boat. Deposits of sediments in hollows will be similar to those of domed terraces, but growth conditions for shrubs such as tamarisk will be somewhat more favorable because of root growth in the more porous crossbedding.

Talus slopes and rock slides occupying 19 percent and 1 percent, respectively, of the shoreline before flooding will be changed. Considerable slumping will have occurred. Some will have slumped away completely and will be part of the bottom sediment, leaving only a scar on the cliff wall. Others will be changed in texture by removal of the silt and sand, leaving cobbles and boulders. Shoreline use of these types will not change, as most were too steeply sloping and unsafe for camping before the flooding occurred. Vegetational succession, principally by shrubs, will not drastically change their appearance.

The sand and alluvium shorelines, each 3 percent at the 3,660-foot contour, will in some areas be smoothed out into gently sloping shores. In deep deposits, wave-cut steps may be left along the shoreline. The surfaces of both types will have an increased percentage of silt and organic matter. This 6 percent of the shoreline at the 3,660-foot contour represents the area which receives almost all of the impact of shoreline use for recreation. These areas are level or gently sloping; they are safe shores on which to land a boat; they provide a means of easily tying up a boat; it is possible to pitch a tent by using stakes in the soil; areas can be found level enough on which to sleep; and for those not using portable toilets, these areas are most frequently used for toiletry. Swimming and other water sports most commonly demand a sloping sandy shoreline, especially for children, and, finally, there is a common attitude that a lake vacation means camping on a sandy shore. The minority--those who sleep on boats, swim only in deep water, or prefer the cleanliness of slickrock surfaces--is recognized.

The 6 percent of the shoreline occupied by sand and alluvium which is the focal point of shoreline recreational use and impact is the same shoreline which is most susceptible to the greatest change by vegetational succession upon its exposure after flooding. The principal concern is the invasion of tamarisk. This process and the potential density of thickets

which may develop have been previously discussed. The perpetuation of
shoreline stands of tamarisk around the lake is partially dependent on the
rate of filling. The faster the increase in water level, the less is the
opportunity for stands of tamarisk to become established and survive to re-
seed additional areas. At the projected rate of filling, it can be anticipated that local stands of tamarisk will be found on sandy shorelines, especially in protected bays and inlets, around the lake when the 3,700-foot
level is attained. Each year that the level is held at that contour, with
the usual seasonal drawdown, the extent of tamarisk establishment along the
shoreline will increase. Also, one can expect a vertical growth of 3 to 6
feet per year and an increase in density of the stands into impenetrable
thickets by the production of shoot sprouts from the horizontal roots.
A visual impression of the potential thickets which can develop along the
sandy-alluvial shoreline can be seen by both pre- and post-dam terrace
stands below Glen Canyon Dam. Not only will tamarisk become established
along and above the high water mark, but during drawdown periods it will
become established in the upper part of the zone and will endure flooding
the next summer. Unlike seasonal stands along a river channel, which are
subject to scouring action, these lakeshore stands are not deleteriously
affected; as illustrated by shoreline stands in 1975, they will start to
grow, flower, and produce seeds a few weeks after emerging from the summer
high water. Each time the thickets are flooded, the woody plants become
covered with a coating of attached algae. Because the principal shoreline
recreational use coincides with the period of high water level, there will
be offshore submerged and partially submerged woody vegetation. The accumulation of organic matter and silt will soon change the nature of the

formerly clean, sandy bottom. These undesirable conditions have not developed in the past because of the continuing rise of water to new high levels. The "honeymoon" period of "virgin" beaches will soon be over. Bringing a boat to shore through a tamarisk thicket, swimming in the submerged stands, and camping in a tamarisk thicket with the resulting populations of insects that soon develop, will not be an enriching, aesthetically pleasing experience.

As the water level is decreased to lower operational levels, one can expect tamarisk to advance downslope as long as sandy or alluvial surfaces are available. The continued survival of stands of tamarisk established at the high levels will depend on the depth of the soil, allowing for progressive root penetration as the water table drops. With its deep taproot, it is reasonable to predict a successful adjustment of this highly adapted phreatophyte species. Where there is a thin mantle over bedrock, upper stands of tamarisk may occupy the entire exposed shoreline slope from 3,700 feet down to the operational level.

The shoreline surface materials of principal concern as sites of tamarisk development are those of sand, alluvium, and domed terrace with a mantle of sand or of alluvium. These types represent the following percentages of the linear shoreline at the 3,660-foot contour: 3, 3, 1, and 11 percent, totaling 18 percent of the shoreline. These percentages represent linear distances of 49, 49, 16, and 180 miles, respectively. The acreages of the above types were determined for the shore between the 3,660- and 3,700-foot levels by use of an electronic planimeter and the computer at Dartmouth College. These areas have been computed by shoreline type and for each tributary containing the types. Tributaries with extensive areas suspect to tamarisk invasion are Wahweap, Bullfrog, Warm Creek, Main Channel, San Juan, and Halls Creek areas which receive the heaviest shoreline use by the public. The total potential area which could likely be invaded by tamarisk is 10,414 acres which is equivalent to over 16 square miles.

There is general agreement that tamarisk does not provide desirable shade or campsite vegetation because of the density of the thickets and the concentration of noxious insects which soon develop. In addition, the role which tamarisk thickets play in the loss of water by evapotranspiration is worthy of consideration. Using a conservative figure of 4 feet per year of water lost from stands of tamarisk, based on work by Turner and Halpenny (1941), Gatewood and coauthors (1950), and Van Hylckama (1970), the potential loss from the 10,400 acres would represent a loss of 41,600 acre-feet per year.

Replacement of tamarisk with another vegetation would not result in complete reduction of water lost by evapotranspiration, but substantial reductions could be attained. Even substitution of a grass sod, as has been reported for Bermuda grass in central Arizona (Decker, Gaylor, and Cole, 1962) reduced losses by 50 percent.

Because of the importance of the limited sandy shoreline for recreational use in Glen Canyon National Recreation Area, it is recommended that control measures for tamarisk be instigated as soon as the maximum water level is reached. The cost is least and the effectiveness greatest when control is started with seedling or young-aged stands. As the stands mature, the increased woody biomass, the increased extensive root system, and the increased capability of root sprout development make control very difficult.

It should be recognized that there is no way to eliminate tamarisk from a reservoir and that control, like maintenance, is a continuing process. A priority ranking of sandy shoreline areas with consideration to their limited availability in certain tributaries and the demand for public use should provide a basis for selection of areas to manage.

Among the species invading the upper drawdown zone have been found seedlings of the native Fremont cottonwood. Occasional mature trees and groves dominated by this species have been desired and heavily used as shoreline campsites. The species offers ideal shade and camping conditions. There is some evidence that its shade provides competition against tamarisk. Both cottonwood and Russian-olive (Elaeagnus angustifolia) can be easily rooted and transplanted. It is suggested that these species be transplanted along selected shorelines of deep sand when the reservoir reaches maximum level. Also, plantings should be made at the summer peak levels during the period of time when the water levels are decreased to lower operational levels. It is suggested that plantings be made in dispersed clumps along the shore. Such plantings would, of course, negate the possibility of widespread herbicide control of tamarisk unless a herbicide is developed which is specific to tamarisk.

To be effective in reducing water loss, mechanical control measures must be repeated frequently during the summer (Campbell, 1966). Rapidly growing shoots transpire heavily. Hughes (1966) reported that chemical spraying to obtain more than a 60 percent mortality is too expensive for effective use on a large scale. A more effective method of combining the use of herbicides and root plowing is not feasible for reservoir shorelines.

There is a need for continued research on the ecology of tamarisk as it relates to the invasion of the Colorado River system and the margin of its reservoirs. Because of its lessening of the recreational value of critical recreational areas, continued investigations of methods of control--chemical, mechanical, and biological--are needed. These may result in varying costs but may be applicable in correspondingly critical areas.

Because much of the Colorado Plateau has similar geology, physiography, climate, and vegetation, the use of a well-studied reservoir, such as Lake Powell, as an example of the features to be expected in other reservoirs in the Upper Colorado River Basin seems justified. The importance of the integrated approach cannot be overstressed. Finally, it is man who created the reservoir, manipulates the water levels which affect the shorelines, uses the reservoir, and impacts on the reservoir. He is a vital part of the shoreline ecosystem and research must be so oriented to include him as both a factor and a benefactor.

References

Campbell, C. J. 1966. Periodic Mowings Suppress Tamarisk Growth, Increasing Forage for Browsing (Washington, D.C., U.S. Department of Agriculture, Forest Service Research Note RM-76).

Cooley, M. E. 1958. "Physiography of the Glen-San Juan Canyon Area," Plateau vol. 31, no. 2, pp. 21-33.

_____. 1959a. "Physiography of the Glen-San Juan Canyon Area, Part II: Physiography of San Juan Canyon," Plateau vol. 31, no. 3, pp. 49-56.

_____. 1959b. "Physiography of the Glen-San Juan Canyon Area, Part III: Physiography of Glen and Cataract Canyons," Plateau vol. 31, no. 4, pp. 75-79.

Decker, J. P., W. G. Gaylor, and F. D. Cole. 1962. "Measuring Transpiration of Undisturbed Tamarisk Shrubs," Plant Physiology vol. 37, no. 3, pp. 393-397.

Dolan, R., A. Howard, and A. Gallenson. 1974. "Man's Impact on the Colorado River in the Grand Canyon," American Scientist vol. 62, pp. 392-401.

Gatewood, J. S., T. W. Robinson, B. R. Colby, J. B. Hem, and L. C. Halpenny. 1950. Use of Water by Bottom-Land Vegetation in Lower Safford Valley, Arizona (Reston, Va., U.S. Geological Survey Water-Supply Paper 1103).

Hansmann, E. W., D. E. Kidd, and E. Gilbert. 1974. "Man's Impact on a Newly Formed Reservoir," Hydrobiologia vol. 45, pp. 185-197.

Horton, J. S. 1964. Notes on the Introduction of Deciduous Tamarisk (Washington, D.C., U.S. Department of Agriculture, Forest Service Research Note RM-16).

Hughes, E. E. 1966. "Single and Combination Mowing and Spraying Treatments for Control of Saltcedar (Tamarix pentandra Pall.)," Weeds vol. 14, pp. 276-278.

Jennings, J. D. 1966. Glen Canyon: A summary. University of Utah Anthropology Papers, No. 81, Glen Canyon Series, No. 31 (Salt Lake City, University of Utah).

Kidd, D. E. 1976. Bacterial Contamination of Lake Powell Waters: An Assessment of the Problem, Lake Powell Research Project Bulletin No. 16, National Science Foundation. (Copies obtainable from: Institute of Geophysics and Planetary Physics, University of California, Los Angeles, California 90024).

Kuchler, A. W. 1964. The Potential Natural Vegetation of the Coterminous United States (New York, American Geographic Society, Special Research Publication No. 36).

Potter, L. D., and Natalie Pattison. 1976. Shoreline Ecology of Lake Powell, Lake Powell Research Project Bulletin No. 29, National Science Foundation. (Copies obtainable from: Institute of Geophysics and Planetary Physics, University of California, Los Angeles, California 90024).

Potter, L. D., D. Kidd, and D. Standiford. 1975. "Mercury Levels in Lake Powell-Bioamplification of Mercury in a Man-Made Desert Reservoir," Environmental Science and Technology vol. 9, pp. 41-46.

Reynolds, R. C., Jr., and N. M. Johnson. 1974. Major Element Geochemistry of Lake Powell, Lake Powell Research Project Bulletin No. 5, National Science Foundation. (Copies obtainable from: Institute of Geophysics and Planetary Physics, University of California, Los Angeles, California 90024).

Turner, S. F., and L. C. Halpenny. 1941. Ground-Water Inventory in the Upper Gila River Valley, New Mexico and Arizona: Scope of Investigation and Methods Used. Transactions of the American Geophysical Union vol. 22, pp. 738-744.

Van Hylckama, T. E. A. 1970. "Water Use by Salt Cedar," Water Resources Research vol. 6, no. 3, pp. 728-735.

Woodbury, A. M., S. D. Durrant, and S. Flowers. 1959. A Survey of Vegetation in Glen Canyon Reservoir Basin. University of Utah Anthropology Papers, No. 36, Glen Canyon Series, No. 5 (Salt Lake City, University of Utah).

THE ECOLOGY OF COLORADO RIVER
RESERVOIR SHORELINES: A DISCUSSION

Robert D. Ohmart[*]

As one of the discussants at this Forum, my task is to evaluate the
completeness and adequacy of the paper presented by Dr. Loren Potter,
"The Ecology of Colorado River Reservoir Shorelines." The following
comments and discussion are based on the assumption that the term "shoreline ecology," as submitted in the original title, is synonymous with
riparian ecology which encompasses the biotic and abiotic components of
the alluvial floodplain adjacent to the Upper Colorado River and its
tributaries above Compact Point at Lee Ferry.

As Dr. Potter clearly points out in his introduction, the paper deals
with a small portion of the ecology of the upper basin; specifically with
the shoreline of Lake Powell from a substrate, vegetational, and aesthetic
viewpoint. Lake Powell is used as an example to illustrate some of the
interactions of the physical and biological environment. The paper does
not attempt to provide an overview of the state of the knowledge of
riparian communities in the upper basin or what types of data are needed
to adequately understand the system for management purposes.

To do this, it would be necessary to review and summarize the published data on the ecology of the upper basin plus applicable data from
other systems, and to assess the present state of knowledge. It would

---

[*]Associate Professor, Department of Zoology, Arizona State University,
Tempe, Arizona.

then be necessary to treat the subject of future research directions and
needs to ensure an adequate data base and tools of analysis for management decisions.

Time and work load have not allowed me to compile the published data
on the upper basin, but a paper by Hayward, Beck, and Tanner (1958) yields
a good overview of ecological conditions in the upper basin.

Geographic provinces of the Upper Colorado River Basin (adapted from
Hayward, Beck, and Tanner, 1958):

 I. Green River Basin

  A. 30,000 sq. mi., chiefly in southwestern Wyoming but
   encompasses portions of northeastern Utah and
   northwestern Colorado.

  B. Topography is rolling hills or flat, gentle
   slopes trenched by broad river valleys.

  C. The parent material is mainly sedimentary of
   Tertiary age.

 II. Uinta Mountain

  A. 14,000 sq. mi. in Utah and Colorado.

  B. Topographically it is a long syncline.

  C. Parent material is of Tertiary age.

 III. Colorado Plateau

  A. 130,000 sq. mi.

  B. Steep slopes and sheer canyon walls formed by period
   of deposition, uplift and erosion.

  C. Sandstone, limestone and shale were deposited during

the Paleozoic and Mesozoic. Horizontal uplifts with much localized folding and warping occurred near the Cretaceous; erosion followed and deposition in the Tertiary with another uplift resulting in increased rainfall and intensified stream erosion.

Biotic communities (Hayward, Beck, and Tanner, 1958):

I.   Cottonwood - willow - tamarisk

II. Northern desert shrub

III. Pinyon-juniper woodland

IV. Hanging gardens

To my knowledge, this is the only attempt to examine the entire basin; other studies of a more limited scope have been conducted at various localities in the area, but these are scattered and have no continuity.

Because the area is so poorly known from an ecological consideration, I have established levels and types of data that are needed to eventually build an adequate data base for proper assessment and management of the Upper Colorado River system.

Level I--Amalgamate all relevant literature and assess for future data needs. In-depth quantitative data are needed for each major plant community type and the wildlife use values therein. If such data are not available from at least a two-year data base, then subsequent levels of effort would have to be undertaken.

Level II--Prepare a vegetation type map of the entire upper basin from existing aerial photographs. Terrestrial communities of 10 hectares (24.7 acres) or more should be defined by major vegetation type. Resolution

of emergent plant communities would have to be more detailed. Areal extent
and distribution of major communities would then be available to assess
potential modifications. Ground truths should follow to ensure map
accuracy.

Level III--Quantify an adequate number of communities of each major
vegetation type. Install permanent line intercepts for species composition and ground cover data, measure foliage profiles and determine
foliage volumes to characterize vegetative structure, conduct tree counts
for density and species composition estimates. Gather information on
phenology, health of communities (age classes represented) and extent of
invasion of the introduced salt cedar (Tamarix chinensis). Data on seed,
leaf, and stem are highly desirable for plant species whose parts are
important food resources (e.g., mesquite) for wildlife.

Level IV--Examine seasonal abundance of insects in each major
community type. These primary consumers and their abundance will dictate
diversity and abundance of secondary consumers.

Level V--Determine avian, mammalian, reptilian, and amphibian
densities and diversities for each major community type on a seasonal
basis. Adequate sampling on a north-south gradient will give insight
into ecological parameters limiting distribution and abundance of each
species.

Level VI--Examine food habits of major vertebrate species (common)
in each dominant community type to determine the relationship of the
secondary and tertiary consumers with primary consumers and autotrophs.
In avian species, data such as foraging rates in one season versus another

may be an excellent index to primary consumer abundance as well as energy
requirements and major avenues of energy flow.

Monthly data resulting from wildlife density data over three or more
years will give seasonal wildlife densities around means which can be
correlated with climate, phenology, primary consumers, etc., to show how
these variables affect wildlife densities. These types of baseline data
are presently not available on the upper basin, but are essential for
proper management as well as for assessment and evaluation of impacts of
proposed modifications. Evaluation of proposed enhancement or modification
is not possible without a data base of this nature.

Methodologies, modes of data storage, and programs for data analysis
are currently in existence and being used on the Lower Colorado River
Basin (Ohmart and Anderson, 1974; Anderson and Ohmart, 1975, 1976a, 1976b,
1976c) in studies being financed by the regional office of the Bureau of
Reclamation in Boulder City, Nevada. A recent paper by Anderson and
coauthors (1977) outlines the methodologies used on the Lower Colorado
River and discusses data applicability for management, operational enhancement, and mitigation.

References

Anderson, B. W., R. W. Engel-Wilson, D. Wells, and R. D. Ohmart. 1977. "Ecological Study of Southwestern Riparian Habitats: Techniques and Data Applicability," in U.S. Department of Agriculture, Importance, Preservation and Management of Riparian Habitat (Washington, D.C., U.S. Department of Agriculture, Forest Service General Technical Report Rm-43) pp. 146-155.

Anderson, B. W., and Robert D. Ohmart. 1975. Annual Report (1975): Vegetation Management Studies (Boulder City, Nev., U.S. Bureau of Reclamation, Contract No. 4-10-01-01-310).

Anderson, B. W., and Robert D. Ohmart. 1976a. Annual Report (1976): Vegetation Management Studies (Boulder City, Nev., U.S. Bureau of Reclamation, Contract No. 14-06-300-2415).

Anderson, B. W., and Robert D. Ohmart. 1976b. Wildlife Use and Densities Report (1976) (Boulder City, Nev., U.S. Bureau of Reclamation, Contract No. 14-06-300-2531).

Anderson, B. W., and Robert D. Ohmart. 1976c. Vegetative Type Maps of the Lower Colorado River (Boulder City, Nev., U.S. Bureau of Reclamation, Contract No. 14-06-300-2415).

Hayward, C. Lynn, D. Elden Beck, and Wilmer W. Tanner. 1958. Zoology of the Upper Colorado River Basin (Provo, Ut., Brigham Young University Science Bulletin).

Ohmart, Robert D., and B. W. Anderson. 1974. Vegetation Management Annual Annual Report (1974) (Boulder City, Nev., U.S. Bureau of Reclamation Contract No. 4-10-01-01-310).

THE ECOLOGY OF COLORADO RIVER
RESERVOIR SHORELINES: A DISCUSSION

Robert D. Curtis*

When I was invited to participate in this Forum, I was asked to dis-

cuss a paper by Dr. Loren D. Potter then entitled "Shoreline Ecology of

the Upper Colorado River Basin." I was specifically requested to cover

elements of the subject which were neglected in the paper. Since that

time, the title of Dr. Potter's paper has been changed to "The Ecology

of Colorado River Reservoir Shorelines."

Because Dr. Potter's paper was limited in scope to the Lake Powell area

of the Upper Colorado River Basin, I was shocked into the realization that

I would have to cover those elements of the subject not covered by Dr. Pot-

ter: the remainder of the upper basin above Lake Powell. Since the re-

mainder of the upper basin occupies in excess of 107,000 square miles as

compared to 300 square miles of Lake Powell and vicinity, I realized that

I had some homework to do.

To acquaint myself with a subject as vast as the 244,000 square mile

watershed of the Colorado River, I reviewed a 1946 publication entitled

The Colorado River (U.S. Department of the Interior, 1946). From this

publication, I learned that the Colorado River rises in the Rocky Mountains

on Mt. Richthofen near Granby, Colorado, and flows generally southwest for

about 1,440 miles to the Gulf of California in Mexico. However, due to

---

*Chief of Wildlife Planning and Development Division, Arizona Game and Fish Department, Phoenix, Arizona.

man's manipulation of the river during the last seventy years, water from the Rockies seldom reaches the Gulf of California.

The main tributary of the Colorado is the Green River which drains some 45,000 square miles of watershed after spawning itself near Gannet Peak (13,785 feet above mean sea level (MSL) in the Wind River Mountains of Wyoming. The Green meanders 750 miles southward to empty into the Colorado in northeastern Utah at elevation 3,876 MSL.

The other major tributary, draining some 39,000 square miles of watershed, is the San Juan River which starts in the lofty peaks of the Continental Divide of southwestern Colorado and northwestern New Mexico and then drops nearly 9,000 feet to join the Colorado in majestic Glen Canyon, some 140 miles south of the confluence of the Green River with the Colorado.

Here is a vast area of this nation that in the last fifty or so years has received widespread attention because of power potential, flood control, and irrigation dams and reservoirs along the Colorado and its tributaries. Here is an area that contains additional energy sources in uranium ore, natural gas, oil shale, coal, and uintaite. Here is an area of magnificent scenery and natural wonders and which has become attractive to a large number of visitors. Here is an area that attracted early explorers such as Escalante in 1776; Jim Bridger, William Ashley, Jedediah Smith, Etienne Provost and Antoine Robidoux in the 1820s and 1830s; J. C. Freemont in 1842; J. W. Gunnison in 1853; and J. W. Powell in 1859-72.

Here is an area so majestic, magnificent and complex with diverse geology, topography, ecology, fauna and flora that it cannot justifiably be discussed in fifteen minutes or fifteen hours.

Consequently, since the subject of discussion is ecology, I remembered my University of Indiana ecology professor's definition of a quarter of a century ago: ecology is the scientific study of the relation of organisms or groups of organisms to their environment (Allee and coauthors, 1949). Odum (1953) defined ecology as the study of the structure and temporal processes of populations, communities, and other ecological systems and of the interrelations of individuals composing these units. A more recent definition by Dasmann (1968) states that "ecology is the study of ecosystems to determine their status and the ways in which they function." An ecosystem is the combination of biotic communities with their physical environment. A biotic community is an assemblage of species of plants and animals in a given area and having impacts on one another. In my opinion Dr. Potter's paper would have been more acceptable if he would have confined his subject to the ecological relationships, ecosystems and biotic communities of Lake Powell.

In any paper on ecology, one would expect to see a discussion of ecosystems, biotic communities, biological productivity, biomass, energy transfer, physical and chemical requirements, biotic pyramids, food chains, food webs, biotic succession, limiting factors, and the interrelation and interaction within the ecosystems. Consequently, before I comment on Dr. Potter's paper and to ensure against the possibility of misconceptions, it should be made clear that I do not profess to be an expert in any special field of biological science. The views and comments expressed are those of a former wildlife research biologist and river basin supervisor. Currently, and for the past thirteen years, I have been a wildlife administrator employed by the Arizona Game and Fish Department to coordinate department responsibilities in federal land and water project evaluations, wildlife planning and development, and environmental studies of any activity having an impact on wildlife resources.

Upon receipt of Dr. Potter's paper, I read the abstract, reviewed the table of contents, and then read the paper in its entirety. Frankly, I was disappointed. It is most difficult for me, as a biologist and a professional wildlife administrator, to accept Dr. Potter's paper as a paper on ecology. Clearly, the paper has value since it does discuss geology, physiography, climate, vegetation, shorelines, shoreline mapping, shoreline vegetation biomass studies, shoreline succession, shoreline submergency, shoreline recreation, and water level management of Lake Powell. But the paper does not discuss ecosystems, total biotic communities, biological productivity, and the multitude of intricate and integrated reactions that take place within ecosystems. Consequently, the paper lacks the scientific ingredients that are necessary in a study of ecology.

The overall subject of this Forum is timely since western water and energy development are on a collision course with environmental and ecological considerations. One of the reasons we are here is to discuss the ecological data deficiencies for analyzing and managing the multiple uses of the vast Upper Colorado River region. From a wildlife biologist's viewpoint, I see the need for additional data on (1) instream requirements, both quantity and quality; (2) shoreline ecosystems; (3) agency and environmentalist reactions to impacts of extensive development of energy resources; (4) riparian and wetland area requirements for wildlife; and (5) protection of threatened and endangered fish and wildlife species.

The study on <u>Critical Water Problems Facing the Eleven Western States</u> by the U.S. Department of the Interior, Bureau of Reclamation (1975), in Chapter IV, Westwide Problem No. 5, concludes that "1. Generally, environmental information...on many water and related land use problems is lacking, incomplete or inconclusive.... 2. ...environmental data necessary for reasonable choices are...basic data on: flows required to maintain... fishery habitat...riparian habitat adjacent to stream channels...to support ...wildlife; wetland habitat requirements...critical factors necessary to maintain...outstanding natural, and terminal lake ecosystems; and habitat required to support big game populations during winter conditions." These problems apply to Lake Powell too and there is a need for continued studies. Earlier studies by Woodbury and Russell (1945), Woodbury (1958), and Hayward, Beck, and Tanner (1958) all provide a basis for the faunal aspect of an ecological study of the shoreline of Lake Powell.

Since we are generally confining our remarks to the shoreline of Lake Powell and its ecology, it is necessary to define shoreline. Shoreline is the demarcation between land and water; it could also be defined as that area of land between the high and low water marks. The amount of marginal slope determines the degree of definiteness of the shore margin of the reservoir. Since Lake Powell is a relatively young body of water, the configuration of the shoreline is in the form of sharp peninsulas, deep angular bays and sinuosities of several kinds, each offering resistance to free movement of water under the influence of wind and gravity. As the reservoir ages, the original shoreline will be modified by shore cutting and shore building and will become more regular and simplified with reduced shoreline length.

Obviously, as physical forces change the shoreline, the ecosystem will change. It is a well known limnological fact that, with the reduction of shoreline, other things being equal, the biological productivity of the reservoir will decline. Currently, since Lake Powell is young with an extensive shoreline of 1,800 miles and high nutrient inflow, the biological productivity is high. The phenomenon of Lake Powell is that it has characteristics of a reservoir with the recycling of nutrients and also the characteristics of a river with the continual introduction of fresh nutrients: dissolved, suspended, and organismal, entering from upstream. Lake Powell illustrates a favorable, but erratic, condition with a high level of biological productivity. The high biological productivity of the reservoir contributes to the productivity of the shoreline. With high biological productivity, the shoreline and flood plain ecosystems of Lake Powell are the most productive inland wildlife habitat in this nation; diversity and relative abundance of species are high.

Prior to the construction of Glen Canyon Dam, the reservoir area behind the current dam site supported a rather limited biotic community. Low rainfall, dry and saline soils, and sparse vegetation of the Great Basin desert scrub type generally created harsh conditions (Brown and Lowe, 1974a and 1974b). Due to these limiting factors, the number of animal species was small except for those that lived in the muddy waters of the Colorado or were able to cling to a fragile existence on the sparse vegetation among the rocks and cliffs.

With the development of Lake Powell, and the efforts of the federal government and the states of Arizona and Utah, the lake has become a mecca

for fishermen and other recreationists. More important than recreation,
as far as the ecology of the area is concerned, are the new habitats
created with resultant diverse biotic communities. These new habitats
have attracted birds, mammals, reptiles, amphibians, and a thousand
less known or less studied species to the area. The vastness of the
Lake Powell area, with new chemical and physical features attracting
new biotic communities, should be a treasure of scientific information
for the serious student of ecology. Universities with their academic
communities should be delighted and enthused with the almost unlimited
areas for scientific research and study.

Post impoundment investigations of Glen Canyon Reservoir have been conducted by the states of Utah and Arizona under funds provided in part by
section 8 of Public Law 84-485, the Colorado River Storage Act of 1956 (U.S.
Congress, 1956). Under section 8 of this act, the Secretary of the Interior
is directed in connection with the project "...to investigate, plan, construct, operate, and maintain (1) public recreational facilities...to conserve the scenery, the natural, historic, and archaeologic objects, and the
wildlife on said lands...and (2) facilities to mitigate losses of, and improve conditions for, the propagation of fish and wildlife."

Under the program developed by the two states, the Utah Department of
Fish and Game was responsible for the investigation of chemical and physical
regimens and fishery development in Lake Powell, while the Arizona Game and
Fish Department was responsible for estimation of fishing pressure and harvest, and the measurement of plankton.

The post-impoundment investigations under section 8 of PL 84-485 began
on Lake Powell in July 1963 and terminated over eight years later, in January

1972. Since 1972, the investigations have continued under state funding
and the Federal Aid to Fishery Restoration Act of 1950 programs.

Chemical and physical collection stations were operated by Utah in fourteen locations from Glen Canyon Dam upstream to the vicinity of Hite (Utah)
near Dirty Devil River (Stone and Miller, 1965). Stations were sampled in
March, June, September, and November to gather information prior to spring
runoff, during spring runoff, in the summer, and in late fall. Samples were
taken from zones of production, the thermocline, the hypolimnion, and the
bottom. Physical and chemical analyses were conducted in accordance with
standard techniques employed in the fisheries field. Results of all studies
appeared in _Glen Canyon Reservoir, Post-Impoundment Investigation Progress
Reports_ (Utah State Department of Fish and Game, 1964).

Creel census and plankton studies were conducted by Arizona from 1964
to 1972 and are reported in reports of the Arizona Game and Fish Department
on reservoir fisheries investigations. The most abundant species of fish
that were creeled from Lake Powell in 1971 were black crappie (_Pomoxis nigro-
maculatus_), largemouth bass (_micropterus salmoides_), bluegill sunfish
(_Lepomis macrochirus_), channel catfish (_Ictalurus punctatus_), and rainbow
trout (_Salmo gairdneri_) (Stone, 1972).

Plankton studies were initiated in July 1964 with samples collected
from a series of thirteen sampling stations arranged throughout the reservoir to provide representative coverage from Glen Canyon Dam upriver to Hite.
Standard sampling techniques as described by Welch (1952) and normally used
in the fisheries field were employed. The genera of phytoplankton most
prevalent were the Diatoms: _Fragilaria_ and _Synedra_; the Desmids: _Gonatozygon_,
Closterium, and _Staurastrum_; the blue-green algae (_Anacystis_) and the green

algae (Pediastrum). The genera of zooplankton most commonly found were crustacea: _Daphia_, _Cyclops_, and _Diaptomus_; rotifers; and protozoa: _Ceratium_ (Stone, 1966).

Earlier, I mentioned the value of Dr. Potter's paper with regard to vegetation and shoreline characteristics. It is a known ecological premise, at least from the point of view of a wildlife biologist, that in defining and describing the biotic communities of Lake Powell, the vegetation probably affords the best basis for comparison, since the nature of plant cover gives us a clue to the physical forces that are important to animal life. The role of vegetation in the occurrence and habits of animals is difficult to evaluate, but there are correlations between the distribution of flora and fauna in biotic communities. Of the several ecological factors (food, water, cover, and space) that are of importance to the general well-being of animals, cover is of utmost importance.

Potter, in his paper, describes the vegetation of the shoreline at Lake Powell as being dominated by willow (_Salix_ sp.), tamarisk or salt cedar (_Tamarix_ pentandra), Emory baccharis (_Baccharis_ emoryi), and arrowed pluchea (_Pluchea_ sericea). On the higher terraces and talus, groves of Gambel oak (_Quercus_ gambellii), netleaf hackberry (_Celtis_ reticulata), and Fremont cottonwood (_Populus_ fremontii) are found.

In the marshy areas, a variety of species of grasses, saltgrass (_Distichles_), muhly (_Muhlenbergia_), common reed (_Phragmites_ communis), and witch grass (_Panicum_), along with horsetail (_Equisetum_), flat sedge (_Cyperus_), sedge (_Carex_), and rush (_Juncus_) form the saline meadows.

According to Hayward, Beck, and Tanner (1958), various species of amphibians, such as the Great Basin spadefoot (_Scaphiopus_ hammondi), red-spotted toad (<u>Bufo</u> <u>punctatus</u>), Rocky Mountain toad (<u>Bufo</u> <u>woodhousei</u>), canyon treefrog (<u>Hyla</u> <u>arenicolor</u>), and western leopard frog (<u>Rana</u> <u>pipiens</u>), are expected to be found in these shoreline habitats. Reptiles expected include leopard lizard (<u>Crotaphytus</u> <u>wislizeni</u>), western chuckwalla (<u>Sauromalus</u> <u>obesus</u>), northern whiptail (<u>Cnemidophorus</u> <u>tigris</u>), and several species of garter snakes (<u>Thamnophis</u> <u>spp.</u>).

These habitats support a wide variety and high populations of bird species: hawks, owls, nighthawks, hummingbirds, woodpeckers, kingbirds and other flycatchers, swallows, crows and ravens, wrens, mockingbirds and thrashers, thrushes, vireos and warblers, meadowlarks and blackbirds, tanagers and grosbeaks, towhees and sparrows, Gambel's quail (<u>Lophortyx</u> <u>gambelii</u>) and mourning dove (<u>Zenaidura</u> <u>macroura</u>). The latter two are game birds with the mourning dove probably the most common species in these habitats. Woodbury and Russell (1945) provide a good checklist of birds of the Glen Canyon area.

Mammals of the shoreline, according to Hayward, Beck, and Tanner (1958), would include, but not be limited to, desert cottontail (<u>Sylvilagus</u> <u>audubonii</u>), blacktailed jack rabbit (<u>Lepus</u> <u>californicus</u>), Colorado chipmunk (<u>Eutamias</u> <u>quadrivittatus</u>), Botta's pocket gopher (<u>Thomomys</u> <u>bottae</u>), Ord's Kangaroo rat (<u>Dipodomys</u> <u>ordii</u>), beaver (<u>Castor</u> <u>canadensis</u>), western harvest mouse (<u>Reithrodontomys</u> <u>megalotis</u>), deer mouse (<u>Peromyscus</u> <u>maniculatus</u>), northern grasshopper mouse (<u>Onychomys</u> <u>leucogaster</u>), coyote (<u>Canis</u> <u>latrans</u>), gray fox (<u>Urocyon</u> <u>cinereoargenteus</u>), western spotted skunk (<u>Spilogale</u> <u>gracilis</u>), river otter (<u>Lutra</u> <u>canadensis</u>), bobcat (<u>Lynx</u> <u>rufus</u>), and mule deer (<u>Odocoileus</u> <u>hemionus</u>).

These species were reported seen in 1958 in the shoreline areas of the Colorado River in the Glen Canyon area. Lake Powell started filling in

1963 and the shorelines of the 1950s are now deep beneath the lake. An interesting comparison could be made between 1958 and today with regard to species composition and relative abundance of each. With the creation of new vegetative habitats, as described by Potter, new species would be expected, particularly mobile species such as birds. In addition, the reservoir would attract deep-water birds such as loons, grebes, pelicans, cormorants, ducks, geese, and swans; marsh birds and shore birds would be attracted to new habitats also.

During the review of Dr. Potter's paper, I became increasingly aware of Dr. Potter's concern with the invasion of tamerisk, or salt cedar, to the newly created shorelines of Lake Powell. Dr. Potter spent a considerable amount of time describing the evils of this species and how it interferes with recreational pursuits and how it may be eliminated or reduced by control methods.

As a wildlife administrator and biologist, I would take strong exception to the elimination of wildlife habitat. Maintenance of wildlife habitat is perhaps the most important tool in wildlife management today. The continuation of wildlife habitat and its protection allows wildlife resources to be classified as renewable. Take away the habitat and you take away the wildlife.

During the last two decades, I have opposed with vigor those developers and agencies that planned to destroy, by removal, wildlife habitat, particularly the riparian habitat. I am disturbed that a biologist would advocate removal programs for salt cedar or any riparian habitat without thorough study and evaluation of its impacts, and without consideration of other resources, particularly wildlife resources. Campbell (1970) stated "Ideally, only riparian vegetation unnecessary for fish and wildlife habitats, control of streambank erosion, or aesthetics should be treated." Campbell went on to say "The ecology of riparian vegetation, particularly in the Southwest, and the significance of these plant communities to the expanding human population, agriculture, and industry is only partially understood."

I do applaud Dr. Potter for his suggestions that "both cottonwood and Russion-olive (Elaeagnus angustifolia)...be transplanted along selected shorelines of deep sand when the reservoir reaches maximum level." This type of management, if successful, would benefit wildlife. But as a wildlife biologist, I feel that wildlife needs more than one vegetative type since the transition area from one vegetation type to another is highly attractive to wildlife. Leopold (1933) formulated a law of dispersion which theorized that the potential density of resident game would be proportional to the sum of the vegetative type boundaries. In short, more edge means more game; this concept applies to most species of wildlife-- anywhere.

Lake Powell is just one of many energy development projects in the West. It is just one of many projects on the Colorado River, but it is important not only as a storage and hydroelectric reservoir, but as a biological entity. As biologists, ecologists, or environmentalists, we must train and educate ourselves as riparian water watchers so that necessary fact-finding can be undertaken. We must make certain that adequate environmental forecasting is a part of all energy development projects. We must try to assess the second- and third-order consequences of river modifications, and we can do this by adequate study of all ecological factors.

Since this Forum is concerned with western water systems, a useful reference should be River Ecology and Man (Oglesby, Carlson, and McCann, 1972). This reference is a series of reports written by scientists, engineers, and social scientists on rivers, and man's interactions with them. This reference should prove valuable to fishery biologists, aquatic ecologists, limnologists, sanitary engineers, hydrologists, economists, political scientists, and administrators concerned with the study and management of freshwater systems. Following review of River Ecology and Man, my faith in the accomplishments of the academic and scientific community was restored.

292

References

Allee, W. C., A. E. Emerson, O. Park, T. Park, and K. P. Schmidt. 1949. _Principles of Animal Ecology_ (Philadelphia, W. B. Saunders).

Brown, David E., and Charles H. Lowe. 1974a. _A Digitized Computer-Compatible Classification for Natural and Potential Vegetation in the Southwest with Particular Reference to Arizona_ (Phoenix, Arizona Academy of Science, Vol. 9, Supplement 2, May).

_____. 1974b. _The Arizona System for Natural and Potential Vegetation_ (Phoenix, Arizona Academy of Science, Vol. 9, Supplement 3, November).

Campbell, C. J. 1970. "Ecological Implications of Riparian Vegetation Management," _Journal of Soil and Water Conservation_ vol. 25, no. 2 (March-April).

Dasmann, Raymond F. 1968. _Environmental Conservation_ (New York, John Wiley and Sons).

Hayward, C. Lynn, D. Elden Beck, and Wilmer W. Tanner. 1958. "Zoology of the Upper Colorado River Basin--1. The Biotic Communities," _Science Bulletin Biological Services_ vol. 1, no. 3 (Provo, Ut., Brigham Young University).

Leopold, A. 1933. _Game Management_ (New York, Chas. Scribner's Sons).

Odum, Eugene P. 1953. _Fundamentals of Ecology_ (Philadelphia, W. B. Saunders).

Oglesby, Ray T., C. C. Carlson, and J. A. McCann. 1972. _River Ecology and Man: Proceedings of an International Symposium on River Ecology and the Impact on Man_ (New York, Academic Press).

Stone, Joseph L. 1966. _Glen Canyon Unit - Colorado River Storage Project, Reservoir Fisheries Investigations, Creel Census and Plankton Studies, February 1, 1965 to January 31, 1966_ (Phoenix, Arizona Game and Fish Department).

_____. 1972. _Glen Canyon Unit - Colorado River Storage Project, Reservoir Fisheries Investigations, Creek Census and Plankton Studies, February 1, 1971 to January 31, 1972_ (Phoenix, Arizona Game and Fish Department).

Stone, Roderick, and Kent Miller. 1965. _Glen Canyon Reservoir Post-Impoundment Investigation._ Progress Report No. 3 (Salt Lake City, Utah State Department of Fish and Game).

U.S. Congress. 1956. Public Law 84-485. 84 Cong. 2 sess.

U.S. Department of the Interior. 1946. _The Colorado River_ (Washington, D.C.).

U.S. Department of the Interior, Bureau of Reclamation. 1975. <u>Westwide Study Report on Critical Water Problems Facing the Eleven Western States</u> (Washington, D.C.).

Utah State Department of Fish and Game. 1964-1972. <u>Glen Canyon Reservoir, Post-Impoundment Investigation Progress Report</u>, Nos. 1-9 (Salt Lake City, Ut.).

Welch, Paul S. 1952. <u>Limnology</u> (New York, McGraw-Hill).

Woodbury, A. M., and H. N. Russell, Jr. 1945. "Birds of the Navajo Country," <u>Biological Series</u> vol. 9, no. 1 (Provo, Ut., University of Utah).

_____. 1958. "Preliminary Report on Biological Resources of the Glen Canyon Reservoir," Anthropological Papers, No. 31, Glen Canyon Series, No. 2 (Provo, Ut., University of Utah).

Chapter 12

THE IMPACTS OF ENERGY DEVELOPMENT ON BIG GAME
IN NORTHWESTERN COLORADO

Bertram D. Baker,* Robert L. Elderkin,** Donald R. Dietz***

Introduction

For purposes of this report, northwestern Colorado is defined as that
part of the state north of the Colorado River and west of the Park and Gore
mountain ranges (see figure 1). In addition to these two ranges, other prominent topographic features located south to north and west to east from the
area's southwestern corner include the following: Roan and White River
plateaus, White River Basin (which includes Douglas and Piceance Creek
drainages), Danforth Hills, Williams Fork River, Williams Fork Mountains,
Yampa River Basin (including the Little Snake and Elkhead rivers), and the
Elkhead Mountains. Moreover, the extreme northwestern corner of Colorado
is marked by Cold Springs Mountain, Browns Park, the Green River and its
Lodore Canyon in Dinosaur National Monument, and Douglas Mountain.

Great climatic, edaphic, and topographic diversity over such a large
plateau and mountainous area produces great habitat diversity, as indicated
by the plant communities of Costello (1954). The western and northern

*Senior Wildlife Biologist, Colorado Division of Wildlife, Fort Collins,
Colorado.
**Environmental Scientist, U.S. Geological Survey, Grand Junction,
Colorado.
***Oil Shale Coordinator, U. S. Fish and Wildlife Service, Grand Junction,
Colorado.

The report is, in part, a contribution of Colorado Pittman-Robertson
and Dingell-Johnson Federal Aid Project FW-21-R.

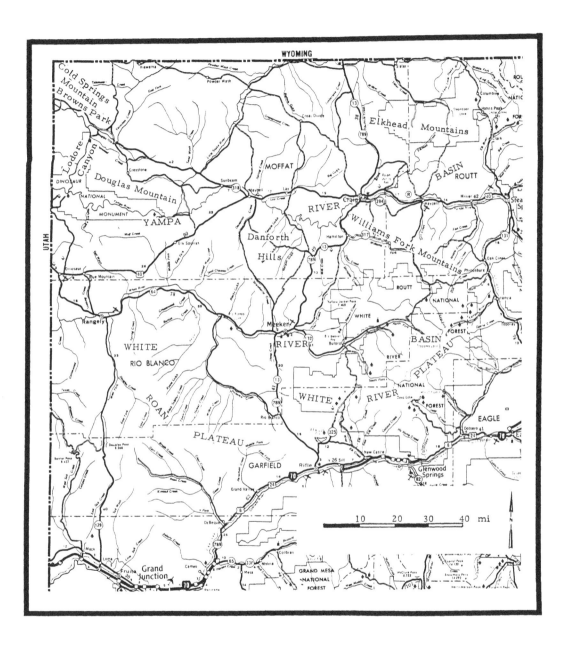

Figure 1. Map of Northwestern Colorado

portions of the region at lower elevations are characterized by the salt-
bush, greasewood, and sagebrush cover types. The next higher elevations,
from about Craig, Meeker, and Glenwood Springs westward, support extensive
pinyon-juniper and mountain or mixed shrub acreages. Coniferous forests
of Douglas-fir, lodgepole pine, and spruce-fir are cover types at the
upper elevations of the mountains and plateaus. Aspens grow on quite
extensive acreages and are generally associated elevationally with the
conifers. Small in total acreages are the alpine and riparian communities,
although drainageway cottonwoods, alder, and willow are of particular
importance because they provide for greater biotic diversity. Alpine
areas, likewise, cannot be overlooked, since they are habitats for species
of restricted tolerances, such as ptarmigan and rosy finches.

Just as it is important to delineate the geographic extent of this
report, so too is it pertinent to limit the biotic scope. A manageable
number of wildlife species with which the authors have experience is the
principal criterion for selecting Colorado's big game (Colorado Department
of Natural Resources, 1975) for discussion. The big game ungulates and
carnivores are discussed in sections to follow. These include: mule
deer, elk, antelope, mountain sheep, bear, and mountain lion. Coverage
of small game, nongame, endangered, and other terrestrial classifications,
though critically needed, is not included here.

Big Game

Mule Deer

Mule deer are found in virtually every habitat and are the most
abundant big game animal in northwestern Colorado. Pooling 1975 harvest

data for the twenty-five big game management units (figure 2) comprising this part of Colorado, we found 14,620 deer were harvested by all means and in all seasons (Denney and coauthors, 1976). The foregoing harvest amounts to 36.14 percent of the 1975 statewide total of 40,449, a very substantial part, particularly since the area is estimated to be only about one-fourth that of the total occupied mountainous range.

Similarly impressive as northwestern Colorado's deer yield in the twenty-five management units is the harvest from a single unit, unit 22. Coincidental with high grade oil shale deposits, unit 22, or Piceance, is also where the still substantial migratory Piceance deer herd winters and is hunted. Over the ten-year, 1966-1975 span, the annual harvest from this unit varied from 2,001 to 5,512 deer for a proportionate statewide share of from 3.05 percent to 8.18 percent, with a mean of 5.15 percent of the average annual statewide kill of 67,753.

The best estimate of the Piceance deer herd population is that given by Bartmann (1976) who stated that the 1975-76 winter herd size consisted of $24,206 \pm 3,724$ deer at the 90 percent confidence level. Despite how one interprets the foregoing estimate, or the harvest indexes previously given, Piceance deer numbers are important by comparison.

Other parts of northwestern Colorado also support substantial deer populations, and several areas coincide with proven and potential fossil fuel deposits. Unit 11 (Strawberry in the Yampa and White River Basins) recorded having 2,003 deer taken, for the second highest harvest of the twenty-five management units in 1975 (Denney and coauthors, 1976). Strawberry also has six fields producing oil and two gas fields, and is scheduled to expand extensively the surface and underground coal extraction

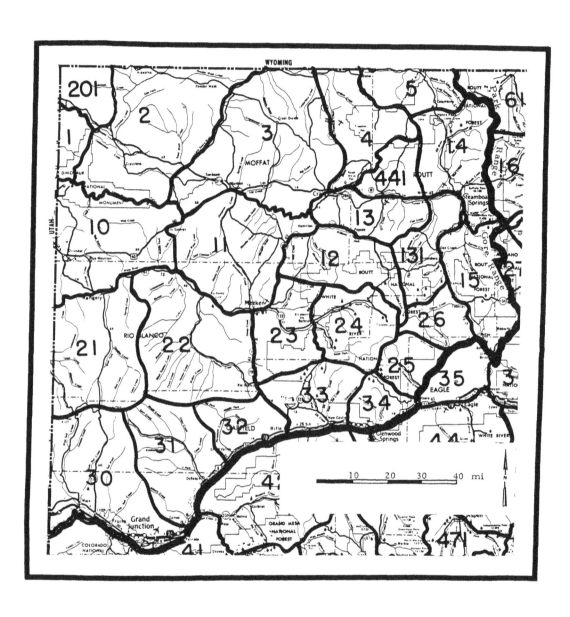

Note: The twenty-five management units discussed in this paper
are located north and west of the heavy black line.

Figure 2. Big Game Management Units in
Northwestern Colorado

soon (U.S. Department of the Interior, Bureau of Land Management, 1976).

Management unit 3 (Craig in the Yampa River Basin), a strong third in the

1975 harvest with a 1,145-head kill, has ten producing gas fields and

tremendous reserves of bituminous and sub-bituminous coal. (See the papers

in the volume by Smith and by Maddock and Matalas, and the discussion by

Steele.) Lastly, of prime importance now and in the future for gas

production is the Cathedral Field (Murray, 1976). Situated mostly in

management unit 21 (Douglas), which ranked fifth in harvest in 1975

(934 deer), gas exploration, development, and transmission activities very

likely will be disadvantageous to deer as well as to other species.

Elk

In the past, a great deal of effort has been devoted by the Colorado

Division of Wildlife to research and management of elk, as well as of deer,

in northwestern Colorado. Justification for this interest in elk rests

upon the high rate of reproduction and the total carrying capacity of the

habitat expressed both in herd numbers and in accompanying annual harvest

yields (Boyd, 1970). Because recent population estimates are not avail-

able, harvests of elk will be expressed as an index of abundance.

First, and starting on the broadest base, northwestern Colorado's

twenty-five big game management units produced a 7,788-head harvest for

1975 (Denney and coauthors, 1976). That yield is equivalent to 34.41 per-

cent of the statewide total kill. Furthermore, eight of those twenty-five

units, collectively called E-6 or the White River elk data analysis unit,

ranked first statewide with a ten-year mean annual harvest of 2,934 elk

(Baker and coauthors, 1974). A more recent ten-year calculation for

1966-1975 harvests showed a mean annual kill of 3,701 elk or 20.72 percent of the statewide total. These harvest data indicate the importance of the White River elk.

It is difficult to assess the extent to which energy development might impact elk. Perhaps a test of big game tolerance to human disturbance is underway right now in northwestern Colorado in the area corresponding with management units 12 (Williams Fork), 13 (Dunkley), and 131 (Oak Creek) in the Yampa River Basin. With 1975 harvests of 1,080 elk in unit 12 and 839 elk in units 13 and 131 combined (Denney and coauthors, 1976), units 12 and 13 are nearly as big producers as unit 23 (Miller Creek) which harvested 1,169 elk in 1975. It is questionable how herds and harvests will hold up in light of action and plans for surface coal mining over massive areas of habitat in the Williams Fork Mountains portion of these two units (U.S. Department of the Interior, Bureau of Land Management, 1976).

Fortunately, management units 23 and 24 (White River), which comprise the solid core of White River elk ranges (Boyd, 1970), do not have appreciable reserves of coal and other fossil fuels (U.S. Department of the Interior, Bureau of Land Management, 1976). Thus, from the standpoint of fossil fuel extraction, these important central elk ranges seem comparatively safe from degradation. Minor Dakota Sandstone uranium deposits are located northeast of Meeker in this region, but currently they are not being mined (U.S. Department of the Interior, Bureau of Land Management, 1976).

Antelope

The pronghorn populations in northwestern Colorado cover six antelope management units: A1 through A5, and A83 (see figure 3). Units A1 through

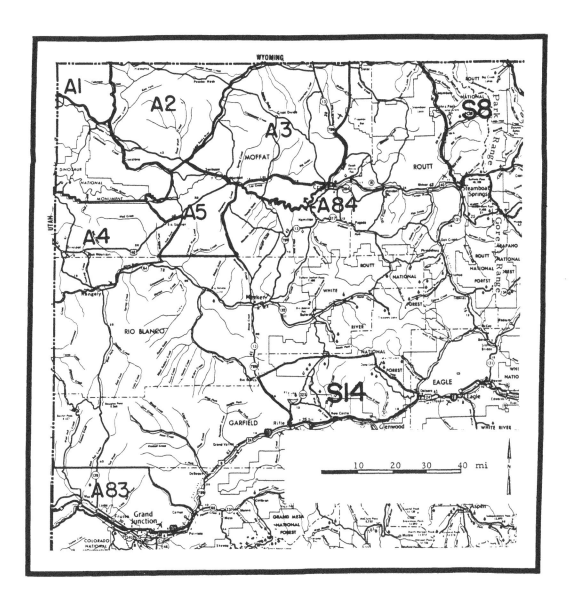

Figure 3. Antelope Management Units (A1-A5, A83, A84)
and Bighorn Sheep Management Units (S8, S14)
in Northwestern Colorado

A5 approximate the northwestern three-fourths of Moffat County, and include that small part of Rio Blanco County north of the White River and west of Wolf Creek. Management unit A83 covers that portion of Mesa County north of the Colorado River and west of Asbury Creek.

Except for comparatively minor boundary differences at southwestern and northeastern corners, management unit 3 (Craig) and antelope unit A3 (Great Divide) are the same territory. Besides being excellent mule deer habitat, this land north and west of Craig and bounded by State Highway 13, the Little Snake River, and U.S. Highway 40, is also very good antelope range. Pojar, using the computer simulation model developed by Gross and coauthors (1973), estimated that the fall 1976 pre-hunt antelope population size in A3 was 6,714.[1] A 1975 harvest of 684 pronghorns was recorded for unit A3 (Denney and coauthors, 1976), in addition to the rather substantial pre-hunt population just mentioned. The 1975 harvest for unit A3 equaled 19.44 percent of the statewide total, and combined with harvests for units A1, A2, A4, and A5, they equal 37.03 percent of the statewide total. Except for unit A83, northwestern Colorado is prime pronghorn territory. Due to extremely low numbers of pronghorns present, unit A83 near Grand Junction has not had an open season in recent years. White estimates that only about thirty pronghorns currently exist in this region.[2]

---

[1] T. M. Pojar, Colorado Division of Wildlife, Fort Collins, Colorado (personal communication, 1976).

[2] C. E. White, Colorado Division of Wildlife, Fort Collins, Colorado personal communication, 1976).

In assessing energy development impacts on pronghorns, attention is called to the earlier section on mule deer and the ten producing gas fields and coal development possibilities in management unit 3 (antelope unit A3). The potential future stresses on deer in this area similarly apply to antelope welfare evaluations.

In addition to the threat of fossil fuel development in management units 3 and A3, unit A2 contains two oil and three gas fields, and unit A4 has the big Rangely and smaller Elk Springs and Winter Valley oil fields. These fields, together with the proposed coal development in both units, promise to apply continuing direct and indirect stress on the pronghorn herds of this area.

Uranium has already been produced from the Browns Park Formation near Lay and Maybell (U.S. Department of the Interior, Bureau of Land Management, 1976). Any sizable revival in the production of this fuel will probably have adverse effects not only on pronghorns and mule deer, but also on the semidesert shrub and pinyon-juniper communities present.

Bighorn or Mountain Sheep

There are two bighorn management units in northwestern Colorado. Unit S8 corresponds with the Mt. Zirkel-Park Range area and animal populations discussed by Bear and Jones (1973). Bear reported that he believes no bighorns presently exist in this unit.[3]

The other bighorn management area is unit S14 which combines herds and areas north of New Castle on Clinetop Mesa and the rims and slopes of Glenwood Canyon north of the Colorado River between Mitchell and French

---

[3]G. D. Bear, Colorado Division of Wildlife, Fort Collins, Colorado (personal communication).

Creeks (Bear and Jones, 1973). Bighorn numbers in this unit have decreased in recent years, and are precariously low at the present time.

Bighorns also exist in Lodore Canyon of the Green River in Dinosaur National Monument (Bear and Jones, 1973). A population estimate of fifty to seventy-five head has been made by Bear for this herd. Hunting is banned in the monument and thus is not a herd drain factor there.

Bear

Of the two big game carnivores in northwestern Colorado, bears are more widely distributed, and apparently more abundant, than are mountain lions. On the basis of license income revenues, bears are definitely the more valuable (Denney and coauthors, 1976).

Harvest data show that a total of 105 bears were killed during the 1975 season, that total coming from nineteen of the twenty-five management units (Denney and coauthors, 1976). Unit 33 (Rifle), which led all units, had eighteen bear killed during 1975. Other units that were fairly important contributors to the harvest were units 30 (Salt Creek) and 14 (Elk River) with kill totals of thirteen and eleven, respectively.

Generally speaking, bears show reasonable tolerance to humans but could expect to lose ground as energy development reaches into preferred brushland and forest habitats. Like the mountain lion, the species deserves more than passing attention in wildlife management.

Mountain Lion

Northwestern Colorado has been well known for its high numbers of lions, gaining extra notoriety when Teddy Roosevelt hunted near Meeker in 1905 (Feltner, 1972). A combination of consistently large numbers of mule deer for food, and great expanses of pigmy woodland and rough terrain for cover, has made this lower White River country better than average lion territory.

Again, using the 1975 harvest data to indicate abundance, a total of forty mountain lions were taken from nine of the twenty-five management units (Denney and coauthors, 1976). Only ninety were taken that year from all of Colorado, so the forty represent 44.4 percent of the total. The Douglas Creek area, defined as management unit 21, had fifteen lions harvested in 1975, the highest for any unit in northwestern Colorado, as well as in the state.

No appreciable impacts on lion populations as a result of the high upward trend in fossil fuel development in northwestern Colorado have yet been observed. The real concern hinges upon how well the mule deer numbers hold up. This is probably the most important factor for the survival of mountain lions in northwest Colorado.

Development Impact Assessment Problems

For the most part, oil shale and coal mining activities present fairly closely related problems and research needs as they relate to wild-life impact assessment. Therefore, except where there are specific differences noted in this section and for the research needs discussed in the next section, comments will apply to both resources.

Uncertainties of mining methods and waste disposal. Coinciding with shale oil industry efforts to find the most economical production-scale method for producing synthetic crude is the attendant problem of gearing vegetation reclamation research to different types of spent and raw shales.

Spent shales from U.S. Bureau of Mines, Tosco II, Union Oil, and Paraho

processes are being tested for suitability as plant growth bases. With

emphasis now shifting from surface retorting and waste disposal to in situ

retorting, research is required to determine how raw shale can best be

stabilized and, concurrently, provide optimum food and cover for major

wildlife species.

In addition to needs for undertaking research in oil shale mining

reclamation, continuing studies on coal spoil banks rehabilitation are also

desirable. Examples of excellent topsoil revegetation experiments current-

ly underway are those at the Colowyo Mine owned by W. R. Grace and Company

in Axial Basin (Phillips and Berg, 1976). Energy Fuels Corporation is

also participating with Agriculture Research Service and the Environment,

Mineral, Rehabilitation Inventory Analysis (EMRIA) program of the Craig

BLM District in reclamation studies at Energy Fuels Corporation's mine south

of Milner. These and other reclamation research programs are necessary to

determine how best to restore coal spoil banks, slopes and micro-reliefs,

and to provide cover most suitable for whatever wildlife species combina-

tions are selected for reestablishment.

Lack of knowledge of wildlife species living requirements. Unfortunate-

ly, we know very little about the habitat requirements of the various wildlife

species. We are not even able to supply a reasonable estimate of what, for

example, one acre of upland sagebrush means in the annual cycle of one mule

deer in Piceance Basin. When it is reported for Prototype Lease Tract C-a

(Rio Blanco Oil Shale Project, 1976) that about 2,417 acres of sagebrush

(42 percent sagebrush of 5,755 acres total) will have been disturbed by

development activities by the year 2015, this is not readily translatable to mean annual numbers of mule deer lost from the Piceance herd. Complexities of this type of evaluation are very great, but answers must be found.

The Tract C-a Detailed Development Plan (Rio Blanco Oil Shale Project, 1976) also states, "Habitat supporting one cottontail will be eliminated for every 1 to 2 ha [hectares] (3 to 4 A) disturbed." No distinction is made as to which of the several vegetative types is disturbed in this formula. Does the "3 to 4 A" rate apply to the entire 5,755 acres by the year 2015? These examples are not given in criticism, but only to demonstrate that the industrial environmental engineer or contract ecologist-consultant is faced with equally perplexing problems as those of us who are making efforts at surveillance.

Lack of easy interchange between environmental staffs. There are continuing difficulties in the exchange of information between government agencies, and between governments and industry, on environmental impact assessment, surveys, and monitoring. Poor communication is a frequent malfunction in all organizations; so, it is not surprising when separate and often opposing agencies are unable to exchange information. There is no simple solution to this difficulty but strong efforts by individuals of all organizations to remain open-minded and amenable to compromise will do much to alleviate this problem.

Research Needs

Whether approached from the standpoint of basic research, applied research, so-called field surveys, or administrative studies, a great

need exists for knowledge upon which to base wildlife impact assessments.
Some needs have been discussed briefly in the previous sections. The
subject will be further expanded here, using outline form, with the order
of presentation an approximate order of priority applicable to development
of both shale oil and coal resources.

1. We need to know the habitat requirements for terrestrial wildlife, concentrating first upon big game. All other species are important,
but would receive subordinate effort.

    A. Food

        (1) By season, studies of food quantities and types
            via nutritional and digestibility requirements
            are needed.

        (2) By season, studies of forage mix and palatability
            demands and factors are needed. Is palatability
            environmental or genetic, and what components
            affect it?

        (3) Do animal food requirements differ by sex and age during
            various physiological activities and if so, to what
            extent?

    B. Cover

        (1) What plant heights, densities, canopy coverages,
            species mixes, and type juxtapositions are optimum
            and minimal?

        (2) Are the relationships of cover, food and water by
            positioning and distances important?

        (3) What are the influences of topography, slope, and exposure?

C. Water

    (1) What are the daily requirements, and are there
seasonal and special demands?

    (2) What are the maximum tolerances to industrial
and mining effluents and leachates?

D. Food, cover, and water combined to produce preferred habitat

    (1) What factors are involved in providing optimum range
by season? Determine and map-delineate different
seasonal ranges.

    (2) Are special habitats needed for special animal
functions such as fawning, calving, nursing, feeding,
escape, critical winter survival, and migration?
Determine and map-delineate these areas.

    (3) Do inherent habitat preferences affect migration and
possible future range expansion?

    (4) What broad changes in vegetative types have occurred
over the last fifty years due to habitat management
practices? How does this correlate with population
increases and declines of deer and elk?

E. Habitat carrying capacity as influenced by:

    (1) People in rural housing subdivisions, snowmobiles,
farm vehicles, bikes, hiking, skiing, and snowshoeing.

    (2) Deer-auto and deer-train collisions. Relative to the
preceding human-caused stress factors is the very
critical need for on-going deer-auto collision studies

on all potential and presently-known important highways and county roads. This will enable integrated planning for con- struction of possible preventive and control structures and fences concurrent with the planning of road upgrading for increased traffic volumes.

(3) Competition with domestic livestock, feral horses, and other wildlife by season.

(4) Stage of vegetative succession. Are important plants being replaced in kind? What are the condition and trend of key species?

(5) Livestock range management practices such as burning, chaining, spraying, plowing, and weed and insect control.

2. We need to know the value of mitigation efforts, by wildlife species--priority, again, as in item 1, with mule deer, elk, pronghorns, and other big game first.

A. Feeding areas

(1) What are the advantages to wildlife, if any, of temporary revegetation crops and cover?

(2) What effects are caused by pinyon-juniper and sagebrush removal, aspen and brush thinning and dozing, and brush burning?

(3) Which plants are best, and where, for replacement or augmentation establishment? Are shrubs best and most practical for all ungulates in mountain areas?

(4) What is the best method for mountain brush rejuvenation?

B. Watering areas

    (1) What are the specific requirements of wildlife species?

    (2) Are springs, seeps, or ponds, or combinations preferred?

    (3) What type of watering area benefits which wildlife species, by vegetative type?

C. Adjacent undisturbed areas

    (1) What are potentials of acquiring long-term, adjacent refuge areas to absorb displaced species?

    (2) What are the population characteristics of portions lost and absorbed? Are they sex and age specific, or related? How important are these factors in species management?

D. Mitigation alternatives

    (1) What is the interplay between forage or cover-type manipulation and watering projects for mitigation? What is the interplay of such mitigations with adjoining undisturbed areas?

    (2) What are the best methods for monitoring mitigation projects?

3. Indicator species of both flora and fauna are needed which can be measured adequately and which are sensitive to such specific parameters as:

A. Airborne and aquatic pollutants

B. Specific soil elements

C. Pivotal species in a change of floral and faunal composition

4. Refinements of wildlife census methodologies are necessary.

   A. We need to develop and refine procedures that yield reasonable
      accuracy for counting highly mobile species such as deer,
      elk, coyotes, bobcats, and mountain lions.

   B. Can procedural refinements be found to accommodate the
      inherent variability in small mammals censusing?

   C. Can present raptor and sohgbird census techniques be improved
      to permit detection of significant population changes by
      vegetative type?

5. We need to know how environmental factors relate to one another.

   A. Example: What effects will there be upon future browse
      production in western Colorado of the severe killing freeze
      of June 1976?

   B. How much change in one factor is necessary before other
      factors become important in the chain? From the preceding
      example, at what level would low moisture during the winter
      of 1976-77 depress 1977 production?

The foregoing presentation of problems and research needs concerning
impact assessment of energy development is incomplete. Nevertheless, it
is presented to draw attention to the ever-growing demands by shale oil
and coal industries, and attendant governmental regulatory agencies, for
obtaining the best environmental baseline, surveillance, and reclamation
data possible.

References

Baker, D. B., R. J. Boyd, J. F. Lipscomb, and T. M. Pojar. 1974. "Deer and
    Elk Management Study" (Fort Collins, Colo., Colorado Division of Wild-
    life, Federal Aid Report W-38-R-28, Work Plan 18, Job 1, Job Progress
    Report, Game Research Report, July).

Bartmann, R. M. 1976. "Piceance Deer Study--Population Density and
    Structure" (Fort Collins, Colo., Colorado Division of Wildlife, Federal
    Aid Report W-38-R-31, Work Plan 16, Job 2, Job Progress Report, Game
    Research Report, July).

Bear, G. D., and G. W. Jones. 1973. "History and Distribution of Bighorn
    Sheep in Colorado" (Fort Collins, Colo., Colorado Division of Wildlife,
    Federal Aid Report W-41-R-22, Work Plan 1, Job 12, Job Final Report,
    January).

Boyd, R. J. 1970. Elk of the White River Plateau, Colorado, Colorado Divi-
    sion of Game, Fish, and Parks Technical Publication No. 25 (Denver,
    Colo., Colorado Department of Natural Resources, Division of Game, Fish,
    and Parks, Big Game Investigations, Project W-38-R, GFP-R-T-25).

Colorado Department of Natural Resources. 1975. Division of Wildlife and
    Division of Parks and Outdoor Recreation Laws, Reprint of Title 33
    of Colorado Revenue Statutes as Amended, DOW-M-I-14-'75 (Fort Collins,
    Colo.).

Costello, D. 1954. "Vegetation Zones in Colorado," in H. D. Harrington,
    Manual of the Plants of Colorado (Denver, Colo., Sage Books).

Denney, R. N., H. D. Riffel, and R. J. Tully. 1976. 1975 Colorado Big Game
    Harvest, (Denver, Colo., Wildlife Management Section, Colorado Division
    of Wildlife).

Feltner, G. 1972. A Look Back. A 75 Year History of the Colorado Game,
    Fish and Parks Division (Denver, Colo., Colorado Department of Natural
    Resources. Division of Game, Fish and Parks, GFP-G-I-14).

Gross, J. E., J. E. Roelle, and G. E. Williams. 1973. "Program ONEPOP and
    Information Processor: A System Modeling and Communication Project,"
    Progress Report, Colorado Coop Wildlife Research Unit (Fort Collins,
    Colo., Colorado State University).

Murray, D. K. 1976. Energy Resource Development Map of Colorado. 1:500,000
    Scale. (Denver, Colo., Colorado Geological Survey).

Phillips, C. M., and W. A. Berg. 1976. Establishment of Native Shrubs on Disturbed Lands in the Mountain Shrub Vegetation Type, Agronomy Report, Colorado State University, and Colorado University Experiment Station (Fort Collins, Colo.).

Rio Blanco Oil Shale Project. 1976. "Detailed Development Plan, Tract C-a," in Vol. 4: Sect. 10 Environmental Assessment, Chapter 5, Terrestrial Ecology (Denver, Colo., Gulf Oil Corp-Standard Oil Co. (Ind.), March) p. 10-5-38.

U. S. Department of the Interior, Bureau of Land Management. 1976. Northwest Colorado Coal: Draft Environmental Impact Statement (Denver, Colo., Bureau of Land Management, Colorado State Office).

THE IMPACTS OF ENERGY DEVELOPMENT ON BIG GAME
IN NORTHWESTERN COLORADO: A DISCUSSION

Duane A. Asherin*

I will begin by confessing that I have no personal knowledge of the flora and fauna in the geographic area this paper addresses, namely the Upper Colorado River Basin in northwestern Colorado. However, I do have some appreciation of the energy development assessment problems and research needs from work we have conducted on hydroelectric projects in the Columbia River Basin, and from knowledge of assessment problems associated with the phosphate mining industry in southeast Idaho. Phosphate mining in this region is of the open-pit type and, therefore, poses assessment problems similar to those of coal and oil shale development.

The authors of this paper are to be complimented in dealing with the major objective of the Forum--to identify specific wildlife-related problems and research needs associated with the development of western energy resources. This paper identifies, in depth, the research needs and priorities which three natural resource agencies--the Colorado Division of Wildlife, the U.S. Fish and Wildlife Service, and the U.S. Geological Survey-- deem necessary to properly assess the impacts on terrestrial wildlife species of coal and shale oil developments in northwestern Colorado. In pursuing this objective, however, I believe that pertinent information has been inadvertently omitted. This information would have been beneficial

*Wildlife Data Analyst, Western Energy and Land Use Team, U.S. Fish and Wildlife Service, Fort Collins, Colorado.

At the time this paper was presented, the author was Research Wildlife Biologist, Idaho Cooperative Wildlife Research Unit, College of Forestry, Wildlife and Range Sciences, University of Idaho, Moscow, Idaho.

to the reader in evaluating the energy development assessment problems and research needs that the authors delineated.

Those unfamiliar with the wildlife resources of northwestern Colorado should find beneficial the presentation of background information on the wildlife, coal, and oil shale resources located in this region. A more in-depth review of pertinent or applicable research and management knowledge available for this region concerning the various habitats and their associated wildlife species would have been extremely helpful. In addition, a time frame or schedule concerning the development of these oil shale and coal resources should have been included. This time schedule should have been worked into the priorities established for the research needs since those areas scheduled for earliest development would require the more immediate research emphasis and baseline data collection efforts. Some of the potential impacts, I suspect, will be site- and species-specific (e.g., on a particular wildlife species or group of species in a given location), while other impacts will probably be of a more region-wide nature (e.g., on particular, widespread plant communities).

Apparently, full-scale development of these oil shale resources will be some time in the future, if at all, while an expansion of coal mining activities appears more immediate. I agree with the authors when they point out that available time should be used judiciously. Representative areas containing representative habitats and wildlife populations should be selected or identified, and baseline field studies performed on these sites prior to the initiation of mining so that a true assessment of impacts is possible. A rather unique opportunity still appears to exist here--to collect data on the quantity and quality of vegetation and wildlife before they are affected by exploratory and pre-mining activities, for example, road building and over-burden removal. Too often in impact assessment analyses we are forced to use control data collected on already impacted resources which generally are not useful for comparison purposes. These baseline studies should then be followed up with studies of the mining and post-mining phases. In this way, we will begin to understand the total extent of the impacts associated with large-scale natural resources developments such as coal and shale oil.

The authors have identified those wildlife species which they feel should receive highest priority regarding the research needs they listed-- namely, big game and more specifically mule deer, elk, and mountain lion. They recognize that other species are also important, but state they should receive subordinate effort. Game species have traditionally recieved this high priority in the establishment and funding of research programs because these are the wildlife species that have provided the operating budgets for state fish and game agencies. I believe the data needs described are relevant for all wildlife. We are all aware of the national trend and public demand for consideration of those other wildlife species. I contend that the energy development assessment problems and research needs in north- western Colorado, and elsewhere, are just as important (perhaps more important, because so little work has been done on them) for these other wildlife species or vertebrate groups. I am aware that the manpower, time, and money do not exist for collecting the missing data on all the hundreds of other wildlife species which have been neglected in the past. However, our impact assess- ments will remain incomplete until we begin to include these species in our baseline, monitoring, and research studies. Perhaps some of the earlier comments at this Forum concerning the need to identify critical river reaches and species in the aquatic environment should be expanded to include representative species of the terrestrial wildlife groups as well. Impact assessments of any kind will only begin to approach completeness when we do.

I am not going to quibble or argue here over the specific research needs identified under the major categories the authors have outlined. These needs have apparently already been encountered by the authors in their impact assessment work with energy development in northwest Colorado. The wildlife impact assessment needs described by the authors under all five major categories should be considered when designing the impact assessment study. Specific information needs will vary with the area to be impacted, but with this type of preplanning, many of the answers to information needs can be obtained during the collection of baseline data; others will have to await the evaluation of follow-up data, when available.

Certain research needs will require more effort than others; some, intensive research efforts. One example of the latter, as indicated under item 4 in the paper, is the need for refinements in wildlife census techniques to yield more accurate estimates of all wildlife populations. Numerous comments have been made during this Forum concerning the need for more complete and accurate data from baseline studies and not simply more data. This problem has plagued us in the past and will continue to plague us in the future, especially when we attempt to evaluate impacts on wildlife other than game species. We must place more research emphasis on developing sound census methodologies.

Certainly, the need exists for more in-depth data on specific life requirements of all terrestrial wildlife species, as outlined by the authors under item 1. Without data identifying habitat requirements, impact assessments will continue to be largely speculative. One item of importance within this major research needs category, which I also believe is very important, is the effect of human-caused stress factors on habitat carrying capacity. Other papers have indicated that we are going to witness an increase in the (human) population of the region due to energy development there, and coupled with this, a substantial increase in recreational demand. (See the paper by Parker in this volume.) Assessment of the impacts of various human-caused stress factors present a real challenge to the researcher. Just the addition of increased road traffic, much less all of the off-road recreational uses and their combined effects, presents potential for tremendous wildlife impacts, and a challenging area for impact quantification.

In addition to the specific information needs described by the authors under item 2, the general need to document the value of mitigation efforts for particular wildlife species, I would add that the economic value of the various wildlife species and their habitats is a question that remains unsettled and certainly in need of better documentation, as well. The economist's and political scientist's roles in these needed evaluations are essential. Under item 3, the authors express a need for identifying indicator species of both flora and fauna that can provide an early warning of the potential damages caused by airborne and aquatic pollutants from energy developments. A summary of the known biological effects of gas and particulate emissions from coal-fired power plants is presented by

Gordon (1975) for the ecosystems of southeastern Montana. This article
discusses specific impacts on those plant species which have been used as
biological indicators of phytotoxic gases, acidic rains, and acidic solutions of coal-fired power plant emissions. This article may have application to impact assessment problems in northwestern Colorado as well.

321

References

Gordon, C. 1975. "Biological Feasibility of Siting Coal-Fired Power Plants at Colstrip, Montana," *Western Wildlands* vol. 2, no. 4, pp. 28-40.

THE IMPACTS OF ENERGY DEVELOPMENT ON BIG GAME
IN NORTHWESTERN COLORADO: A DISCUSSION

Jack Howerton*

Energy development in the Upper Colorado River Basin differs somewhat
from that of the state of Washington; however, as we will see, the environmental problems associated with such development are similar.

Major energy developments in Washington State are hydroelectric and
nuclear. There are also several sites along the Cascade Mountain range
within the state that have potential for geothermal development. And we
are faced with the real possibility of supertankers transporting oil on
Puget Sound. We already have refineries on the sound, and proximity to
Alaskan oil fields increases the possibility of further petrochemical
development.

As one can see, we are attempting to deal with many kinds of energy
development on very different kinds of terrain, in vegetation types from
rain forest to upper sonoran desert and including marine and fresh-water
ecosystems. Each project is unique; consequently, we must design studies
to fit each project.

Before we get to the point of assessing impacts, however, there are
other obstacles that must be overcome. First, we must convince project
initiators and sponsors, public decision makers, and the public that impacts will occur and sponsors must be willing to accept responsibility for
damages to wildlife caused by their projects. The Bureau of Reclamation,

*Wildlife Biologist, Washington State Department of Game, Olympia,
Washington.

the Corps of Engineers, and three public utility districts have an agreement regarding the operation of Columbia River dams to maximize power production, but they make no effort to determine how these coordinated operations affect wildlife.

Closer coordination and cooperation are also needed among state resource agencies. In Washington, various agencies have responsibilities that relate to the environment: the Department of Natural Resources administers the Surface Mine Reclamation Act of 1971; the Department of Ecology has responsibility in the areas of air and water quality; and the Washington Game Department is responsible for preserving, protecting, and perpetuating wildlife. But unfortunately, there is little cooperation among these agencies.

Economics is another problem that must be addressed. No one has yet devised an acceptable means of determining the values of wildlife. Economists tell us that gross expenditures by recreationists for wildlife-oriented recreation are not acceptable. They tell us that the net economic value based on a demand curve, assuming that it can be determined, is the route to go. Yet, we find that wildlife does not fit the traditional supply and demand curves. Methods must be devised to obtain the information we need regarding the social value of wildlife. We must have values in order to compute cost-benefit ratios, and for use by our legislators and other decision makers.

Wildlife and development agencies and biologists must begin to deal with total impacts of development projects. I believe few biologists today realize the full impact on wildlife of an energy development project. Traditionally, we have studied the direct and initial impacts of energy development on game species, and we have provided this to some development agencies. When we attempt to include other wildlife in our assessments, we often receive negative responses.

We must begin to consider the total ecosystem rather than selecting, out of context, segments of that ecosystem for isolated study. Granted, this will be difficult, especially when too often we do not even understand the relationships of game species to their habitat well enough to accurately determine impacts. Let me give an example. On Wells Hydroelectric Project on the Columbia River in eastern Washington, Oliver and Barnett (1966) conducted pre-project studies and estimated 13,860 units of game would be lost through inundation of 4,680 acres of land. After the project was completed, Oliver conducted post-project studies and found actual losses of more than 200,000 units of wildlife.[1] This did not include wildlife other than game species.

Let me take another example. Two years ago, most biologists in eastern Washington considered bitterbrush the major food plant of good mule deer range. A recent study, however, showed that nutrient content determined the use of plants. Nutrient levels in plants change throughout the year (McArthur, 1975). Researchers on that study found that mule deer browsing followed nutrient peaks. Deer selected plants which provided the highest food value to the animal. Clearly, we need to learn more about the relationships of wildlife to habitat. Food, water, and cover are no longer enough to consider--wildlife need places to feed, breed, brood, loaf,

[1]Wendell H. Oliver, Washington Department of Game, Olympia (personal communication).

325

sleep, hide, keep warm or cool, and play, and must have unrestricted movement between these areas.

Let us take a closer look at hydroelectric projects. Hydroelectric
projects in Washington have inundated nearly one-half million acres. If
we extrapolate figures from other areas, we estimate that one-half million
acres would support annual fall populations of 3,000 elk, 15,000 deer,
250,000 pheasant, 750,000 quail, 50,000 chukar, and 30,000 grouse (Oliver,
1966). I wonder how many waterfowl, fur animals, songbirds, shore birds,
raptors, reptiles, and amphibians were affected. We have lost more than
one million game animals of those species for which calculations were done,
and we have not considered the secondary and tertiary development impacts,
such as those discussed in an earlier paper at this Forum (see Parker's
paper). Who is responsible for these impacts? Impacts on recreation, and
present and future impacts on management programs of wildlife agencies have
also been ignored. Problems associated with hydroelectric projects at
times seem insurmountable. They do not necessarily end with construction of
the projects and flooding of riparian habitats. Additional powerhouses may
be constructed. Operations may be changed to meet peak power demands, which
result in vastly different water level fluctuations than those experienced
before the construction. At Grand Coulee Dam, a third powerhouse was added.
With power peaking, we expect 28-foot fluctuations in tailwater. These
developments are further complicated by irrigation withdrawals and other
water uses.

We do have laws and regulations which provide for compensation of wildlife losses caused by hydroelectric projects. Nuclear, coal, geothermal,
and petrochemical development must meet the requirements of the National

Environmental Policy Act and the State Environmental Policy Act, but nothing says developers of these energy developments must compensate for wildlife losses. Changes in laws and regulations could help in this area.

In conclusion, I am sure there are problems and needs we have not considered here, but if we resolve the major problems which were discussed, we will have come a long way. We have a long way to go, and for every hour it takes to get there, we lose habitat and wildlife.

References

McArthur, M. B. 1975. Food Habits of Deer and Elk on Colockum Wildlife
    Recreation Area (Olympia, Wash., Washington Department of Game).

Oliver, W. H. 1966. "The Effects of River Basins Development on Wildlife."
    Paper presented at the Washington Chapter Meeting of the Wildlife
    Society. [Copies can be obtained from the author, P. O. Box 187,
    Bueno, Washington 98921.]

Oliver, W. H., and D. C. Barnett. 1966. Wildlife Studies in the Wells
    Hydroelectric Project Area, Federal Power Commission License No. 2149,
    Columbia River--Washington (Olympia, Wash., Washington Department of
    Game).

PART III. FISH AND WILDLIFE MANAGEMENT IN THE UPPER
          COLORADO RIVER BASIN: PLANNING AND
          EVALUATION CONSIDERATIONS AND PROCEDURES

Chapter 13

THE POTENTIAL IMPACTS OF ENERGY DEVELOPMENT
ON WATER RESOURCES IN THE YAMPA RIVER BASIN

Thomas Maddock III,* N. C. Matalas**

Introduction

Efforts to formulate a national energy policy have focused considerable attention on coal development. The coal resources in the western states are substantial and offer the potential for making an important contribution to energy self-sufficiency. To what extent this potential can be realized depends in part on the available water resources for supporting a coal-based industry. On an aggregate scale, the nation's water resources could support very large-scale exploitation of coal. But on a more local scale where coal mining and conversion would actually be located, water-related problems might arise. This might well be the case in those locales where there would be conflicting water interests such as water for energy versus water for agriculture. Also, fish and wildlife constitute a resource whose exploitation or preservation may pose additional conflicting demands on the uses of water.

To examine some of the problems inherent in water resources planning with respect to large-scale energy development, the Yampa River Basin, situated in the northwestern part of Colorado, is considered. Above

*Hydrologist, U.S. Geological Survey, National Center, Reston, Virginia, and Professor of Hydrology and Water Resources, University of Arizona, Tucson, Arizona.
**Hydrologist, U.S. Geological Survey, National Center, Reston, Virginia.

Maybell, Colorado, the Yampa Basin has a drainage area of approximately 3,400 square miles. The basin contains large reserves of high quality coal, located principally in the areas of the Williams Fork Mountains and the Danforth Hills. The coal reserves are characterized by low contents of ash, approximately 5 percent by weight (air dried), and sulphur, about 0.4 percent by weight (air dried). Total reserves are estimated to be 5.7 billion tons (Landis and Cone, 1971) of which Speltz (1974) estimates there is a strippable resource of 750 million tons in the Yampa coalfield and 164 million tons in the Danforth Hills coalfield.

Coal mining itself is not a water intensive operation although surface mining together with the accompanying land reclamation will place some demand on water. Coal conversion, however, such as the production of thermal electric power or synthetic fuels, is likely to be water intensive, particularly if water is used for cooling. It is noted that in the production of high Btu synthetic gas, water is used for hydrogen production. In a given region, mine-mouth coal conversion operations might impinge on the region's water resources and the more so if the water resources are largely committed to other purposes.

In the following few sections, an assessment of water use in coal mining and conversion is presented. Following that, a brief account is given of the potential of the water resources in the Yampa Basin to support coal-based operations there.

Coal Development Scenarios

A recent environmental study of the Yampa Basin by the Bureau of Land Management (U.S. Department of the Interior, Bureau of Land Management, 1975) was based on the production of 33 million tons of coal per year by 1990, with 24 million tons per year being produced by surface min-ing. At present, thermal electrical power plants in the basin have a combined capacity of about 1,170 MW. It was assumed that by 1990 addition-al capacity would be built, bringing the total to about 2,930 MW. The coal not utilized by the plant would be shipped to markets outside the basin. The Bureau of Land Management study provides a framework for assessing the water resources of the Yampa Basin.

For the purposes at hand, two scenarios were considered (table 1). For the first scenario, coal production is by surface mining alone, such that by the year 1990 the annual production rate is 24 million tons per year with thermal electric power capacity being 2,000 MW. The coal not utilized in the power production is shipped (by means other than a water intensive method of transport) to markets outside the Yampa Basin. The second scenario is similar to the first scenario but also includes the production of high Btu synthetic gas (SNG). It is noted that the produc-tion of synthetic gas has not been seriously contemplated in the Yampa Basin. However, the production of synthetic fuels has not been ruled out as a possible component of future energy policies and is thereby consid-ered to illustrate its water impacts relative to those of thermal electric power production.

Under both scenarios, water use is assessed in the case of no re-strictions (base case) placed on discharging residuals and in the case where restrictions (constrained case) are placed to meet a standard of zero discharge of liquid residuals for other than tar condensates (in the case of synthetic gas production), and to meet the Colorado ambient air quality standard of 150 ppm of SO_2.

Table 1. Coal Development Scenarios for the Yampa River Basin (1990)

Scenario	Residuals management level	Number of power plants[a]	Number of gasification plants	Coal converted to electric power (million ton/yr)	Coal converted to SNG (million ton/yr)	Coal transported (million ton/yr)	Total coal mined (million ton/yr)
1	Base case	2	0	5.52	0	18.48	24
	Constrained case	2	0	5.65	0	18.35	24
2	Base case	2	1	5.52	6.25	12.23	24
	Constrained case	2	1	5.56	6.25	12.10	24

[a]Each plant has a design capacity of 1,000 MW.

Only the direct effects of mining and production of electrical power and synthetic gas on the basin's water are considered. Realization of either of the two scenarios would be accompanied by an increase in population, which at present in the basin is about 16,000, and development of supportive industries and services. Thus, the assessment of water use described in this report does not fully illustrate the stress on the basin's resources. Also, estimates of the capital and operating costs relative to the scenarios are not presented.

With respect to surface mining, four levels of land reclamation, denoted as $\emptyset$, A, B, and C, were considered (see also table 1 in Smith's paper in this volume). Level $\emptyset$ implies no reclamation and was included to provide a base for residuals generation comparisons with the other three land reclamation levels. Level $\emptyset$ consists only of those activities which provide for efficient mining. For example, among the activities would be road wetting for dust control and the diversion of surface water away from strip pits.

Reclamation level A involves striking off spoil ridges to a 15-foot width, covering final coal seams, and seeding. Level A is based on the Colorado Open Mining Land Reclamation Act of 1973. Reclamation level B requires a high degree of effort in that the land surface is returned to its original contours, implying a substantial amount of spoil flattening. Under this level, it was assumed that the land would be revegetated for a return to grazing. One year of irrigation was assumed to enhance the early establishment of vegetation.

Reclamation level C seeks a rolling landscape and revegetation suitable for game habitat with an emphasis on the reduction of wind- and waterborne sediment. The rapid vegetation required is aided by mulching, three years of irrigation, and heavier applications of soil amendments and fertilizers than required for level B.

Thus, for the two scenarios, water use is assessed in terms of two levels of residuals management with respect to the production of electrical power and synthetic gas and four levels of land reclamation with respect to mining. A more detailed account of the material presented in this report is given in a recent study by the U.S. Geological Survey on water and energy use in coal conversion (James and coauthors, 1977).

Coal Development Operations

A brief description is first given for the three basic operations --surface mining, the production of thermal electric power, and the production of synthetic natural gas. Water use under alternative levels of residuals management is then discussed in the following section.

Surface Mining. No specific sites are assumed for the mining operations. However, it is assumed that coal is of the quality typical of that found in the Williams Fork Mountains, that is, with a heat value of 11,460 Btu/lb and characteristics as shown in table 2.

A mine site is assumed to have characteristics representative of those at some of the better resource locations in the basin. A single coal seam of 10-foot thickness with an average overburden thickness of 53.5 feet is assumed. The overburden sequence measured upward from the top of the coal is assumed to consist of 0.5 feet of shaley and bony coal, 6 feet of soft, easily weathered, organic shale, and the remainder interbedded shale and sandstone. With a recovery ratio of 0.90, the stripping

Table 2. Williams Fork Coal Quality

Composition	Percent weight
Hydrogen (H_2)	4.84
Carbon (C_2)	61.74
Nitrogen (N_2)	1.54
Oxygen (O_2)	15.15
Sulphur (S_2)	0.40
Water vapor (H_2O)	11.00
Ash	5.35
Total	100

ratio, that is the overburden per ton of mined coal, was computed to be 5.30 cubic yards per ton (yd^3/ton).

For a given mine, the mining rate is taken to be 3.125 million tons/year. Thus, for the 1990 mining rate of 24 million tons/year, approximately eight mines would be required. At a mining rate of 200 acres of coal per mine per year, each mine, assuming production commences this year, will encompass approximately 2,800 acres of land by 1990. The eight mines collectively will require 22,400 acres by 1990. Relative to each mine, it is assumed that 17,188,000 yd^3/year of overburden overlying the coal will be removed.

Thermal Electric Power Plant. The design capacity of the thermal electric power plant is taken to be 2,000 MW. It is assumed that the design capacity is provided by two plants, each of 1,000 MW capacity, sharing a common stack. The plant operated with no restrictions on residuals discharge is referred to as the "base" plant, and the plant operated to meet the residuals discharge constraints noted above is referred to as the "constrained" plant. The heat rate values were varied from 8,700 Btu/kWh for a plant using a cooling pond to 8,910 Btu/kWh for a complete plant using an evaporative natural draft cooling tower. Correspondingly, the raw coal input rates varied from 5.52 million tons/year to 5.65 million tons/year. In either case, the power plant's coal input rate is somewhat less than the coal produced by two of the surface mines. The excess mining output is assumed to be shipped to markets outside the Yampa Basin. No doubt there would be some markets within the basin. Because the Williams Fork coal is low in sulphur content and the ash may have a high melting or slagging point, the power plant is assumed to use a dry bottom furnace rather than a cyclone furnace.

With a dry bottom furnace, coal preparation consists of pulverizing the raw coal to the consistency of a very fine power, and to improve combustion efficiency the pulverized coal is fed into the furnace under high temperatures (over 500° F). With a dry bottom furnace, approximately 90 percent of the coal ash is entrained in the flue gas, and the remaining 10 percent is collected as bottom ash (Delson and Frankel, 1972).

Coal Gasification Plant. It is assumed that the coal gasification plant uses the SYNTHANE process (Forney and McGee, 1972) and has a capacity of 250 million standard cubic feet per day (SCFD) with a 0.90 plant availability factor (24 hrs/day; 330 operating days/yr). The plant includes a thermal electric power plant operated on the generated tar and char and also an oxygen plant. The raw coal input rate is 6.25 million tons/yr (790 tons/hr), a rate equal to the output of two of the surface mines. Thus, for the second scenario which consists of surface mining 24 million tons of coal per year and producing 2,000 MW of electrical power and 250 million SCFD of pipeline quality synthetic gas, approximately 50 percent of the mined coal would be available for export.

The SYNTHANE process involves six principal operations: (1) coal preparation, (2) gasification, (3) dust and tar removal, (4) shift conversion, (5) gas purification, and (6) methanation. The last operation is to upgrade the heat content of the gas. For the Williams Fork coal, the product gas has a heat value of 926 Btu/SCF.

Residuals Generation

Mining activities and the coal conversion processes invariably lead
to the generation of residuals--liquid, solid and gaseous--that either
must be reused or modified and disposed of as wastes. The types and quantities of the residuals generated depend on the coal mining and conversion
technologies, and are affected by efforts to meet environmental standards.
For the constrained case, the Colorado ambient air quality standard of 150
ppm of SO_2 was considered together with zero discharge of liquid residuals
for other than tar condensates in synthetic gas production. A summary of the
liquid, solid, and gaseous residuals generated is given in table 3.

In the case of surface mining, the principal residual is waterborne
sediment. This is particularly high under reclamation level ϕ. The waterborne sediment load is substantially reduced under reclamation level C and,
in fact, was found to be negative indicating a slight reduction in natural
sediment yields. Relative to reclamation level ϕ (base case), reclamation
level C (constrained case) results in a substantial reduction in the total
quantity of residuals generated, principally the liquid residuals.

For the residuals generated in the production of thermal electric
power and synthetic pipeline gas, no account was taken of possible trace
elements in Williams Fork coal (table 2). For the most part, the
residual streams for the power plant are in the gaseous phase (i.e.,
SO_x and fly ash), and the solids entrained in the bottom ash. In the
case of the SYNTHANE plant, the principal residual streams include
process condensates (from dust and tar removal, and from heat exchang-
ing operations), gaseous effluents from the on-site power plant and gas
knock-out drum, and bottom ash. With both types of plants, the cooling

Table 3. Residuals Generation

(lbs/ton of coal mined)

Residual management level	Residual type	Mining[a] (M)	Coal conversion process		Scenario 1 M+P	Scenario 2 M+P+G
			Electric power plant (P)	SYNTHANE plant (G)		
Base case	Liquid	59	0	1,870	59	1,930
	Solid	0.1	11.2	10.6	11.3	21.9
	Gaseous	2.9	4,930	12,800	4,980	17,700
	Total	62.0	4,940	14,700	5,000	19,700
Constrained case	Liquid	0.02	0	86.5[b]	0.02	86.5
	Solid	0.09	129	127	129	256
	Gaseous	1.0	4,930	13,100	4,930	18,000
	Total	1.11	5,060	13,300	5,060	18,300

[a] Reclamation level ∅ in base case; level C in constrained case.

[b] tar (may be considered a by-product).

341

and boiler systems contribute liquid residuals in the form of blowdown. The liquid residuals generated in the SYNTHANE plant result primarily from purification processes that are necessary to maintain production and are not a result of environmental standards. Combustion is assumed to be complete in the power plant such that residuals are either entrained in the gas or end up in a solid form as ash.

For the SYNTHANE plant, the production of by-products, such as ammonia and tar, is possible whereas for the power plant there are no by-products. By-product recovery reduces the quantity of residuals to be disposed of in the SYNTHANE process. On the other hand, because there is no by-product recovery and because materials are added in residuals modification (treatment) processes, the power plant yields a net increase in the quantity of residuals to be handled and disposed of.

The residuals generated in the mining and conversion processes must ultimately be discharged to the environment--atmosphere, water courses, and land--but not necessarily on site. The impacts of these residuals on the environmental media as well as on the terrestrial and aquatic eco-systems of the Yampa Basin are not considered in this paper, although some of these impacts are addressed in other papers in this volume. A quantitative assessment of the environmental impacts requires an esti-mate of the extent of energy development in the basin as well as de-velopment of environmental models for predicting the associated environ-mental impacts. The paper by Sawyer and coauthors in this volume pre-sents a range of possible energy development scenarios for the Four Corners states and for the Upper Colorado River Basin. The paper by

;mith in this volume addresses the impact of surface mining on the trout
fisheries in the Yampa Basin.

We will confine the remainder of our analysis to the impact of
energy development on Yampa River flows and to the necessity of
providing additional storage for surface water in the Yampa River
Basin.

Water Availability in the Yampa River Basin and Its Use

For the period 1917 through 1974, the mean annual flow of the Yampa
River near Maybell, Colorado, was 1,130,000 acre-ft/yr (1,560 cfs), with
September being the low-flow month. The mean monthly flow for September
based on the same period was 174,500 acre-ft/yr (241 cfs). The cumulative
probability distribution of September flows for the fifty-seven-year period
of record is depicted in figure 1.

To ascertain whether or not the Yampa flow can support the coal
development scenarios outlined above, it is necessary to consider the
requisite water intake and circulation values for the coal conversion
processes which are given in tables 4 and 5, respectively. Relative to
the coal conversion processes, the water use for mining is small and is
not taken into account in the subsequent discussions. For the coal conversion processes, water is withdrawn from the river and is circulated
through the plants where it is either consumed, recycled, or discharged
(figure 2) depending on the production technology and on the environmental
standards that are to be met.

The minimum and maximum water withdrawals (intakes) for the two
coal development scenarios are given in table 6. For each of the

344

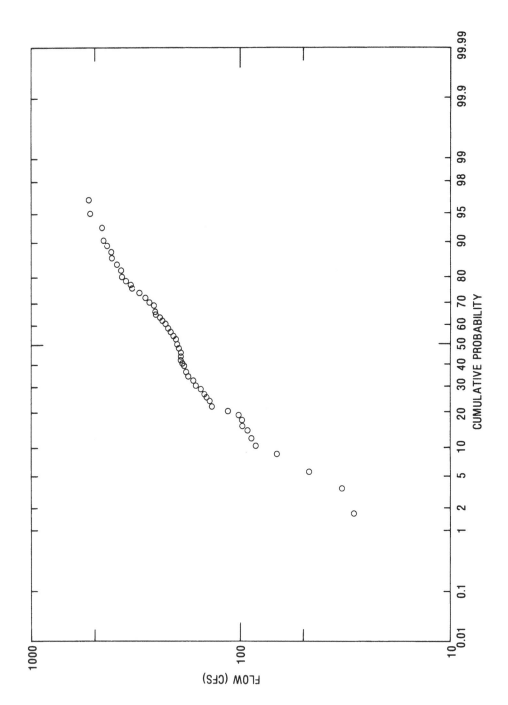

Figure 1. Probability Distribution: September Flows, Yampa River Near Maybell, Colorado

345

Table 4. Water Intake for Coal Conversion Processes for the Year 1990

(acre-feet/yr)[a]

Residual management level	Cooling system	Coal conversion process[b]			Cooling			Total		
		P	G	P+G	P	G	P+G	P	G	P+G
Base	A	190	5,470	5,660	675,000	509,000	1,180,000	675,000	514,000	1,190,000
	B	190	5,470	5,660	31,900	14,700	46,600	32,100	20,200	52,300
	C	190	5,470	5,660	25,800	17,800	43,600	26,000	23,300	49,300
	D	190	5,470	5,660	25,800	17,800	43,600	26,000	23,300	49,300
Constrained	A	990	445	1,440	718,000	571,000	1,290,000	719,000	571,000	1,290,000
	B	990	445	1,440	33,500	16,500	50,000	34,500	16,900	51,400
	C	990	445	1,440	26,100	20,000	46,100	27,100	20,400	47,500
	D	990	445	1,440	25,900	20,000	46,100	26,900	20,400	47,500

A = Once through-cooling

B = Cooling pond

C = Natural draft wet towers

D = Mechanical draft wet towers

P = Power plant

G = SYNTHANE plant

[a] Three significant figures.

[b] Includes net boiler feed for the SYNTHANE plant.

Table 5. Water Circulation for Coal Conversion Processes for the Year 1990

(acre-feet/yr)^a

Residual management level	Cooling system	Coal conversion process			Cooling			Total		
		P	G	P+T	P	G	P+G	P	G	P+G
Base	A	37,500	22,500	60,000	675,000	509,000	1,180,000	713,000	532,000	1,240,000
	B	37,500	22,500	60,000	1,110,000	509,000	1,620,000	1,150,000	532,000	1,680,000
	C	37,500	22,500	60,000	1,540,000	509,000	2,050,000	1,580,000	532,000	2,110,000
	D	37,500	22,500	60,000	1,070,000	509,000	1,580,000	1,110,000	532,000	1,640,000
Constrained	A	37,500	29,400	66,900	718,000	571,000	1,290,000	756,000	600,000	1,360,000
	B	37,500	29,400	66,900	1,220,000	571,000	1,290,000	1,260,000	600,000	1,860,000
	C	37,500	29,400	66,900	1,640,000	571,000	2,210,000	1,680,000	600,000	1,280,000
	D	37,500	29,400	66,900	1,180,000	571,000	1,750,000	1,220,000	600,000	1,820,000

A = Once-through cooling

B = Cooling pond

C = Natural draft wet towers

D = Mechanical draft wet towers

P = Power plant

G = SYNTHANE plant

^{a}Three significant figures.

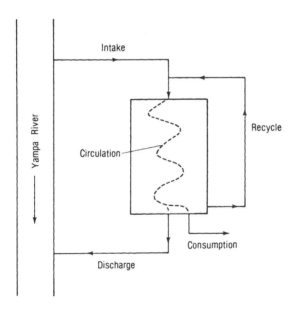

Figure 2.   Schematic Water Use Pattern for a Coal Conversion Process

Table 6. Minimum and Maximum Water Intakes for the Coal Conversion Processes for the Year 1990

(acre-feet/yr)

Coal development scenario	Residual management level	Minimum (Mechanical or natural wet cooling)		Maximum (Once-through cooling)	
		Intake	% of mean flow[a]	Intake	% of mean flow[a]
1	Base	26,000	2.30	675,000	59.73
	Constrained	26,900	2.38	719,000	63.63
2	Base	49,300	4.36	1,190,000	105.31
	Constrained	47,500	4.20	1,290,000	114.16

[a]Mean annual flow of 1,130,000 acre-feet/year (1,560 cfs).

scenarios, the minimum and maximum withdrawals do not vary appreciably
with the residual management level. Water withdrawal with the addition
of a coal gasification plant (in scenario 2) is somewhat less than if
two additional power plants had been considered.

As shown in table 6, once-through cooling in scenario 2 is not
feasible without the development of storage. In this case, the annual
water withdrawal exceeds the mean annual flow and even if there were no
variability at all in the flow of the Yampa River, some storage would
be required for recirculation. But the river flow from month to month
and year to year is variable and thus storage is likely to be necessary
for scenario 1 as well. In the case of natural or mechanical draft wet
cooling, the water intake is substantially less. However, even in this
case, there remains a finite probability that the Yampa flow, particularly during the drier months, would be insufficient to sustain the coal
conversion processes. Relative to the two coal development scenarios,
the probability of a flow during the month of September being less than
the required withdrawal is in the range of 0.05 to 0.10. Thus, with the
use of mechanical or natural draft wet cooling, the withdrawals could be
rather reliably met by the natural flow of the Yampa River assuming that
there were no other demands for water in the basin. However, coal development is not the only source of water demand. Existing demands,
as well as the additional demands of supportive activities for the coal
conversion processes under scenarios 1 and 2, would need to be taken
into account to determine the extent to which coal development could be
supported by the Yampa flows.

No attempt is made here to determine the total water demands for either coal development scenario. However, to gain some insight as to storage requirements for different levels of demand, storage capacities are derived.

Storage Estimates

The historical flows of the Yampa River are unlikely to repeat themselves during the economic life of the coal development activities. The flows are assumed to be random and though future flows cannot be predicted, it is possible to generate synthetic flow sequences where the sequences may be regarded as equally likely realizations of future flows. With respect to each synthetic flow sequence, the storage requirement for various levels of demand can be estimated. Using a series of synthetic flow sequences and the associated storage requirements, the expected storage requirements for various levels of demand can be determined.

A multiseason (twelve months) single site (near Maybell, Colorado) synthetic flow generator (Matalas, 1967) was used to generate twenty, fifty-year, monthly sequences of flow. For each of the flow sequences, the storage capacities required to meet various levels of demand were estimated by means of the Thomas "sequent peak algorithm" described by Fiering (1967). Three levels of demand were considered. The smallest level was taken to be 47,500 acre-ft/yr which corresponds to the constrained plants in scenario 2 using natural or mechanical draft wet cooling (table 4). The two additional levels were taken to be 362,000 acre-ft/yr and 734,000 acre-ft/yr. These three demands are approximately equal to 4 percent, 32 percent, and 64 percent of the mean annual flow of the Yampa River near Maybell.

From the twenty synthetic flow sequences, the mean and standard deviation of the storage capacities were estimated and are presented in table 7.
Thus, the storage requirement for the coal conversion processes would be
quite small if there were no other water demands in the basin. However,
there would be additional demands which could greatly increase the storage requirements. If once-through cooling had been considered, storage
requirements would have greatly increased, and because the maximum withdrawals would have exceeded the mean annual flow, an algorithm other than
the "sequent peak algorithm" would have had to be used to estimate the required storage capacities.

Table 7. Storage Capacities (Mean and Standard Deviation)

Water demand (acre-ft/yr)	Storage capacity (acre-ft)	
	Mean	Standard deviation
47,500	9,240	8,620
362,000	1,830,000	221,000
734,000	6,110,000	1,330,000

The above analysis is meant to be illustrative only, and the estimates
of the storage capacities are rough. The analysis is not intended to provide
precise estimates of the storage requirements. To ascertain these requirements, a much fuller account of the region's coal and water development objectives and of the hydrology, outside the scope of this paper, would be needed.

Summary and Conclusions

Two coal development scenarios and two residual management levels for each were considered for the Yampa River Basin. Scenario 1 is based on surface mining of coal, part of which is used for the production of electrical power and the remainder is transported to markets outside of the basin using non-water intensive transport systems. Scenario 2 is similar to scenario 1 but also includes the production of synthetic natural gas.

In the base case, it is assumed that no environmental standards concerning residuals discharges have to be met. A constrained case is based on meeting a standard of zero discharge for liquid residuals and the state of Colorado standards for air quality.

Based on plant level analyses of the mining and coal conversion activities, water use patterns and the types and quantities of residuals generated were determined for various types of cooling systems. The minimum water withdrawals were determined to be 26,000 acre-ft/yr under scenario 1 (residuals management base case) and 47,500 acre-ft/yr under scenario 2 (residuals management constrained case). The flow of the Yampa River during the low-flow month of the river (September) was estimated to be less than these demands with a probability in the range of 0.05 to 0.10.

Synthetic flow sequences were used to estimate the mean and variance of storage requirements for three levels of water demand. The demand levels varied from 47,500 acre-ft/yr to 734,000 acre-ft/yr with corresponding mean storage capacities ranging from about 9,000 acre-ft to 6,000,000 acre-ft. The estimates of storage capacity are not to be taken

as firm estimates, but rather serve to indicate that the flow of the
Yampa River cannot support the coal development scenarios discussed
herein without providing reservoir storage. Based on more detailed
information concerning the types and quantities of residuals generated
and the patterns of water use provided by plant level coal mining and
conversion studies, in conjunction with a regional economic assessment
of alternative water demands, the feasibility of these and other coal
development scenarios could be evaluated in a more exacting manner.

References

Delson, J. K., and R. J. Frankel. 1972. "Residuals Management in the Coal-Electric Industry." Incomplete draft manuscript, Resources for the Future, Washington, D.C.

Fiering, M. B. 1967. Streamflow Synthesis (Cambridge, Mass., Harvard University Press).

Forney, A. J., and J. P. McGee. 1972. "The Synthane Process--Research Results and Prototype Plant Design," Proceedings of the 4th Synthane Pipeline Gas Symposium (Chicago, Ill., October) pp. 51-71.

James, I. C. II, E. D. Attanasi, Thomas Maddock III, S. H. Chiang, B. T. Bower, and N. C. Matalas. 1977. "Water and Energy Use in Coal Conversion," U.S. Geological Survey Open-File Report (in draft) (Reston, Va., U.S. Geological Survey, Water Resources Division).

Landis, E. R., and G. C. Cone. 1971. "Coal Reserves of Colorado Tabulated by Bed," U.S. Geological Survey Open-File Report (Reston, Va., U.S. Geological Survey).

Matalas, N. C. 1967. "Mathematical Assessment of Synthetic Hydrology," Water Resources Research vol. 3, no. 4.

Speltz, Charles N. 1974. "Strippable Coal Resources of Colorado," U.S. Bureau of Mines Preliminary Report 195 (Washington, D.C., U.S. Department of the Interior).

U.S. Department of the Interior, Bureau of Land Management. 1975. Environmental Statement, Northwest Colorado Coal (Washington, D.C.).

THE POTENTIAL IMPACTS OF ENERGY DEVELOPMENT
ON WATER RESOURCES IN THE YAMPA RIVER BASIN:
A DISCUSSION

Myron B. Fiering*

Introduction

The paper by Maddock and Matalas can be divided into two major sections:
the first specifies water requirements imposed by various coal mining and
gasification scenarios, and the second examines the adequacy of water availability to meet such requirements in the Yampa River Basin and the storage
requirements for overcoming potential deficits. Their hydrologic analysis
shows, for a highly simplified case, shortfalls which might be expected and
the storage capacity needed to meet these deficits at a prescribed level of
reliability. This discussion focuses on the second portion of their study,
and in particular suggests the form and extent of the analysis that would
have to be undertaken to promote their storage study from a simple didactic
model to a working tool suitable for implementation by agencies or institutions interested in monitoring and managing water supply and quality.
This discussion is not intended as a criticism of the analysis suggested
by Maddock and Matalas because it is understood that their work is merely
illustrative of the form which a complete analysis should take.

Principal Shortcomings of the Proposed Approach

The sequent peak algorithm is featured in the Maddock-Matalas paper
as a method for calculating the storage required to meet a set of target

*Gordon McKay Professor of Engineering and Applied Mathematics,
Harvard University, Cambridge, Massachusetts.

flows at a given level of reliability. It should be noted that this algorithm gives the same storage requirement as the much older Rippl, or mass curve, analysis; its principal value lies in the fact that it can be shown to produce an optimal strategy for releasing (wasting) water under any demand pattern. The sequent peak algorithm was never intended to supplant the mass curve analysis for calculating storages.

The concept of reliability requires further elaboration because there are many ways of utilizing synthetic streamflows to estimate reliability of system performance. For example, it might be sufficient simply to count the number of failure periods (where a failure occurs whenever the target is not met) and to calculate the ratio of the number of failures to the total number of time periods in a hydrologic simulation run; this ratio subtracted from 1.0 gives a measure of system reliability, but it should be noted in using this method that small deficits penalize system performance just as heavily as large ones.

Another measure of system reliability assumes a linear loss (or penalty) function, accumulates the deficits for all time periods in which the target is not met, and then divides the sum by the nominal target over all time periods; this produces a failure ratio which represents the extent of all failures and which could easily be modified to represent the average failure. More elaborate transformations, such as quadratic or piecewise linear, of the physical deficit could be utilized to penalize the system more heavily, or more gently, to include the proximity in time of one failure to another (that is, by incorporating a measure of persistence among failures), and so forth.

Perhaps the most important shortcoming of the Maddock-Matalas analysis, and it is well recognized by the authors, is its reliance on September flows alone for establishing the cumulative flow deficit to be met by releases from storage. The storage requirements are based on a sequence of flows drawn only from the most critical month of the year so the reservoir cannot refill during periods of high flow. The study also ignores other actual and potential uses of water in the Yampa River Basin, other than for coal production and energy development, so the required storage is reduced because no other uses are incorporated.

How to Structure a More Comprehensive Water Allocation Model

A more comprehensive and satisfactory model from which to develop and assess water policy options for a given river basin, including the storage requirements, would contain at least seven important components:

(i) a statistically valid representation of basin hydrology, with appropriate temporal and spatial correlations calculated for inclusion in a stochastic model for generating "seasonal" flows;

(ii) a realistic set of demands (price-quantity relationships) for water diversion and use;

(iii) an unambiguous operating policy under which the proposed storage configuration would be operated;

(iv) an algorithm for assessing the costs (in terms of the opportunity costs) associated with failure to meet the several water demands placed on the system;

(v) a set of environmental standards whereby the conflicting interests can admit or reject the various policy options;

(vi) an administrative structure for identifying the conflicting interests, for articulating the several objectives and potential negotiating positions, and for introducing proposed solutions and compromises; and

(vii) institutions for implementing compromise solutions along the negotiating frontier (the locus of Pareto-optimal strategies).

These components are discussed in turn below.

(i) <u>Basin Hydrology</u>

When selecting a model for synthetic streamflow generation, it is tempting to use a stochastic process for which the parameter estimates are reasonably stable. For example, it has been difficult to object to the use of a normal density function because we know that the sampling errors of its first two moments are manageably small. However, it is probably more appropriate in studies of water shortage to be less concerned about the moments and other measures of central tendency than about quantiles, because it is at the <u>tails</u> of the flow distribution, in periods of protracted low flow, that the system tends to fail. Preservation of the lower moments of the probability distribution is not sufficient in studies which turn on extrema. Therefore, it is more appropriate to select distributions and fitting techniques which give rise to robust estimates of <u>quantile</u> flows near the tails and to deemphasize stability of the parameters themselves. Recent work by Thomas and Houghton which is summarized in Houghton (1977), Matalas and Wallis (1978), and others suggests that the Wakeby distribution, particularly when suitable parameter estimating techniques such as "probability-weighted moments" are used, produces remarkably stable estimates of extreme quantile-flows even though the parameters themselves are highly variable. The reason for this robustness of the tails might inhere in the fact that the Wakeby's five parameters provide enough flexibility, that is, enough overlap between high and low values (or right-hand and left-hand tails), to force a functional form for which estimates well beyond the limit of observations are very robust.

Whether the Wakeby or some other distribution is imposed, it is clear that traditional reliance on lower moments alone, as Maddock and Matalas did in their paper, should be replaced by techniques which are robust in their extrema, and that suitable temporal and spatial dependence should be built into generating algorithms for extreme flows. Currently available multisite flow generators do not use the Wakeby or similar (Tukey-type) distributions; work is now underway to produce such generators.

(ii) Incorporation of Water Demands

To save computer time, it is appropriate to divide the year into seasons of unequal length so that high-flow periods, during which there is presumably no problem in meeting demands imposed by energy, and other natural resource, development projects can be represented by coarse time steps, while low-flow periods can be represented by months, weeks, or days to obtain a resolution appropriate to the problem.

This discussion does not deal with demand patterns for several competing uses of a water resource; however, if these demands were defined and ranked in terms of priority, they could be used as inputs to the simulation program whose operating policy treats them in order.

(iii) The Operating Policy

There have been many efforts to use programming techniques to identify optimal operating policies, but according to Loucks[1] these methods cannot

---

*Daniel P. Loucks, Professor of Water Resources Engineering, Cornell University (personal communication, April 1977).

readily handle multiple-reservoir systems. Thus, the work of Young (1967),
Buras (1965), and Jacoby and Loucks (1972) is not generally applicable to
large, multi-reservoir storage systems unless it can be modified to accommodate multiple storage sites. An algorithm developed recently by
Fiering and Loucks (1977) is a fast approximation to the Space Rule developed earlier in the Harvard Water Program (Hufschmidt and Fiering, 1966).
This rule allocates storage, rather than releases, so as to equalize the
probability of spill among all reservoirs in the system; the new algorithm
shows promise of being widely adopted in simulation studies.

The identification of an optimal storage policy cannot proceed inde-
pendently of the calculation of costs and benefits (or penalties). That
is, an optimal policy purports to promote beneficial uses or to reduce
nonbeneficial uses; without a clear statement as to what constitutes such
uses and how multiple uses are to be evaluated and compared, it is impossible to promulgate an optimal policy. But it is mathematically awkward to
include realistic benefit and loss functions in most operating policy
formulations; the resulting dynamic programming techniques have been applied
to significant planning problems of this sort but they characteristically
consume so much computer time and storage that on balance their utility is
suspect. The burden of obtaining optimal solutions essentially precludes
the use of the technique for all but large water resource systems in which
it is anticipated that even modest improvements in system operations would
justify the incremental expense of the analysis.

(iv) Alternative Analyses of Reservoir Storage-Yield

A computer simulation program for storage design operates in one of
two modes. If a storage-yield (using a sequent peak or Rippl curve)

analysis is to be made, the synthetic hydrologic sequences are divided into N specified lengths or design periods of length T for each of which a storage-yield analysis produces a requisite storage capability for the entire system. These N storage values are then arrayed and averaged over all design periods of equal length T. The distribution and moments can be assessed with respect to cost, risk, and so on. By changing the water demands, design period, operating policy, and the like, a range of requisite storages (or their distributions) can be calculated, as Maddock and Matalas have done in their study. This form of analysis is not properly a simulation because the analysis produces a solution; a true simulation makes a design choice and asks: "What are the consequences of this decision?"

The alternative form of analysis utilizes a very long synthetic sequence under a presumed system design and system operating policy, and from the calculated indexes of system performance a plot is made on a contour map of the index of system reliability (note the several definitions given above) as a function of the design decisions (including system targets or demands). In the simple one-reservoir case, this is analogous to a storage-yield function, or input-output curve. However, the input-output curve is not a deterministic locus, but a family of curves or contours relating storage (input), "firm" yield (output), and the associated level of reliability which is tantamount to defining just how "firm" the "firm yield" really is.

When dealing with more than one storage facility, the analysis is slightly more complicated because many combinations of storage could produce the same yield and reliability levels, which complicates the

definition of the abscissa. For management purposes, the quantity to plot
is the least-cost combination of storages which produce the given yield
and reliability; this is similar to the so-called equivalent storage or
cheapest single downstream storage whose performance characteristics are
identical to some upstream combination (the work of Moen (1977) has been
important in this regard).

(v)   The Use of Standards for Reducing the Policy Options

Assume we have before us a simulation model which is driven by a
reasonably comprehensive operating policy, subject to appropriately selected
historical, or to generated synthetic records, or both, and augmented by a
set of functions for calculating benefits and costs of the water resource
system. The assessment of these economic outcomes has historically been
taken to be derived directly from some set of functions and values which
are appropriate and applicable to all the parties interested in the project.
It should be noted that almost all important projects, and certainly those
that produce undesirable and controversial side effects such as water pollution, destruction of wild and scenic regions, upstream-downstream con-
troversies with regard to flood control, and the like, engender powerful
interest groups whose perceptions of the value of specified physical outcomes vary enormously. It is naive to believe that conflicting interests
will place the same values on the outcomes of a given project. Further, it
is naive to believe that any set of criteria or standards will be uniformly
acceptable, given our institutional arrangements, to all the parties who
profess vested interests in the project. There are several alternatives
to the inevitable chaos, including administrative fiat which establishes
an objective function by executive action (as might occur in the private

sector); some form of plebiscite or referendum under which the various interests are given a range of weights; and some set of guidelines for reducing the potential chaos by compressing the total number of options to an irreducible minimum which might then be sorted and ranked through a variety of negotiating and political processes.

Most decision makers would agree that executive order is the least complicated because there is no appeal, no disagreement, and no conflict. But given the realities of our political and institutional infrastructure, it seems clear that some limited negotiation must take place, that incentives or side payments must be arranged, and that interest groups must be offered compromises for resolving the conflict.

In order to reduce the number of options for consideration at the final stages of planning and negotiation, it is useful to establish constraints which then dictate that certain combinations or alternatives are not feasible; we call these constraints standards. While it is conceptually attractive to develop such limitations to the technologically possible alternatives, it is operationally a difficult task. For example, if environmental standards are to be based on toxicological evidence concerning long-term cumulative effects of pollutants, the statistical evidence often leads to contradictory judgments concerning allowable levels of exposure. Even more basic, the notion of exposure is ill-defined because duration, frequency, and intensity are all intimately involved in calculating the cumulative insult. Is long-term, but slight, violation of some toxicological standard equally as dangerous as periodic violations which produce the same total pollutant flux in the environment? Is either of these equally as dangerous as an infrequent or short-term exposure to a greater intensity?

If a standard is written in terms of the seven-day, ten-year return period level (for example, low flows or pollutant concentrations), how do we account for a protracted period, say fourteen days, which is in violation? Is that to be counted as one violation or more than one?

Thus, the establishment of standards that deal with extreme events becomes a particularly slippery statistical matter, whereupon it becomes essential in establishing such standards further to suggest the use of probability estimating procedures which are robust with respect to their ability to estimate the tails of the distributions of violations. It has previously been noted in this discussion that the Wakeby distribution, a member of the generalized Tukey family of λ-densities, is particularly resilient and robust with respect to the tails. This remarkable property of the Wakeby, owing in part to the richness and flexibility imparted by its five parameters, allows for adjustments and adaptations to small-scale sampling fluctuations, and even to some outliers, so that important advances in the establishment of standards can be made through the use of similar densities. By application of such standards, and by careful definition of the vectorial aspects of a "violation," a substantial number of options can be excluded from consideration because they do not meet the conditions for further consideration.

(vi) The Analysis of Policy Options: the Negotiation Frontier

Utilization of the Yampa River for process water in the Maddock-Matalas coal development model poses potential conflicts between proponents of the scheme and those who support competitive uses of the water resource. Such conflict is typical rather than rare because it only infrequently happens that all the parties impacted by a proposed development can agree on the need for, and assessment of, a resource project. The relationships among users are not necessarily symmetrical; upstream withdrawals are made at the expense of downstream users, but there is little that the downstream users can do physically to alter the upstream supply. However, there may be institutional arrangements and incentives which might allow the laws of economics and political science somehow to drive the disparate interests toward a compromise.

For purposes of illustration, we assume that the interests of the various parties are not antagonistic; that is, no party desires the impoverishment of the others. Note that this differs from the usual zero-sum game in which losses to one player represent gains to another so that the players compete against each other for those assets which change hands. In the application here, and in the model used symbolically to represent it, interests of the opposing parties may conveniently be thought of as being independent (in mathematical terms, orthogonal). Consider only two interest groups, A and B. Group A does not care if Group B does very well so long as it does not cost them anything, and conversely. The best alternative (in that neither group can do any better without harming the other) is not necessarily the same as that which optimizes a benefit/cost or net benefit criterion function, unless side payments and other compensations are allowed. Optimization of a single criterion function is more appropriate in the private sector, in the military services, and in a setting in which the objective function can be established by decree, whereas the negotiated or compromise optimum, known by economists as the Pareto-optimal solution, is more appropriate in the public sector and in those resource allocation problems whose solutions must ultimately

serve several vested interests whose objectives are not mutually congruent.

A graphic way to display Pareto-optimality is to consider a decision
which impacts two groups, A and B, whose interests are independent. Limitation to two groups is suggested because it requires only simple graphi-
cal techniques to demonstrate the arguments. In the general case, any
number of groups or vested interests can be represented, but the dimension-
ality of the space required to visualize the process increases with the
number of such groups.

Each group is asked to consider a set of decisions or alternatives.
In fact, there are an infinite number of such alternatives because design
decisions are generally make along a continuum. But in practice it is
reasonable to reduce this to a set of decisions on a discrete domain so
that the total number of options is manageable. It should be emphasized
that the art of resource-system design is contained in generating and
sorting a sufficiently small number of alternatives to allow systematic
evaluation by simulation or other form of analysis. This is not a trivial
task and involves intuition, judicious application of standards and criteria
which preclude as infeasible large portions of the decision space, con-
sultation with clearly identified interest groups, articulation of objectives and objections for each, and so on. The participants are asked
to rank or to evaluate (without communication with one another) the options.
Groups A and B derive their benefits and calculate their costs using different algorithms. In any event, the point (A_i, B_i) can be plotted on
Cartesian coordinates for which A_i is the benefit perceived by Player A
for the ith project and B_i is the benefit perceived by Player B for the ith

alternative. Clearly if there were more players, the point would have to be plotted in n-dimensional hyperspace and would have coordinates $(A_i, B_i, C_i, \ldots)$.

Under our assumed rules of behavior, any solution point which lies to the north or east of point i is said to dominate it because at least one party is made better off and no party is penalized. Similarly, any point which lies to the south or west is said to be dominated by point i, for which all players do at least as well and at least one player does better.

Thus, the northeastern boundary of all points can be thought of as dividing the (A,B)-plane into three portions: the dominated solutions which lie to the south and west of some other point(s), the infeasible zone which lies to the north and east of the undominated solutions and represents that portion of the plane which cannot be reached because of technological or institutional limits, and the boundary which separates these regions. This boundary is called the <u>negotiation frontier</u>.

If Pareto-optimality is to be achieved, the solution must lie along this frontier. Identification of the alternatives which define the frontier is an extremely useful task because it reduces the number of solutions which remain as serious competitors in the process and because it hints at side payments or incentives which might be proposed if the individuals agree to some solution. It is also interesting to consider the effects of having one player or another change his evaluation function without changing the ranking; that is, of making a monotonic transformation of the response frontier. Consider, for example, that the analysis is plotted on an elastic sheet rather than a piece of graph paper and that the sheet is stretched differentially in each of the coordinate directions. While the slope of the negotiation locus will change, indicating different levels of incentives required to promote movement between alternatives, the undominated strategies remain unchanged. Thus, each group does not have to inquire, and should not even care, how the other group evaluates the outcomes; it is essential only that the evaluation function produce a monotonic or transitive ranking which is based on multi-dimensional preference expressed for each of the alternatives.

(vii)   The Role of Public Institutions in Public Policy Decisions

Having identified the (hopefully) small undominated set of strategies for final negotiation, it remains to implement the necessary incentive programs and to provide for transfer of articulated benefits and costs so that the interest groups can move toward a compromise solution with appropriate guidelines from the program administrator. It would be naive to think that every conflict could be so peacefully resolved, that every difference could be reduced to monetary terms, or that we could rapidly become adept at arranging side payments--a mechanism which we in the United States are historically loathe to adopt and which our institutions are not well designed to accomplish. But the alternatives to accommodating such payments are not attractive. One is to do nothing, to continue the debate by authorizing additional studies and restudies which promise to be no more illuminating or convergent than their predecessors. Another alternative is to initiate a series of lawsuits and legal actions which force the decision-making process into the judicial system--perhaps the worst possible place for it to be, with all of its rigidity and bias. A final alternative is to throw the whole matter to the executive or legislative branch in the hope that there will emerge a politically acceptable decision by which all parties would have to abide.

Compared to these alternatives, Paretian analysis is benign. Experience dictates that incrementally negotiated decisions will be made, that small improvements will be initiated, and that as this approach to resolving conflicts grows in popularity and acceptance, more substantial problems might be solved through its application.

The proper role of the administration or government concerns identification of the parties with vested interests, generation and validation of the data bases on which the several parties establish their technological and economic responses, provision of a forum within which the parties can articulate their objectives and preferences, analysis of alternatives whereby the vector of physical responses is assessed, and, finally, presentation of the multi-dimensional scoresheet or plot from which the various trade-offs and negotiations can be assessed and compared and from which the parties can be encouraged to move toward a Pareto-optimal resolution. Thus, government's role is to facilitate, and to provide the basic data for, negotiations along the negotiation frontier. This is not a customary role for American bureaucracy which is more accustomed to developing, interpreting, and executing federal regulations and to constructing the actual systems.

The role of government as a mediator rather than as a participatory agent represents an important shift in emphasis and involvement which should be encouraged. The important point is that decisions under uniformly agreeable objective functions reducible to monetary terms are virtually never made in the public sector except under emergency conditions. In

the absence of these conditions, some form of negotiation must be undertaken, and this can best be done in an atmosphere which encourages articulation of the various objectives, values, and criterion functions.

Conclusions

An appropriate systems analysis of a water resource problem has components which deal with the following issues: (1) statistical description
of the hydrologic inputs with particular reference to distribution of extrema and classification of the stochastic meteorological processes (particularly precipitation and streamflow) in the area; (2) an analytical or
numerical model for calculating the (physical) criterion function, which
model would include a system operating policy, a vector of system targets,
and a set of system constraints; (3) rules for identifying the various
interest groups and for assessing the system performance and system costs
for each; (4) an algorithm for identifying the Pareto-optimal solution or
for conducting an appropriately thorough search of the multi-dimensional
response space to be comfortable with the solution; and (5) a formalism
for recognizing that the best solution represents a consensus or compromise
among conflicting interests and for encouraging the institutions to recognize that most water resources decisions are made under conditions of
conflict of interest.

Maddock and Matalas deal with only a small portion of this list; their
objective was not to describe the details of a comprehensive reservoir
simulation and storage study but rather to show, by excising a small piece
of the total package, how different levels of demand for water supply to
serve a coal gasification project exact a wide range of storage requirements

on a typical river system. The more complete type of study described in this addendum would be required for actual decision making.

References

Buras, Nathan. 1965. "The Optimization of Large Scale Water Resource
    Systems: Operational Aspects." Unpublished memorandum (Los Angeles,
    Calif., University of California, Department of Engineering,
    September).

Fiering, Myron B., and Daniel P. Loucks. 1977. Unpublished memorandum on
    work in progress as reported to the U.S. Department of the Interior,
    Office of Water Research and Technology (Cambridge, Mass., Harvard
    University, Division of Applied Sciences, 30 June).

Houghton, John. 1977. "Robust Estimation of the Frequency of Extreme
    Events in a Flood Frequency Context" (Ph.D. dissertation, Harvard
    University, Division of Applied Sciences, Cambridge, Mass.).

Hufschmidt, Maynard, and Myron B. Fiering. 1966. Simulation Techniques for
    Design of Water Resource Systems (Cambridge, Mass., Harvard University
    Press).

Jacoby, Henry D., and Daniel P. Loucks. 1972. "Combined Use of Optimiza-
    tion and Simulation Models in River Basin Planning," Water Resources
    Research vol. 8, no. 6 (December) pp. 1401-1414.

Matalas, N. C., and James Wallis. 1978. Unpublished memorandum on work
    in progress as reported to the Environmental Systems Program, Harvard
    University (Reston, Va., U.S. Geological Survey, Water Resources
    Division, 27 April).

Moen, Terje. 1977. "Development of Mathematical Models for Preliminary
    Screening in Water Resources Design" (Ph.D. dissertation, Harvard
    University, Division of Applied Sciences, Cambridge, Mass.).

Young, George K., Jr. 1967. "Finding Reservoir Operating Rules,"
    American Society of Civil Engineers, Journal of Hydraulics Division
    HY6.

THE POTENTIAL IMPACTS OF ENERGY DEVELOPMENT
ON WATER RESOURCES IN THE YAMPA RIVER BASIN:
A DISCUSSION

Timothy Doak Steele*

Introduction

The analyses presented in the paper by Thomas Maddock III and N. C.
Matalas describe possible direct effects on the flow and quality of the Yampa
River of mining and conversion of coal, based upon two assumed development
alternatives, each with two levels of environmental controls. The purpose
of this discussion is to supplement the material contained in the Maddock-
Matalas paper with a brief overview of several recent assessment studies of
the Yampa River Basin. These assessments and studies were supported and
undertaken by the U.S. Geological Survey's Colorado district office in Denver
by a project known as the Yampa River Basin Assessment (Steele and coauthors,
1976a; 1976b).

The Yampa River Basin Assessment Project

The Yampa River Basin Assessment is a three-year project supported by
the U.S. Geological Survey designed to evaluate the water resources related
implications of the region's economic development which is rapidly increasing
due to increased mining, conversion, and transport of coal. For purposes of
the basin assessment analyses, the entire Yampa River Basin east of Dinosaur
National Monument (figure 1) was included which encompasses an area of

---

*Hydrologist, U.S. Geological Survey, Lakewood, Colorado.

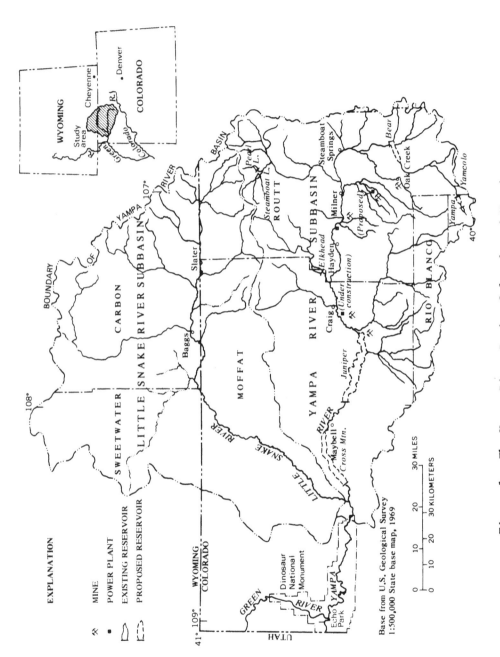

Figure 1. The Yampa River Basin, Colorado and Wyoming

<u>Source:</u> See following page.

Figure 1. (continued)

Source: I. C. James II and T. D. Steele, "Applications of Residuals Management Techniques for Assessing the Impacts of Alternative Coal-Development Plans on Regional Water Resources," Proceedings, Third International Symposium in Hydrology (Fort Collins, Colo., Colorado State University, 1977).

approximately 8,080 square miles (21,900 km^2) in northwestern Colorado and south-central Wyoming drained by the Little Snake and Yampa rivers. Several physical based models and qualitative analyses conducted as part of the basin assessment project have been coordinated with regional residuals-environmental quality management and water-use studies conducted by a systems analysis group of the U.S. Geological Survey in Reston, Virginia. These joint undertakings are discussed in James and Steele (1977).

As noted in the Maddock-Matalas paper, projections of water consumed and residuals generated by coal resource development alternatives provide information for evaluating the water-related environmental consequences of each alternative. Projected water requirements for various scenarios are compared with estimates of water availability in both a physical and legal sense (Knudsen and Danielson, 1977). As a consequence of coal resource and associated economic development, major surface water impoundments have been proposed in the basin (Steele and coauthors, 1977; Steele, 1978). A total of thirty-five reservoirs have been proposed under nineteen projects by the U.S. Geological Survey (table 1). At least two planned transbasin diversions exporting water from tributary streams of the Yampa and Little Snake rivers indicate additional competition for uses of available surface waters in the basin (Steele and coauthors, in press). These are the Hog Park diversion of tributary stream waters of the Little Snake River to the city of Cheyenne, Wyoming, and the Vidler Tunnel diversion of headwater tributary streams of the Yampa River, as well as other west-slope waters, to the city of Golden, Colorado. Both planned diversions are for municipal and industrial uses. Two small existing diversions have a negligible effect on downstream flow conditions.

Table 1. Proposed Major Surface Water Impoundments

Reservoir	River or creek	Total capacity (acre-feet)[a]	Project	Project purpose	Remarks
		Carbon County, Wyoming			
Savery	Savery	(19,000)	SP	I	Alternative to Sandstone site (USBR EIS).
Sandstone	Savery	15,500	SP	I	
		Moffat County, Colorado			
Pothook	Slater	60,000	SP	I	Water rights application of 73,580 acre-feet.
Juniper	Yampa	1,079,990	LY, JC	I,P	Original water right of 844,290 acre-feet plus enlargement of 235,700 acre-feet.
Cross Mountain	Yampa	142,000	LY,JC	I,P	
Jubb	Jubb	2,250	LY	I	Application for water rights not found.
Thornburg	Milk	36,000	YJ	I	Water rights application of 31,810 acre-feet.
Craig	Yampa	44,490	UI	N,D	
Rampart	Fortification	12,330	LY,GN	I	

Table 1. (continued)

Reservoir	River or creek	Total capacity (acre-feet)[a]	Project	Project purpose	Remarks
		Routt County, Colorado			
California Park	Elkhead	36,540	LY,GN	I	
Hayden (Mesa)	Sage	8,620	CU	I,D	
Trout	Trout	23,340	PM	I,N,D	
Childress	Trout	24,160	OCP	M,N,D	
Upper Middle	Middle	102,200	OCP	P,N	Water rights application of 17,000 acre-feet.
Lower Middle	Middle	25,150	OCP	P,N	Water rights application of 17,000 acre-feet.
Twenty Mile	Fish	15,300	JEL	I	
Dunckley	Fish	57,090	UY,HM	I,D,S	
Hinman Park	Elk	44,040	PSC	P	
Big Creek	Big	6,900	b	b	
Grouse Mountain	Willow	79,260	b	b	
Pleasant Valley	Yampa	(43,220)	CRWCD	I	
Woodchuck	Yampa	(40,000)	CRWCD	I	

Table 1. (continued)

Reservoir	River or creek	Total capacity (acre-feet)[a]	Project	Project purpose	Remarks
		Routt County, Colorado			
Lake Catamount	Yampa	7,400	PVI	R,M	
Yampa	Yampa	(151,120)	OCP	P	Water rights conditional decrees; not included in Oak Creek Water & Power Project.
Blacktail	Yampa	229,250	YGC	P	
Lower Green	Green	99,600	YGC	b	Water rights application of 45,000 acre-feet.
Main Green	Green	(103,230)	YGC	b	Water rights conditional decrees; not included in Oak Creek Water & Power Project.
Yampa	Yampa	(32,500)	VTW	X	Overlapping sites, Blacktail Reservoir.
Morrison	Morrison	(12,500)	VTW	X	
Service	Service	(22,000)	YGC	b	Water rights conditional decrees; not included in Oak Creek Water & Power Project.
Wren	Fish	2,160	b	I,R,D	
Allen Basin	Middle Hunt	2,250	b	I,D	

Table 1. (continued)

Reservoir	River or creek	Total capacity (acre-feet) [a]	Project	Project purpose	Remarks
		Routt County, Colorado			
Bear	Yampa	11,610	UY,W	I	
Bear	Yampa	(30,000)	CU	N	
Yamcolo	Bear	9,000	UY,T	L,N,D	Water rights application of 6,530 acre-feet.
Total		2,176,430			

Project codes:

CRWCD, Colorado River Water Conservation District
CU, Colorado-Ute Association
GN, Great Northern Project (CRWCD)
HM, Hayden-Mesa Project (CRWCD)
JC, Juniper-Cross Mountain Project
JEL, J. E. Lutrell
LY, Lower Yampa Project (CRWCD)
OCP, Oak Creek Power Company
PM, Pittsburg & Midway Coal Company
PSC, Public Service Company of Colorado
PVI, Pleasant Valley Investment Corporation

SP, Savery-Pothook Project (USBR)
T, Toponas Project (CRWCD)
UI, Utah International, Inc.
USBR, U.S. Bureau of Reclamation
UY, Upper Yampa Projects (CRWCD)
VTW, Vidler Tunnel Water Company
    (Sheephorn Project)
W, Wessels Project (CRWCD)
YGC, Yampa-Green Corporation
YJ, Yellow Jacket Project (USBR)

Project purposes:

D, Domestic; I, Irrigation; M, Municipal; N, Industrial; O, Other; P, Power; R, Recreation; S, Stock supply; X, Export (transbasin diversions)

Table 1. (continued)

Source: T. D. Steele, D. P. Bauer, D. A. Wentz, and J. W. Warner, "The Yampa Basin, Colorado and Wyoming—A Preview to Expanded Coal-Resource Development and Its Impacts on Regional Water-Resources," Yampa River Basin Assessment Project Report, in review (Denver, Colo., U.S. Geological Survey, in press).

[a]Capacities in parentheses indicate reservoirs competing for the same sites.

[b]Information not available.

For purposes of evaluating water resources related implications of coal development in the Yampa River Basin, uncertainties in the projections of coal production and in the forms of utilization of future mined coal have been considered (Udis and coauthors, 1977; Steele and coauthors, in press). In 1975, only one coal-fired power plant operated in the basin, near Hayden, Colorado (figure 1); the bulk of the mined coal was transported out of the basin by unit train to markets in the Front Range of Colorado and in midwestern states. An additional coal-fired power plant currently is under construction near Craig, Colorado (figure 1). Proposed are at least one more power plant as well as possibly a coal gasification plant and a coal slurry pipeline. For the same amount of mined coal, various considerations of these facilities would have a range of implications in terms of water use and discharged residuals.

Effects of these uncertainties are transferred when projecting population growth due primarily to the anticipated increase in coal production as shown in figure 2. The population growth projections made by the Bureau of Reclamation in its environmental statement on northwest Colorado coal (U.S. Department of the Interior, 1976) considered primarily effects associated with mining of coal on federally controlled leases. The range of population projections made as part of the Yampa River Basin Assessment (Udis and co-authors, 1977) considered a range of in-basin utilization of coal mined in Moffat and Routt counties in Colorado (figure 1). In both cases, a baseline growth trend is assumed to occur in spite of projections of mined coal. The "alternative 5" projection is a "rapid growth" option assuming heavy in-basin conversion of coal to electricity.

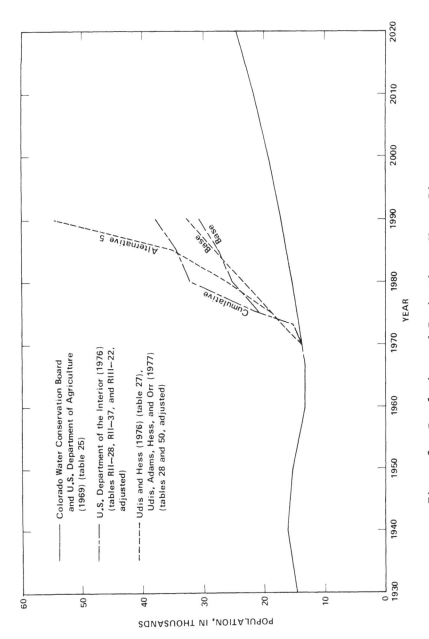

Figure 2. Population and Projections, Yampa River Basin, 1930-2020

Source: See following page.

Figure 2. (continued)

Source: T. D. Steele, D. P. Bauer, D. A. Wentz, and J. W. Warner, "The Yampa River Basin, Colorado and Wyoming--A Preview to Expanded Coal-Resource Development and Its Impacts on Regional Water Resources," Yampa River Basin Assessment Project Report, in review (Denver, Colo., U.S..Geological Survey, in press).

———— Colorado Water Conservation Board and U.S. Department of Agriculture, Water and Related Land Resources, Yampa River Basin, Colorado and Wyoming (Denver, Colo., 1969) Table 25.

— — U.S. Department of the Interior, Northwest Colorado Coal--Environmental Statement (Denver, Colo., Bureau of Reclamation, December 1976) Tables R11-28, R11-37, and R11-22, adjusted.

– – – – Bernard Udis and R. C. Hess, Input-Output Structure of the Economy of that Part of the Yampa River Basin in Colorado--1975. Final Report for the U.S. Geological Survey contract P.O. 12166 (Boulder, Colo., December 1976); and Bernard Udis, T. H. Adams, R. C. Hess, and D. V. Orr, Coal Energy Development in Moffat and Routt Counties of the Yampa River Basin in Colorado--Projected Primary and Secondary Economic Impacts Resulting from Several Coal-Development Futures. Final Report for U.S. Geological Survey contract P.O. 12185 (Boulder, Colo., June 1977).

As noted in the Maddock-Matalas paper, the indirect (population-related and support services) effects of regional economic development, as well as those of agricultural and recreational activities, need to be considered along with the direct effects of coal mining, conversion, and transport on regional estimates of water use and residuals discharges. Using plant process and surface mining models, the preliminary analyses undertaken by Maddock and Matalas considered only the direct effects. For several residuals, such as dissolved solids and nutrients, the effects of indirect and other sources may exceed those associated with coal development. Based upon a set of seven coal resource development scenarios assumed for the Yampa River Basin (Udis and coauthors, 1977; Steele and coauthors, in press), the projected water uses and residuals discharges shown in table 2 provide a range of possible conditions for 1990. Both direct and indirect effects (Hirsch and coauthors, in press) are considered in the results tabulated in this table.

Projections of population, water use, and residuals discharges serve as input data to several modeling and assessment techniques for evaluating the environmental consequences of the anticipated development under study (Steele, 1978; Bauer and coauthors, 1978). Tabular summaries of these modeling and assessment techniques are provided in project planning reports (Steele and coauthors, 1976a; 1976b). In the Yampa River Basin assessment studies, one alternative included possible export of coal by slurry pipeline (Palmer and coauthors, 1977); estimates of consumptive water use by this alternative range from one-fifth to one-fourth of the water consumed by alternatives considered in the Maddock-Matalas analysis, which are predominantly cooling-water uses.

Table 2. Water Use and Residuals Projections for the Yampa River Basin, Colorado, 1990

| | 1975 Base | Slow growth | Coal resource development alternative | | | | | |
| | | | Moderate growth | | | | Rapid growth | |
		7	1	2	3	4	5	6
Water use(acre-feet per year):								
Withdrawal	406,785	420,135	425,960	453,490	442,493	449,906	536,965	506,639
Consumption	142,267	154,804	159,486	185,612	175,818	183,183	267,979	237,489
Total employment	8,709	10,281	12,021	15,776	12,576	12,688	17,437	17,814
Total population	17,897	21,128	24,703	32,419	25,844	26,075	35,834	36,607
Residuals discharged (tons per year):								
Ash	73,926	214,331	214,331	863,441	436,024	436,024	1,618,383	1,351,758
Other solids	12,601	36,535	36,535	182,065	74,325	74,325	275,870	247,862
Total suspended particulates (TSP)	122,225	126,077	133,076	132,546	134,448	133,218	150,856	145,425
Hydrocarbons (HC)	5,682	6,103	6,311	6,983	6,813	6,765	9,785	8,715
Nitrogen oxides (NO_x)	22,193	48,915	49,224	74,562	91,182	91,108	315,393	216,075
Sulfur oxides (SO_x)	11,635	20,992	21,167	26,936	35,834	35,765	114,334	78,022

Table 2. (continued)

	1975 Base	Coal resource development alternative						
		Slow growth	Moderate growth				Rapid growth	
		7	1	2	3	4	5	6
Carbon monoxide (CO)	7,790	8,926	9,147	10,361	10,804	10,729	19,969	16,068
Biochemical oxygen demand (BOD)	2,152	2,169	2,188	2,228	2,194	2,195	2,245	2,250
Chemical oxygen demand (COD)	7,145	7,149	7,155	7,166	7,156	7,157	7,170	7,171
Nitrogen (N)	130	136	145	162	147	148	169	171
Phosphorus (P)	199	201	204	210	205	205	213	214
Suspended solids (SS)	13,276	20,636	36,661	36,712	36,668	36,670	54,016	54,021
Dissolved solids (DS)	62,790	74,991	89,604	89,949	89,655	89,655	105,708	105,744

Note: No baseline growth trend is assumed. Alternative code designation and composition of alternative is given in Bernard Udis; T. H. Adams, R. C. Hess, and D. V. Orr, Coal Energy Development in Moffat and Routt Counties of the Yampa River Basin in Colorado--Projected Primary and Secondary Economic Impacts Resulting from Several Coal-Development Futures. Final report for U.S. Geological Survey contract P.O. 12185 (Boulder, Colo., June 1977). Slow growth, 10 million tons per year; moderate growth, 20 million tons per year; rapid growth, 30.8 million tons per year.

388

Table 2. (continued)

Source: Adapted from water use and residuals discharge projections made by J. E. Schefter and R. M. Hirsch of the U.S. Geological Survey Systems Analysis Group. The method used is described in R. M. Hirsch, I. C. James II, and J. E. Schefter, "Residuals Management--A Tool in River Quality Assessment Applied to Coal Development in the Yampa River Basin, Colorado," Symposium on River Quality Assessment, Proceedings (Minneapolis, Minn., American Water Resources Association, in press).

If additional water demands for energy development as well as other industrial and municipal purposes are superimposed on the traditional seasonal use of surface water for irrigation, additional impoundment of surface water will be required in the basin. As many as thirty-five major reservoirs proposed under nineteen projects in the Yampa River Basin (table 1) may attempt to impound an estimated 141 percent of the long-term mean-annual streamflow discharge from the basin at the Dinosaur National Monument (figure 1) (Steele and coauthors, 1977).

The downstream effects of one possible configuration of four reservoirs in the Yampa River Basin (figure 1) on the monthly patterns of streamflow and dissolved solids concentrations are depicted in figure 3. This and other reservoir configurations are currently being evaluated in some detail (Adams and coauthors, 1978).

Summary

The analyses reported in the paper by Maddock and Matalas indicate some of the direct surface water related effects of coal resource development in the Yampa River Basin for four hypothetical coal mining and conversion alternatives. Several environmental assessment studies recently have been completed which evaluate in greater detail the water resources related consequences of projected development. The Maddock-Matalas paper considered a fixed amount of mined coal, used either for conversion to electricity or gas products in the basin. In contrast, the U.S. Geological Survey's Yampa River Basin assessment studies consider some uncertainty in the coal mining level projected for 1990 (table 2), and in the mixes of the utilization of coal for conversion to other energy forms in the basin or for transport to out-of-basin markets (Udis and coauthors, 1977).

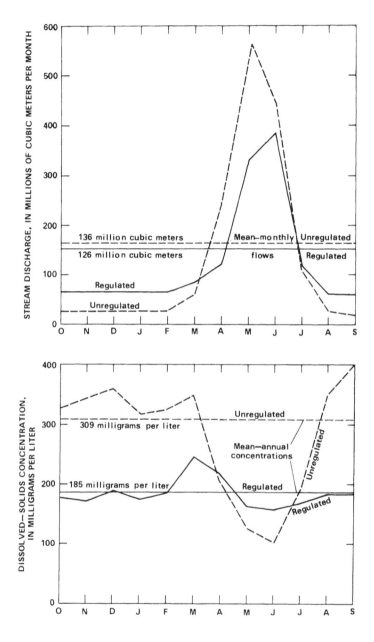

Figure 3. Estimated Downstream Effects of Selected Reservoir
Development on Streamflow and Dissolved Solids
Concentration of the Yampa River at Echo Park,
Dinosaur National Monument, Colorado

Source: See following page.

Figure 3. (continued)

Source: T. D. Steele, "Assessment Techniques for Modeling Water Quality in a River Basin Affected by Coal-Resource Development," Proceedings, International Symposium on Modelling the Water Quality of the Hydrological Cycle (Baden, Austria, International Association of Hydrological Sciences, in press); D. B. Adams, R. H. Dale, D. P. Bauer, and T. D. Steele, "Reservoir and Salinity Modeling in the Yampa River Basin, Colorado and Wyoming," Yampa River Basin Assessment Project Report, in review (Denver, Colo., U.S. Geological Survey, 1978).

Environmental consequences depend in part on siting assumptions. Unfortunately, the Yampa River Basin assessment studies did not progress as far as initially intended in linking projected residuals loadings with various environmental models. Particularly in the water quality area, several linkages are not available in an operational sense, and data on ambient levels are lacking in necessary detail. Effects of upstream water withdrawals and consumptive use on downstream flows depend upon assumed configurations of impoundments and upon what specified senior (mostly irrigation) water rights are preserved under the appropriative doctrine (Knudsen and Danielson, 1977). Assuming certain hypothetical residuals loadings, some site-specific effects are being evaluated (Bauer and coauthors, 1978; Andrews, in press; Warner and Dale, 1978). Richard A. Smith s analysis (see Smith paper in this volume) has looked at possible effects on fish and wildlife, and Andrews' (in press) regional sediment appraisal could lead to an extension of Smith's work. Ongoing cooperative studies between the U.S. Geological Survey and the U.S. Fish and Wildlife Service (Region 6 and Instream Flow Group) involve assessing travel-time and flow-regime changes caused by proposed reservoir development in the Yampa River Basin. Other potential environmental changes, and assessment possible changes, remain topics of future follow-up studies.

The additional information provided by these studies should assist environmental and water resource planners and managers both in anticipating and in dealing with a range of potential environmental changes resulting from such development. Regional "post-audit" studies (Langbein, 1974) would enable further evaluation of the relevance and adequacy of the design and execution of such physical based studies.

References

Adams, D. B., R. H. Dale, D. P. Bauer, and T. D. Steele. 1978. "Reservoir and Salinity Modeling in the Yampa River Basin, Colorado and Wyoming" (Denver, Colo., U.S. Geological Survey Water-Resources Investigations, in review).

Andrews, E. D. In press. "Present and Potential Sediment Yields in the Yampa River Basin, Colorado and Wyoming" (Denver, Colo., U.S. Geological Survey Water-Resources Investigations 78-105).

Bauer, D. P., T. D. Steele, and R. D. Anderson. 1978. "Analysis of Waste-Load Assimilative Capacity of the Yampa River, Steamboat Springs to Hayden, Routt County, Colorado" (Denver, Colo., U.S. Geological Survey Water-Resources Investigations 77-119).

Hirsch, R. M., I. C. James II, and J. E. Schefter. In press. "Residuals Management--A Tool in River Quality Assessment Applied to Coal Development in the Yampa River Basin, Colorado," Symposium on River Quality Assessment, Proceedings (Minneapolis, Minn. American Water Resources Association).

James, I. C. II, and T. D. Steele. 1977. "Applications of Residuals Management Techniques for Assessing the Impacts of Alternative Coal-Development Plans on Regional Water Resources," Proceedings, Third International Symposium in Hydrology (Fort Collins, Colo., Colorado State University).

Knudsen, W. I., Jr., and J. A. Danielson. 1977. A Discussion of Legal and Institutional Constraints on Energy-Related Water Development in the Yampa River Basin, Northwestern Colorado. Final report for U.S. Geological Survey contract no. 14-08-0001-15075 (Denver, Colo., U.S. Geological Survey, December).

Langbein, W. B. 1974. "The Case for Case Histories," Memorandum to Chief Hydrologist, U.S. Geological Survey (5 July 1974).

Palmer, R. N., I. C. James II, and R. M. Hirsch. 1977. "Comparative Assessment of Water Use and Environmental Implications of Coal Slurry Pipelines," Symposium on Critical Water Problems and Slurry Pipelines, U.S. Geological Survey Open-File Report 77-698 (Reston, Va., U.S. Geological Survey).

Steele, T. D. 1978. "Assessment Techniques for Modeling Water Quality in a River Basin Affected by Coal-Resource Development," Proceedings, International Symposium on Modelling the Water Quality of the Hydrological Cycle (Baden, Austria, International Association of Hydrologica Sciences, Publication No. 125) p. 322-332.

Steele, T. D., D. P. Bauer, D. A. Wentz, and J. W. Warner. 1976a. "An Environmental Assessment of Impacts of Coal Development on the Water Resources of the Yampa River Basin, Colorado and Wyoming--Phase I Work
Plan," U.S. Geological Survey Open-File Report 76-367 (Lakewood, Colo.,
U.S. Geological Survey).

_____. In press. 'The Yampa Basin, Colorado and Wyoming--
A Preview to Expanded Coal-Resource Development and its Impact on
Regional Water-Resources" (Denver, Colo., U.S. Geological Survey
Water-Resource Investigations 78-126).

Steele, T. D., R. H. Dale, D. B. Adams, and D. P. Bauer. 1977. "Flow and
Salinity Modeling of Proposed Reservoirs in the Yampa River Basin, Colorado and Wyoming" (abstract), Symposium on Lake Water Quality and Quantity Management, Transactions of the American Geophysical Union vol. 58,
no. 12 (December) p. 1132.

Steele, T. D., I. C. James II, D. P. Bauer, and Others. 1976b. "An Environmental
Assessment of Impacts of Coal Development on the Water Resources of the
Yampa River Basin, Colorado and Wyoming--Phase II Work Plan," U.S.
Geological Survey Open-File Report 76-368 (Lakewood, Colo., U.S. Geolog-
ical Survey).

Udis, Bernard, T. H. Adams, R. C. Hess, and D. V. Orr. 1977. Coal Energy
Development in Moffat and Routt Counties of the Yampa River Basin in
Colorado--Projected Primary and Secondary Economic Impacts Resulting
from Several Coal-Development Futures. Final report for U.S. Geological
Survey contract P.O. 12185 (Boulder, Colo., June). (Copies can be obtained
from the Bureau of Economic Research, University of Colorado, Boulder,
Colo.).

Udis, Bernard, and R. C. Hess. 1976. Input-Output Structure of the Economy
of that Part of the Yampa River Basin in Colorado--1975. Final report
for U.S. Geological Survey contract P. O. 12166 (Boulder, Colo., December).
(Copies can be obtained from the Bureau of Economic Research, University
of Colorado, Boulder, Colo.).

U.S. Department of the Interior. 1976. Northwest Colorado Coal--Environ-
mental Statement (Denver, Colo., Bureau of Reclamation, December).

Warner, J. W., and R. H. Dale. 1978. "Digital-Transport Model Study of
Probable Effects of Coal-Resource Development on the Ground-Water
System in the Yampa River Subbasin, Moffat and Routt Counties, Colorado" (Denver, Colo., U.S. Geological Survey Water-Resources Investigations, in review).

Chapter 14

PREDICTING THE IMPACTS OF
SURFACE COAL MINING ON TROUT POPULATIONS
IN THE YAMPA RIVER BASIN

Richard A. Smith*

Introduction

Coal development has gained much attention in recent years as the
nation has become more and more concerned with establishing a comprehensive
energy policy. The abundant coal resources in the western states are a
critical element in the effort to achieve national energy self-sufficiency.
But integral to consideration of western coal development is consideration
of that development's potential impacts on western water resources, alternative uses of the land, and more broadly on the aesthetic qualities of the
environment--impacts that ultimately will affect the people, fish, and
wildlife of the region. As more demands are made upon the region's limited
water supply, the decisions regarding the allocation of that supply become
more difficult. In the arid West, where mining and the actual processing
of coal would occur, conflicts already exist among various water interests
including energy, agriculture, land development, and fish and wildlife.
The purpose of this paper is to examine the potential impacts on fish,
trout in particular, of changing land use patterns resulting from surface
coal mining in the Yampa River Basin.

---

*U.S. Geological Survey, National Center, Reston, Virginia.

The Yampa River Basin, located in the northwestern part of Colorado, is considered here in examining a potential widespread fisheries problem that could arise from large-scale surface coal mining. This is the same basin that Maddock and Matalas (paper in this volume) used to explore the impacts on streamflows and on surface water storage requirements of large-scale surface coal mining and conversion.

The Yampa River enters the Green River in the Dinosaur National Monument, Colorado. Above Maybell, Colorado, the Yampa Basin has a drainage area of approximately 3,400 square miles. The basin contains large reserves of high quality coal, located principally in the areas of the Williams Fork Mountains and the Danforth Hills. The coal reserves are characterized by low contents of ash (approximately 5 percent by weight, air dried), and sulfur (about 0.4 percent by weight, air dried). Total reserves are estimated to be 5.7 billion tons (Landis and Cone, 1971) of which Speltz (1974) estimates there is a strippable resource of 750 million tons in the Yampa coalfield and 164 million tons in the Danforth Hills coalfield. A recent environmental study of the Yampa River Basin by the Bureau of Land Management (U.S. Department of the Interior, Bureau of Land Management, 1976) was based on the production of 33 million tons of coal per year by 1990, with 24 million tons per year being produced by surface mining (figure 1).

Given the projected levels of surface coal mining in the Yampa River Basin, it is reasonable to hypothesize that one of the more serious environmental problems will be the impact of increased suspended sediments on stream quality and biota. The fact that the per acre yield of sediment from a drainage basin commonly increases, at least temporarily, in the wake of strip mining has been well documented (see review by Environmental Studies Board, 1974).

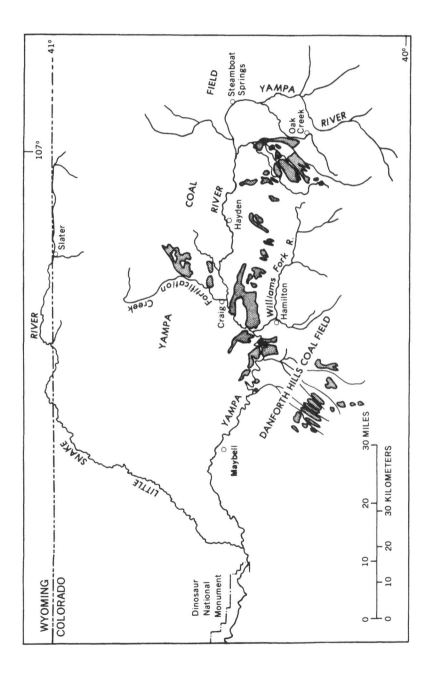

Figure 1. Location of Strippable Coal Deposits
in the Yampa River Basin

Increased sediment yields can be expected to be most immediately evidenced by increased suspended sediment concentrations in the small ephemeral streams draining mine sites, and depending on the sediment-carrying characteristics of the larger tributaries, could result in higher sediment loads elsewhere in the basin.

Perhaps the most important ramification of increased sediment loads in streams in the Yampa River Basin would be potential effects on aquatic organisms. Despite their tolerance for short-term pulses in sediment concentrations, the communities of aquatic organisms in the basin today can with very few exceptions be described as comprising clean water populations, and virtually all species can be considered vulnerable, either directly or indirectly, to a general increase in turbidity and sedimentation.

Fish populations are of particular importance among those species potentially affected by increased sediment yields. Their significance with respect to the prospects for major coal development in the basin is twofold. First, a number of fish species, the salmonids (trout) in particular, are important in the recreation economy of the area. Second, many of these same species are quite sensitive both to the direct physiological effects of suspended solids as well as to the indirect effects of turbidity and sedimentation on their food supply.

Among the many factors determining the extent of future sediment problems in the Yampa River Basin, and in regions of coal development in general, will be the nature of reclamation practices in mined areas. Reclamation methods will influence not only total quantities of sediment carried from mine spoils but also will determine the time course of increased sediment yields as well. Thus, for example, in the case of topsoil application prior to reseeding, erosion may carry large quantities of soil solids into streams initially. However, as a more stable vegetative cover develops in later years, sediment yields will actually be reduced below natural levels. Clearly such qualitative differences in sediment yield patterns will influence the nature of the biological impacts.

This paper describes a method for estimating the impact of surface coal mining and different land reclamation practices on trout populations in streams draining proposed strip mining areas in the Yampa River Basin. The method involves the joint use of two models: first, a stochastic simulation model of stream sediment concentrations under different land reclamation levels; and second, a trout population response model. Whereas the stream sediment model is developed specifically for watersheds in the Yampa Basin, owing to a lack of trout life table statistics for Yampa Basin populations, the ecological response model is based on published data for other watersheds. Thus, while the methodology discussed here is general and may be easily adapted to other areas, it is emphasized that the model has not been calibrated for the Yampa subbasins, and that before applying it to any of the subbasins, the model should be calibrated and verified with field data.

Further information on sediment yield and transport, as well as at least some baseline information on fish and invertebrate populations, should be forthcoming from an ongoing study (Steele and coauthors, 1976) of the effects of increased coal development on water resources in the region.

Methods

Stream Sediment Model

Five "trout" streams in the Yampa River Basin were selected for the study on the basis of availability of data on suspended sediment. The five streams are: Fortification Creek above Craig, Colorado; Little Snake River near Slater, Colorado; Williams Fork at Hamilton, Colorado; and two locations on the Yampa River, one at Steamboat Springs, Colorado, and the other near Oak Creek, Colorado. The five are all classified cold water fisheries (B1) by the Colorado State Division of Wildlife, and were selected to be generally representative of different types of trout streams in the basin. However, they were not selected on the basis that they now support a natural trout population of known size.

Measurements of suspended sediment concentrations in the five streams were made periodically between 1952 and 1958 by Iorns and coauthors (1964). Streamflow measurements for the Yampa River near Maybell, Colorado, on the same dates sediment samples were collected were also available (U.S. Geological Survey, 1952-1958). A least squares regression of sediment concentration on Yampa River streamflow at Maybell was performed combining the sediment data for the five streams. The results are shown in figure 2. Using the regression equation

$$S(T) = 0.28Q(T)^{0.96} \tag{1}$$

is is possible to translate a given time series of Yampa River flows, $Q(T)[ft^3/sec]$, into a corresponding series of estimated sediment concentrations, $S(T)[mg/l]$, for the five respresentative "trout streams" in the basin.

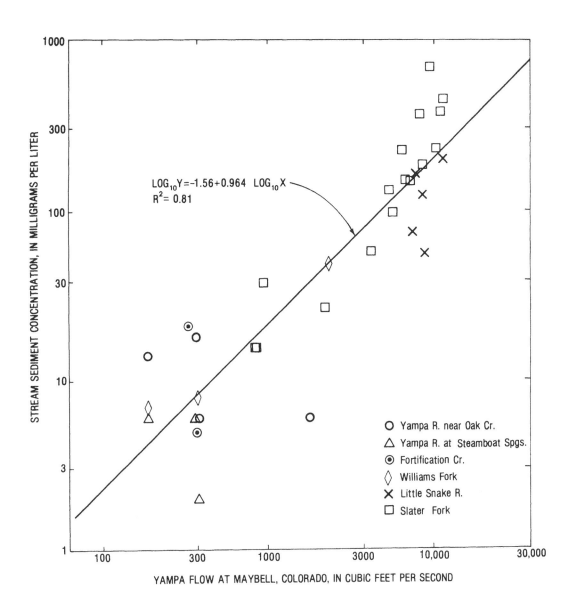

Figure 2. Regression of Suspended Sediment Concentrations
in Five Streams in the Yampa River Basin on the
Flow in the Yampa River at Maybell, Colorado

Since data used in the regression were collected prior to the onset
of surface mining activity, equation 1 gives baseline (natural) estimates
of sediment concentration.   In an analysis of sediment loss due to future
surface mining in the Yampa River Basin, James and coauthors (1977) assumed
a baseline yield of 0.5 tons per acre per year and estimated the 100-year
history of annual sediment generation from a hypothetical 12,000-acre site
mined at 400 acres per year.   Sediment yield calculations were made using
the universal soil loss equation (U.S. Department of Agriculture, 1961),
and were repeated for four land reclamation levels based on estimated rates
of revegetation.   Level $\emptyset$ indicates no reclamation and consists only of
those activities which provide for efficient mining.   For example, among
these activities would be road wetting for dust control and division of
surface water away from strip pits.   Land reclamation level A is based
on the Colorado open Mining Land Reclamation Act (1973), and involves
striking off spoil ridges to a 15-foot width, covering final coal
seams, and seeding.   Level B involves a high degree of spoil flattening
and returns the land to its original contours.   Under this reclamation level,
it is assumed that the land is revegetated for a return to grazing.   One
year of irrigation is included to enhance the early establishment of vegetation.   Land reclamation level C provides a rolling landscape and revegetation
suitable for game habitat with an emphasis on the reduction of wind- and
waterborne sediments.   Rapid revegetation is obtained through mulching,
three years of irrigation, and heavier applications of both soil amendments
and fertilizers than those required for level B.   The four reclamation
levels, the assumed sediment yields (tons per acre over natural yields,
and the assumed costs of reclamation are summarized in table 1.

Table 1. Sediment Yields from, and Costs of, the Four Land Reclamation Levels[a]

Reclamation level	Description	Water requirements for irrigation (acre-feet/acre/year)	Net sediment yields (above-natural yield[b]) (tons/acre/year)	Mining costs[c] (dollars/ton of coal mined)	Reclamation costs[c] dollars per acre	Reclamation costs[c] dollars per ton of coal mined[d]	Increased costs of mining due to reclamation (percent)
Ø	No reclamation	0	26.3	$ 4.92	0	0	0
A	Spoil ridges are struck off to a 15 foot width, final coal seams are covered, and land surface is seeded.	0	17.0	4.96	$ 625	$.04	0.8
B	Land surface is returned to its original contours, land is revegetated for a return to grazing, one year of irrigation is provided to enhance the early establishment of revegetation.	1.0	1.5	5.57	10,156	.65	13.2
C	A rolling landscape and revegetation suitable for game habitat is provided, and mulching, three years of irrigation, and fertilizers are applied.	2.25	0.47	5.66	11,562	.74	15.0

[a]Based on data taken from I. James, E. D. Attanasi, T. Maddock, S. H. Chiang, B. T. Bower, and N. C. Matalas, "Water and Energy Use in Coal Conversion," U.S. Geological Survey Open-File Report (in draft) (Reston, Va., U.S. Geological Survey, Water Resources Division, 1977).

[b]The natural sediment yield is assumed to be 0.5 tons per acre per year.

[c]Costs in 1975 dollars, based on annualized capital costs (12 percent capital recovery factor) and annual operating and maintenance costs.

[d]Based on 15,625 tons of coal mined per acre and 400 acres developed per year for a total of 6.25 million tons of coal per year mined.

The ratio of post-mining sediment yield to estimated baseline yield
provides a scaling factor which can be applied to baseline sediment concentration values generated by equation 1. This provides stream concentration
estimates for the post-mining period. Baseline sediment concentrations were
scaled to give post-mining concentrations according to the following relation:

$$S_M = S_B \left[\frac{(A_T - A_M) Y_B + A_M \cdot Y_M \cdot D_M}{A_T \cdot Y_B} \right] \qquad (2)$$

where S_M and S_B are post-mining and baseline (natural) stream sediment concentrations, respectively; Y_M and Y_B are post-mining and baseline yields
(per acre), respectively; A_M and A_T are mine area and total basin area, respectively; and D_M is the sediment delivery ratio between the mine site and
the stream location of interest. Inherent in this procedure of scaling
stream sediment concentrations on the basis of estimated changes in annual
yield is the assumption that sediment delivery ratios are either known or
can be determined. In the study reported here, stream sediment concentrations
have been modeled for a range of assumed delivery ratios so as to determine
the sensitivity of model outputs to the relative location of mine sites and
streams.

The five streams which supplied sediment data for the regression analysis range in size of drainage area from 604 square miles to less than 30
square miles. This study focuses on a hypothetical 100-square mile watershed
in the Yampa River Basin containing a 12,000-acre surface mine site.

As implied by the strong correlation with streamflow, sediment concentrations are quite variable both seasonally as well as over shorter time
periods. The timing and duration of these sediment pulses in relation to

trout spawning periods is an important aspect of the problem at hand. Equations 1 and 2 can be used to obtain realistic estimates of pre- and post-mining stream sediment concentrations provided that the input time series of streamflow values displays the temporal variability that is characteristic of the Yampa at Maybell, Colorado. To accomplish this, a synthetic streamflow generator was constructed using the fifty-eight-year record of monthly mean streamflow values for that stream gaging station. The generator produces log normally distributed flows and is autoregressive with respect to consecutive months and consecutive years (see Matalas (1967) for detailed description of synthetic streamflow generators). The decision to use monthly sediment concentration values was based on experimental observations of trout mortality as a function of exposure period (Cordone and Kelley, 1961) and is discussed in detail in the next section.

To summarize, the stream sediment model described in this section is used to produce a time series of monthly sediment concentrations which display both seasonal and shorter-term variability characteristics of trout streams in the Yampa River Basin. This is accomplished by generating a synthetic series of streamflow values for the Yampa at Maybell and then translating this series into baseline sediment concentrations for specific locations upstream and in certain tributaries using equation 1. Finally, baseline sediment concentrations from equation 1 are scaled, using equation 2, on the basis of estimated changes in annual sediment yield for the four different levels of land reclamation effort.

Trout Response Model

To some extent the task of modeling the effects of suspended sediment levels on trout has been facilitated by good agreement between the results of laboratory and field studies. Alabaster (1972) finds that a reported 100 ppm threshold for measurable effects on salmonids of prolonged exposures is consistent with field observations correlating sediment concentrations and fish abundance in a number of European rivers. There are shortcomings, however, in relying only on field observations of fish densities in assessing the effects of sediments. The consequences of increased mortalities among eggs and larvae, stages that have been found to be particularly sensitive to sediments (Cordone and Kelley, 1961), continue to be felt in a population for some time after the incidence, in terms of subsequent reductions in spawning population. These losses tend to be incompletely accounted for in field surveys that are limited in time to one or even a few years, or in surveys limited in scope to the more readily observable adult individuals.

Furthermore, even where field efforts are extended to monitor population recovery times (Saunders and Smith, 1964), observations are confounded by reinvasion of individuals from neighboring populations. To be sure, rates of immigration are important in reestablishing a fishery in any one area. The need to make decisions on land reclamation practices to be employed basinwide, however, suggests that some consideration be given to sediment levels that are simultaneously tolerable to all populations of a watershed, and therefore requires information on recovery times of populations denied immigrants from other stocks.

The response and recovery of a fish population under the influence of increased sediment concentrations can be examined on a theoretical basis without the interference of reinvasion encountered in the field. For this purpose, a trout model has been developed using published life table statistics for the trout population and employing empirical data relating relevant parameters to concentrations of suspended sediments.

The Leslie Matrix and Hunt Creek Trout

Given an age distribution of a population at time t, the age distribution for that population at time t+1 can be computed in the following way:

$$\sum_x F_x n_{x,t} = n_{0,t+1}$$

$$P_0 n_{0,t} = n_{1,t+1}$$

$$P_1 n_{1,t} = n_{2,t+1}$$

$$P_2 n_{2,t} = n_{3,t+1}$$

$$\vdots \qquad \vdots$$

$$P_{m-1} n_{m-1,t} = n_{m,t+1}$$

where: $n_{x,t}$ is the number of individuals of age x to x+1 at time t.

F_x is the number of offspring produced over the interval t to t+1 per individual of age x.

P_x is the probability that an individual of age x at time t will be alive at time t+1.

A concise statement of all the above equations can be achieved in matrix form (Leslie, 1945) by defining the age distribution vector

$$\underline{n} = \begin{pmatrix} n_0 \\ n_1 \\ n_2 \\ \cdot \\ \cdot \\ \cdot \\ n_m \end{pmatrix}$$

and the transition matrix

$$\underline{M} = \begin{pmatrix} F_0 & F_1 & F_2 & F_3 & \cdots & F_{m-1} & F_m \\ P_0 & 0 & 0 & 0 & \cdots & 0 & 0 \\ 0 & P_1 & 0 & 0 & \cdots & 0 & 0 \\ 0 & 0 & P_2 & 0 & \cdots & 0 & 0 \\ 0 & 0 & 0 & P_3 & \cdots & 0 & 0 \\ \cdot & & & & & & \\ \cdot & & & & & & \\ \cdot & & & & & & \\ 0 & 0 & 0 & 0 & \cdots & P_{m-1} & 0 \end{pmatrix}$$

and writing, in matrix notation

$$\underline{M} \cdot \underline{n}_t = \underline{n}_{t+1} \tag{3}$$

Also, at times subsequent to t ≠ 0, the population age distribution will be given by

$$\underline{M}^t \cdot \underline{n}_0 = \underline{n}_t \tag{4}$$

Values of P_x and F_x for the fish population under consideration here
(table 2) were calculated from life tables assembled by McFadden and coauthors
(1967) for the brook trout (<u>Salvelinus fontinalis</u>) population of Hunt Creek,
Michigan. This particularly extensive study followed the age class transi-
tions of eleven consecutive cohorts and has been used elsewhere in estimating
the effects of mortality on productivity (Jensen, 1971).

In repeated iterations of equation 3, oscillations that occur in the
relative sizes of the age classes are progressively damped towards stable age
proportions, but total numbers of individuals continue to increase infinitely.
In a natural population, age specific survival and fecundity values (P_x, F_x)
would not remain constant with time. They can be expected to vary both with
density-related factors, as population size increases, as well as with density
independent factors, such as suspended sediment concentrations. Ricker (1954)
proposed that density dependent factors in fish populations operate most
effectively on young individuals, particularly eggs and larvae, with the
explanation that suitable spawning and nursery areas are generally limiting.
That theory is strongly supported by the Hunt Creek study. There the authors
traced a tendency toward a constant population size to the dynamics of the
zero age class alone. In figure 3, values of P_0 calculated from the Hunt
Creek data are plotted against n_0 for each of the eleven years of the study.
The curve in the same figure represents the regression of those data using
an equation of the form

$$P_0 = A \cdot e^{-bn_0}.$$

$$(5)$$

Table 2. Values of Age-Specific Survival and
Fecundity for Hunt Creek Trout

Age (x) (Years)	Probability of Survival from year t to t+1 (P_x)	Fecundity (F_x) (Offspring per individual per year)
0	a	0
1	0.405	0
2	0.176	38.4
3	0.093	129.5
4	0.018	303.1
5	0	500.0

[a] $P_0 = f(n_0)$.

$= 0.132e^{-1.95(n_0 \times 10^{-5})}$ See figure 3.

411

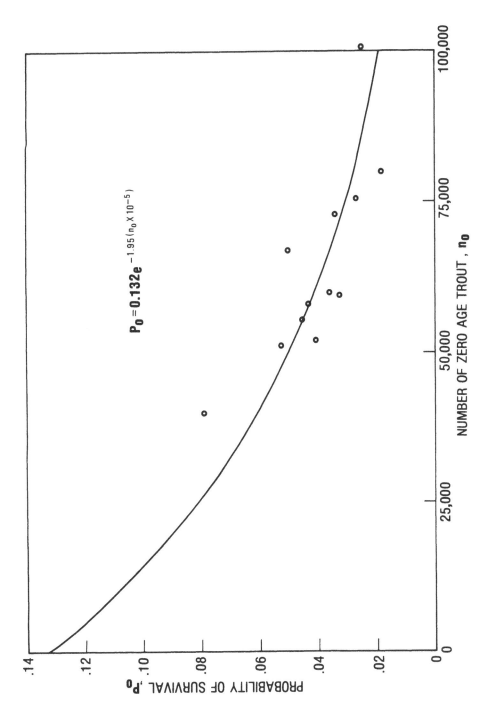

$$P_0 = 0.132e^{-1.95(n_0 \times 10^{-5})}$$

PROBABILITY OF SURVIVAL, P_0

NUMBER OF ZERO AGE TROUT, n_0

Figure 3. Probability of Survival of Zero-Age Trout as a Function of Size of That Year Class

This formulation has two significant properties:

1. As the number of zero age class individuals decreases, P_0 increases and takes the value "A" at $n_0 = 0$.

2. The probability of survival falls exponentially with increasing n_0 and therefore never reaches zero within finite limits to population size.

A density dependent form of matrix $\underline{M}$ can be made explicit for the Hunt Creek trout population with values of P and F from table 2, and with $P_0 = f(n_0) = 0.132e^{-1.95(n_0 \times 10^{-5})}$. Beginning with any initial population, repeated iterations of equation 3 with this density dependent term for P_0 ultimately give rise to the stable population shown in table 3. Mean values of n_x observed in the actual Hunt Creek population are included for comparison; the close agreement is, of course, not surprising since survival and fertility values in the model were estimated from the Hunt Creek life tables.

Effects of Suspended Solids on Trout Survival

The relationship between trout survival and suspended sediment concentration has been investigated in several laboratory studies with the general conclusion that survival remains high even at extremely high concentrations (e.g., 5,000 ppm) when exposure time is kept short (see Griffin, 1938; Wallen, 1951; review in Cordone and Kelley, 1961). This might be expected from the fact that even in normally clear trout streams, sediment concentrations are periodically high, typically so, for example, after storms. By contrast, effects of much lower concentrations, less than 1,000 ppm, for long periods were found to be considerable (Herbert and Merkens, 1961), and

Table 3. Stable Population Distribution Generated from Equation 3 and
Observed Mean Hunt Creek Population

Age (x)	Stable population distribution (n_x)	Observed mean population distribution[a]
0	64,906	62,954
1	2,417	2,502
2	978	1,013
3	172	183
4	16	18
5	1	1

[a]James T. McFadden, Gaylord R. Alaxander, and David S. Shetter,
"Numerical Changes and Population Regulation in Brook Trout Salvelinus
fontinalis," Journal of Fisheries Research Board of Canada vol. 24, 1967.

these prolonged conditions are more pertinent to the objectives of setting
sediment standards. In the latter study, percent mortality of rainbow trout
(Salmo gairdneri) increased with increasing contact time at various concentrations, but reached an apparently stable value after one to two months'
continuous exposure. Those stable values are plotted against suspended
solids concentrations in figure 4, along with two fitted, empirical curves.
The curve crosses the abscissa at 30 ppm in accordance with the conclusion
reached in the study that no significant mortality occurred up to that level.

The porous gravel nests or "redds" that provide protection for developing trout eggs must allow for sufficient water flow to maintain critical
oxygen concentrations if spawning is to be successful. The tendency for
sediments to clog redd interstitial areas and reduce egg survival was investigated in a hybrid lab-field experiment with rainbow trout in Bluewater
Creek, Montana (Peters, 1962). A series of artificial redds along a 15-mile
reach were filled with eggs and exposed to the natural gradient in suspended
load. Monitoring of the sediment concentration at each station for the duration of the embryonic development period was followed by examination of the
redds for hatching frequency. Results of the study appear in figure 4.with
a regression of egg mortality on mean suspended sediment concentration.

The response of the Hunt Creek trout population to a given schedule of
suspended sediment concentrations can be simulated with a series of iterations of equation 3 by adjusting the survival and fecundity values in each
iteration, according to the regression equations in figure 4. Adjustments
to P_x and F_x values are calculated in the following way:

$$P_x = P_x^{0}(1-m_f)$$

and

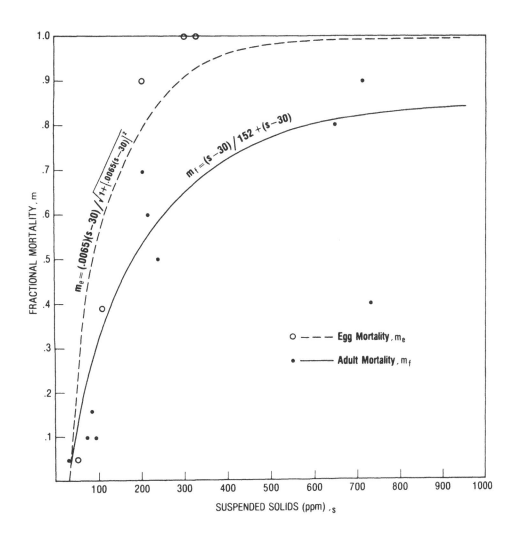

Note: Data for adult fish are from D. W. M. Herbert and J. C. Merkins, "The Effect of Suspended Mineral Solids on the Survival of Trout," International Journal of Air and Water Pollution vol. 5, no. 1, 1961. Data for eggs are from John C. Peters, "The Effects of Stream Sedimentation on Trout Embryo Survival," Paper presented at the Third Seminar on Biological Problems in Water Pollution, Cincinnati, Ohio, 16 August 1962 (Mimeographed).

Figure 4. Fractional Trout Mortality as a Function of Sediment Concentration

$$F_x = F_x^0 (1-m_e)$$

where P_x and F_x are adjusted survival and fecundity values, respectively.

P_x^0 and F_x^0 are the standard, unadjusted survival and fecundity

values given in table 2.

m_f and m_e are fractional mortality values of fish and eggs,

respectively, obtained from the equations in

figure 4 evaluated at a given sediment concentration.

Standard, unadjusted survival and fecundity values are used when the given

sediment concentrations are less than the 30 ppm threshold.

Simulation of Population Response

Simulations of trout population dynamics that result from various sediment concentration schedules appear in figures 5, 6, and 7. In all cases, it is assumed that stated sediment concentrations persist for a sufficient portion of a given year to cause the "stabilized" mortality values discussed above. The results of Herbert and Merkins (1961) indicate those mortalities occur when sediment exposure persists for at least one to two months.

In figure 5, trout population size (expressed as a fraction of initial population) is projected annually for seven years showing the influence of a hypothetical 300 ppm sediment concentration continuously applied during the second year. The initial population is the stable age distribution (table 3), grouped as juveniles and adults in the illustration. The erratic recovery, requiring five years under these conditions, results from altered age class proportions. The relatively few eggs and fingerlings that survive the exposure period give rise two years later to reduced adult numbers,

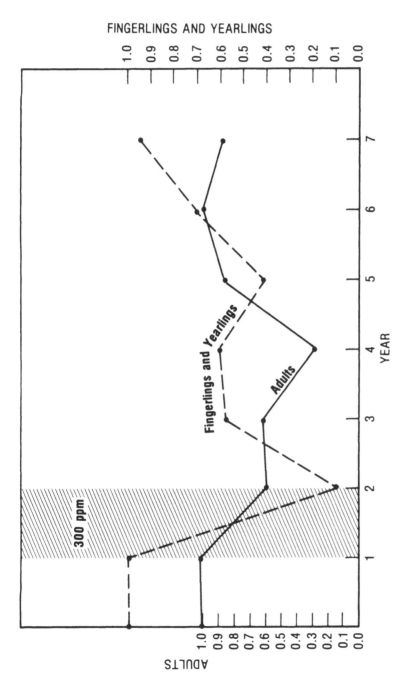

Note: The simulation assumes sediment exposure occurs during the egg development period and persists for at least 2 months of the year. Trout numbers are expressed as a fraction of initial population.

Figure 5. Simulated Effect of One-Year Exposure to 300 ppm Suspended Sediment Concentration on Trout Population Dynamics

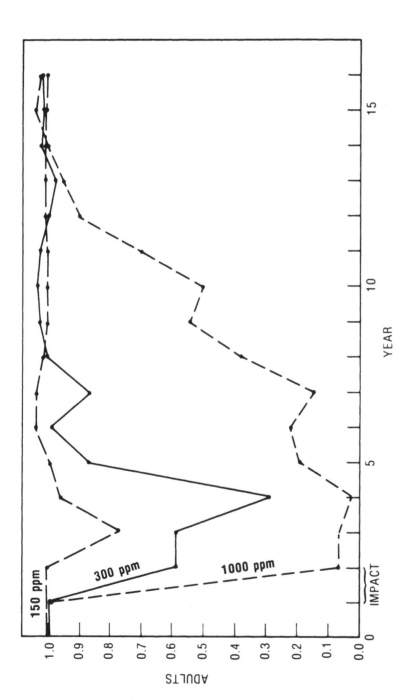

Note: The simulation assumes sediment exposure occurs during the egg development period and persists for at least 2 months of the year. Trout numbers are expressed as a fraction of initial population.

Figure 6. Comparison of the Effects of Different Suspended Sediment Concentrations on Trout Population Dynamics

419

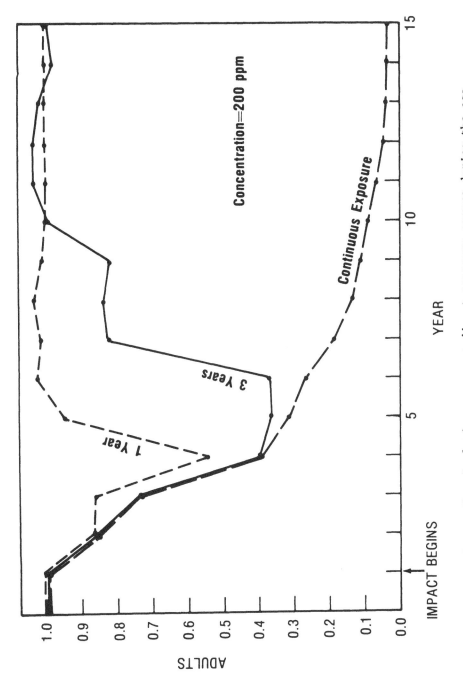

Note: The simulation assumes sediment exposure occurs during the egg development period and persists for at least 2 months of the year. Trout numbers are expressed as a fraction of initial population.

Figure 7. Effect of Varying Exposure Time on Simulated Trout Population Dynamics

which, still later, produce fewer young. The cycle continues but density dependent factors come into play and gradually damp the oscillations.

Comparisons of the effects of different concentrations and different exposure times appear in figures 6 and 7, respectively. It is apparent from figure 6 that when a severe sediment load reduces the population to a small fraction of its normal size, recovery time, in the absence of reinvasion from other populations, may require ten to twenty years. Saunders and Smith (1964), on the other hand, found that a normal population of brook trout returned in one season in the wake of heavy silt loads. Moreover, it is likely that trout, and fish in general, will avoid lethal concentrations of suspended material provided that they have some place to go. The task of setting land reclamation standards, however, requires some analysis of sediment concentrations which are tolerable region-wide, making no assumptions concerning the availability of refuges for affected populations or the existence of neighboring stocks to repopulate the watershed. For this purpose, a model provides a valuable accompaniment to field observations.

It is reasonable to expect that sediment concentrations have indirect effects on fish populations such as reducing the availability of benthic food organisms. Results of several investigations indicate that these dietary effects are exhibited more in the size and weight of trout individuals than in the mortality statistics (Cooper, 1953; Benson, 1954; Powell, 1958). Nevertheless, terms that include the effects of reduced food supply would be a valuable addition to the model.

It is apparent from figure 7 that continuous exposure to even moderate concentrations (i.e., 200 ppm) will drive the population to extinction after fifteen or twenty years. Continuous exposures to less severe concentrations, however, can be tolerated at a reduced population size; slightly increased mortalities due to sediment, in this instance, are balanced with the increased survival probability (P_0) of smaller populations. The relationship between stable population size and sediment concentration is given in figure 8, and indicates that non-zero stable populations can exist up to concentrations of just under 100 ppm.

Field data in the review of Alabaster (1972) show a fairly clear threshold in sediment concentration of about 100 ppm separating English streams with healthy fisheries and ones nearly void of fish, and thus lend at least some degree of credibility to the model outputs.

Joint Operation of the Trout and Sediment Models

Continuous exposure to a constant sediment load is a very unlikely circumstance in the Yampa River Basin or in any western watershed experiencing a rapid increase in surface mining. Variability in stream sediment concentrations is to be expected first because of the great seasonal variation in hydrologic forces which carry sediment through a watershed and, more to the point of the present investigation, because the yield of sediment from mined areas is likely to change radically in the course of surface mining and land reclamation as vegetation on the site is first removed and later reestablished. The details of alternative reclamation efforts, such as replanting and topsoil application, will largely determine the time course of sediment loss, and thus need to be evaluated in relation to the capacity of affected fish populations to withstand increased concentrations and recover from periodic heavy loads. Accordingly, the stream sediment model and trout population model were designed to work in tandem, using the output time series of

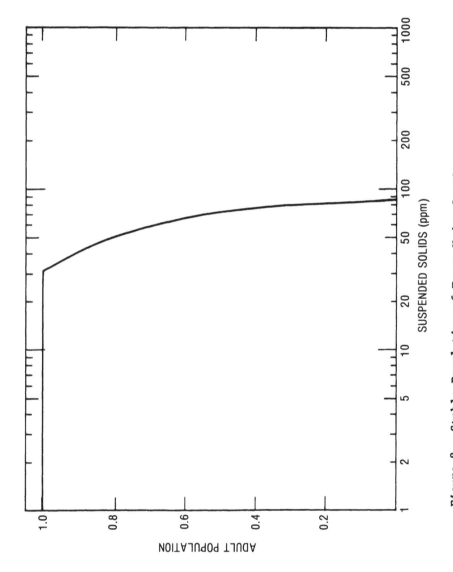

Figure 8. Stable Population of Trout Under Continuous Exposure

sediment concentrations for a specified land reclamation level as input to the population response model.

One thousand years of synthetic monthly streamflow records of the Yampa River at Maybell, Colorado, were generated, using a Markovian model that preserved the mean and standard deviation of the flows.[1] These flows were then translated into ten 100-year time series of monthly baseline stream sediment concentrations using equation 1. The ten synthetic sediment records are each considered equally likely flow sequences. Using estimated changes in annual sediment yield for different levels of land reclamation (see table 1), these baseline sediment records were then scaled (using equation 2) to represent sediment concentrations over the 100-year post-mining period for each of the four alternative reclamation levels discussed above. Annual sediment yields associated with surface mining were estimated assuming a baseline yield of 0.5 tons per acre per year in a 100-square mile drainage basin, and mining at the rate of 400 acres per year for thirty years (that is, a total of 12,000 acres mined).

The forty 100-year sediment records (i.e., ten records for each reclamation level) were then used as input to the trout population model, initializing the population in each 100-year simulation at the stable age distribution (table 3). The spawning and embryonic period was considered to be November through March, and annual egg mortality was computed on the basis of the average sediment concentration over that period. Adult mortalities were computed on the basis of the highest two-month average sediment concentration in each year.

---

[1]For more information on the streamflow generator, see Matalas (1967).

Results, Discussion, and Research Needs

Results

Figures 9 through 13 present example 100-year simulations for the
natural sediment level (baseline conditions) and for the four reclamation
levels. Dashed lines indicate sediment concentrations used to compute
annual adult mortality (that is, maximum two-month averages). In addition
to year-to-year variability, stream sediment concentrations show a gradual
rise and decline in the years following the commencement of mining (assumed
to begin the first year). Not shown, but of great importance in determining
egg mortality, is the tremendous seasonal variation in sediment concentration which typically reaches a maximum in the period March to May, just
after the close of the embryonic period.

The solid lines in figures 9 to 13 show adult trout population size
expressed as a fraction of initial value. Model results suggest that, under
natural conditions (figure 9), there would be relatively infrequent episodes
of significant trout mortality due to suspended sediment. Those that would
occur correspond in time to periods of unusually early snowmelt or to periods
of record precipitation. It is important to note, however, that only mortality due to sediment is dealt with stochastically in the model. In reality,
trout numbers would be somewhat more variable in time as a result of year-to-
year variation in other natural and man-related factors which also bear on
the trout population.

Figures 10 through 13 indicate that there would be considerable differences in the effectiveness of the four land reclamation levels in mitigating
the effects of mining operations on the trout fishery. Reclamation level 0,

425

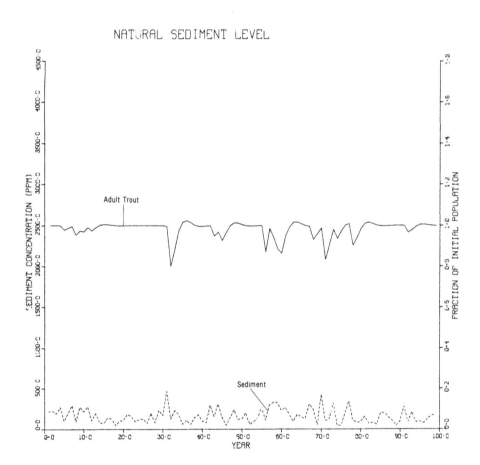

NATURAL SEDIMENT LEVEL

Assumption: Basin size: 100 square miles

Figure 9.   One-Hundred-Year Simulation of the Trout Population
Response to Stochastic Variation in Suspended Sediment
Yields in a Basin Experiencing Surface Mining Activity
--Natural Sediment Level

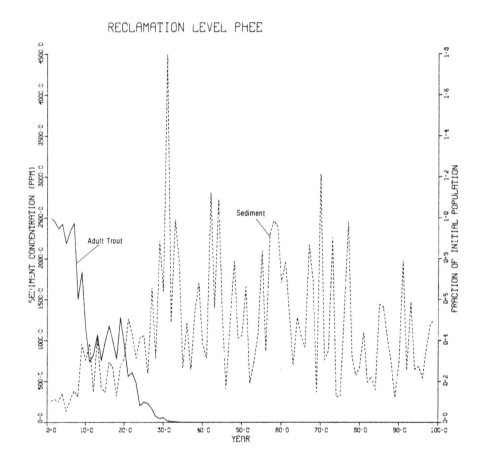

RECLAMATION LEVEL PHEE

Assumptions: Basin size: 100 square miles
              Acres mined: 12,000 (in 30 years)
              Sediment delivery ratio: 1.0

Figure 10. One-Hundred-Year Simulation of the Trout Population
            Response to Stochastic Variation in Suspended Sediment
            Yields in a Basin Experiencing Surface Mining Activity
            --No Reclamation of Strip-Mined Land

427

RECLAMATION LEVEL A

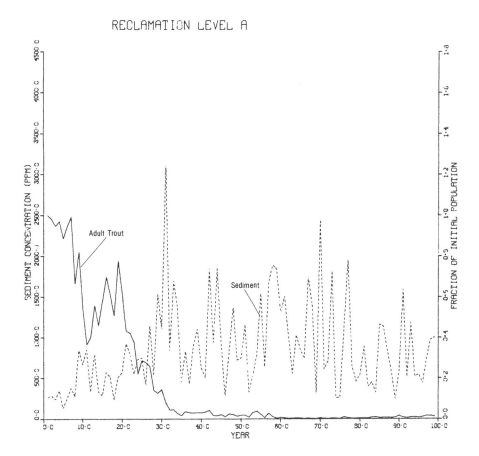

Assumptions: Basin size: 100 square miles
Acres mined: 12,000 (in 30 years)
Sediment delivery ratio: 1.0

Figure 11.   One-Hundred-Year Simulation of the Trout Population
Response to Stochastic Variation in Suspended Sediment
Yields in a Basin Experiencing Surface Mining Activity
--Reclamation Level A

428

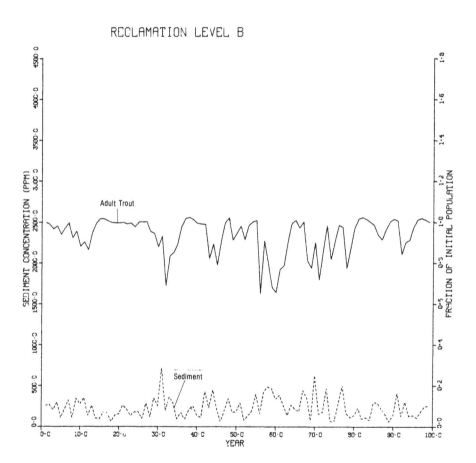

Assumptions: Basin size: 100 square miles
              Acres mined: 12,000 (in 30 years)
              Sediment delivery ratio: 1.0

Figure 12. One-Hundred-Year Simulation of the Trout Population
            Response to Stochastic Variation in Suspended Sediment
            Yields in a Basin Experiencing Surface Mining Activity
            --Reclamation Level B

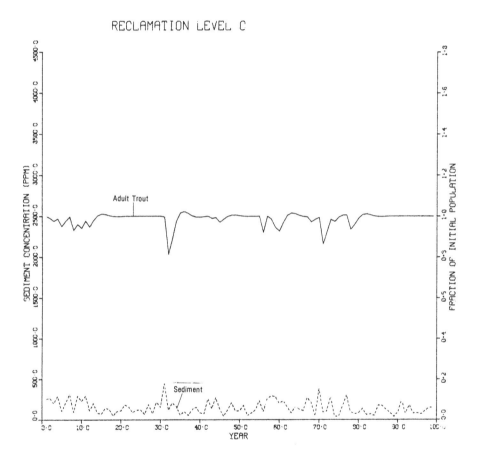

RECLAMATION LEVEL C

Assumptions: Basin size: 100 square miles
Acres mined: 12,000 (in 30 years)
Sediment delivery ratio: 1.0

Figure 13. One-Hundred-Year Simulation of the Trout Population
Response to Stochastic Variation in Suspended Sediment
Yields in a Basin Experiencing Surface Mining Activity
--Reclamation Level C

no reclamation (figure 10), and level A (figure 11), which do not include specific efforts to reestablish vegetation on the site, lead to a precipitous decline in trout numbers and eventual elimination of the fishery. By contrast, the revegetation and soil stabilization achieved in reclamation levels B and C (figures 11 and 12, respectively), which increase mining costs by roughly 13 and 15 percent, respectively (table 1), is sufficient to maintain a viable population with only periodic declines. Further, the declines associated with level C are similar to those associated with the natural sediment levels.

Results depicted in figures 10 through 13 are based on the assumption of 12,000 mined areas in total, and a sediment delivery ratio of 1.0. In other words, sediment concentrations in the trout stream reflect the total yield of sediment from mine spoils assuming no deposition of solids between the mine site and the stream segment in question. Except for stream segments immediately downstream of mined areas, sediment delivery ratios would likely be somewhat less than unity.[2] In this case, partial deposition of strip mine generated sediments would continue over long periods interrupted only by infrequent, short episodes of resuspension and transport during major storms (see Trimble, 1975).

Similarly, the assumption of a 100-square mile watershed and 12,000 mined acres represents only one of a number of reasonable scenarios for

---

[2]It is also worth noting that several trout streams in the Yampa River Basin extend upstream of anticipated mining activity and would provide some possible refuge despite the assumption made here that populations are isolated. This, of course, is a critical assumption in the analysis as refuge areas would allow for some survival which would assist in reestablishing the trout population.

coal development in a typical watershed in the Yampa River Basin. Thus,
it is worth examining the sensitivity of model outputs to changes in these
basin characteristics. For this purpose model runs were conducted for
several other representative cases, and the results are presented in table 4.
First, locations farther downstream from the mine site (e.g., sediment
delivery ratios of 0.5 and 0.1) are considered; then the sensitivity of
results to an increase or decrease in the size of the watershed is examined; and finally, the effect of reducing the rate of strip mining to 200
acres per year (for thirty years) is examined. Results are expressed as
the expected frequency of population decline below 50 percent of the equilibrium (initial) population during the hundred years following the onset of
mining. Frequency estimates are based on results of the ten 100-year simulations generated for each of the four land reclamation levels.[3]

Discussion and Research Needs

In view of the numerous assumptions made in developing both the stream
sediment model and the trout population model, and in the absence of field
verification of either, there are stringent limitations to the conclusions
which can be drawn from the results shown in table 4. Moreover, it is in
keeping with the purpose of this paper, which is to present a modeling

---

[3]For each 100-year simulation, the number of population declines below
50 percent of the initial population is recorded. The expected frequency
of population decline is determined by the following formulation:

$$\bar{N} = \frac{1}{10} \sum_{i=1}^{10} N_i \, ,$$

where N_i is the number of declines in the ith simulation run.

Table 4. Estimated Frequency of Population (Mean Annual Adult) Decline Below
50 Percent of Pre-Mining Size (Number of Occurrences in 100 Years)

	No mining	Reclamation level[a]			
		∅	A	B	C
Acres mined: 12,000 (400 acres per year for 30 years)					
Basin size = 100 sq. mi.; sediment delivery ratio = 1.0	b	88[c]	86[c]	b	b
Basin size = 100 sq. mi.; sediment delivery ratio = 0.5	b	64	27	b	b
Basin size = 100 sq. mi.; sediment delivery ratio = 0.1	b	b	b	b	b
Basin size = 500 sq. mi.; sediment delivery ratio = 0.5	b	b	b	b	b
Basin size = 50 sq. mi.; sediment delivery ratio = 0.5	b	88	84	b	b
Acres mined: 6,000 (200 acres per year for 30 years)					
Basin size = 100 sq. mi.; sediment delivery ratio = 0.5	b	9	2	b	b

[a]See text for description of land reclamation levels.

[b]Frequency less than once in 100 years.

[c]The trout population is eliminated.

technique for estimating the potential impact on a fishery of possible future strip mining and land reclamation scenarios, that the results be used to illuminate data needs rather than to make predictions.

Model outputs are quite responsive to changes in the sediment delivery ratio indicating that efforts to analyze sediment transport in more detail would be worthwhile. There is some reason to suspect, for example, that sediment delivery ratios might actually increase with increasing reclamation effort since solids escaping from retention ponds would be considerably finer and more mobile than those entering streams directly from the mine spoils. More thorough sediment transport studies would also permit development of separate regression relations between river flow and concentration for each trout stream in question.

There are numerous data relevant to trout populations in the Yampa River Basin which would improve the predictive capacity of the fishery response model described here. Necessary information would include life table statistics based on regular sampling, sediment induced mortality as a function of exposure time and sediment concentration, and, as discussed above, effects of sedimentation on food supply. An important aspect of the latter topic might be the influence of turbidity on predatory behavior of trout in the Yampa River system.

Because life table statistics are fundamental to the task of estimating the loss of fishery production over time due to sediments or, indeed, any environmental factor, priority should be assigned to obtaining these data. Perhaps most important, the level of compensatory or density dependent mortality in the population needs quantification since this factor determines the tendency toward population equilibrium. In the model reported here, density dependence is expressed as the dependence of P_0 on n_0 (equation 5).

It seems likely that in the future, as questions regarding land reclamation policy become more specific, the availability of information for predicting the effects of surface mining on stream fisheries will continue to fall short of "needs," and model refinement will continue to be necessary. Thus, it is through the iterative process of gathering data, constructing models which illuminate information shortages and needs, and returning to collect more data, that reliable information upon which to base public policy can be most efficiently generated.

References

Alabaster, J. S. 1972. "Suspended Solids and Fisheries," Proceedings of the Royal Society, London vol. 180, pp. 407-419.

Benson, Norman G. 1954. "Seasonal Fluctuations in the Feeding of Brook Trout in the Pigeon River, Michigan," Transactions of the American Fisheries Society vol. 83, pp. 76-83.

Colorado Open Mining Land Reclamation Act. 1973. (Copies available from the Colorado Division of Mines, 1845 Sherman Street, Denver, Colorado 80203.)

Cooper, Edwin L. 1953. "Periodicity of Growth and Change of Condition of Brook Trout (Salvelinus fontinalis) in Three Michigan Trout Streams," Copeia vol. 2, pp. 107-114.

Cordone, Alma J., and Don W. Kelley. 1961. "The Influences of Inorganic Sediment on the Aquatic Life of Streams," California Fish and Game vol. 47, no. 2, pp. 187-228.

Environmental Studies Board. 1974. Rehabilitation Potential of Western Coal Lands: National Academy of Sciences, Study Committee on the Potential for Rehabilitating Lands Surface Mined for Coal in the Western United States (Cambridge, Mass., Ballinger).

Griffin, L. E. 1938. "Experiments on the Tolerance of Young Trout and Salmon for Suspended Sediment in Water," in Henry Baldwin Ward, Placer Mining on the Rogue River, Oregon, in its Relation to the Fish and Fishing in that Stream (Salem, Oreg., Oregon Department of Geology and Mineral Industries, Bulletin 10) Appendix B, pp. 28-31.

Herbert, D. W. M., and J. C. Merkins. 1961. "The Effect of Suspended Mineral Solids on the Survival of Trout," International Journal of Air and Water Pollution vol. 5, no. 1, pp. 46-55.

Iorns, W. V., C. H. Hembree, D. A. Phoenix, and G. I. Oakland. 1964. "Water Resources of the Upper Colorado River Basin--Basic Data," U.S. Geological Survey Professional Paper 442 (Reston, Va., U.S. Geological Survey).

James, I. C. II, E. D. Attanasi, Thomas Maddock III, S. H. Chiang, B. T. Bower, and N. C. Matalas. 1977. "Water and Energy Use in Coal Conversion," U.S. Geological Survey Open-File Report (in draft) (Reston, Va., U.S. Geological Survey, Water Resources Division).

Jensen, A. L. 1971. "The Effect of Increased Mortality on the Young in a Population of Brook Trout, A Theoretical Analysis," Transactions of the American Fisheries Society vol. 3, p. 456-459.

Landis, E. R., and G. C. Cone. 1971. "Coal Reserves of Colorado Tabulated by Bed," U.S. Geological Survey Open-File Report (Reston, Va., U.S. Geological Survey).

Leslie, P. H. 1945. "On the Use of Matrices in Certain Population Mathematics," Biometrika vol 33, part III, pp. 183-212.

Matalas, N. C. 1967. "Mathematical Assessment of Synthetic Hydrology," Water Resources Research vol. 3, no. 4, pp. 937-945.

McFadden, James T., Gaylord R. Alexander, and David S. Shetter. 1967. "Numerical Changes and Population Regulation in Brook Trout Salvelinus fontinalis," Journal of Fisheries Research Board of Canada vol. 24, pp. 1425-1445.

Peters, John C. 1962. "The Effects of Stream Sedimentation on Trout Embryo Survival," in U.S. Public Health Service, Biological Problems in Water Pollution (Washington, D.C., U.S. Department of Health, Education, and Welfare, Public Health Service Publication No. 999-WP-25) pp. 275-279.

Powell, Guy C. 1958. "Evaluation of the Effects of a Power Dam Water Release Pattern Upon the Downstream Fishery," Quarterly Report vol. 4 (Denver, Colo., Colorado Cooperative Fish Research Unit) pp. 31-37.

Ricker, W. E. 1954. "Stock and Recruitment," Journal of Fisheries Research Board of Canada vol. 11, pp. 559-586; 609-623.

Saunders, J. W., and M. W. Smith. 1964. "Changes in a Stream Population of Trout Associated with Increased Silt," Journal of Fisheries Research Board of Canada vol. 22, pp. 395-404.

Speltz, Charles N. 1974. "Strippable Coal Resources of Colorado," U.S. Bureau of Mines Preliminary Report 195 (Washington, D.C., U.S. Department of the Interior).

Steele, Timothy Doak, Ivan C. James II, Daniel P. Bauer. 1976. "An Environmental Assessment of Impacts of Coal Development on the Water Resources of the Yampa River Basin, Colorado and Wyoming, Phase II Work Plan," U.S. Geological Survey Open-File Report 76-368 (Reston, Va., U.S. Geological Survey).

Trimble, Stanley W. 1975. "Denudation Studies: Can We Assume Stream Steady State?" Science vol 189, no. 4194, pp. 1207-1208.

U.S. Department of Agriculture. 1961. "A Universal Equation for Predicting Rainfall-Erosion Losses," Agricultural Research Service Special Report, Soil and Water Conservation Division (Washington, D.C.).

U.S. Department of the Interior, Bureau of Land Management. 1976. _Final Environmental Statement: Northwest Colorado Coal_ (Washington, D.C.).

U.S. Geological Survey. 1952-58. "Streamflow Records for Colorado," Water Supply Papers (Reston, Va.).

Wallen, Eugene I. 1951. "The Direct Effect of Turbidity on Fishes," _Biology Series_ vol. 48, no. 2 (Oklahoma Agricultural and Mechanical College, Arts and Sciences Studies).

PREDICTING THE IMPACTS OF SURFACE COAL MINING
ON TROUT POPULATIONS IN THE YAMPA RIVER BASIN:
A DISCUSSION

Gordon C. Jacoby, Jr.*

This discussion consists of two parts. The first portion reviews
the paper by Richard A. Smith which is a study of one small area within
the Upper Colorado River Basin. The second portion consists of general
comments based on considerations of the Yampa River and how they may be
extrapolated to cover certain other problems of the basin as a whole.

The paper by Smith analyzes the impact of an aspect of energy development, specifically surface coal mining, on the water quality and trout
populations of the Yampa River Basin. The author correctly identifies an
increase in the amount of suspended sediments as one of the serious future
problems that may affect the trout fishery in the Yampa River. The paper
is a fusion of suspended sediment analyses and the effect of suspended
sediment on trout populations. The combining of these two types of models
serves to produce better insights into one of the effects of surface mining
in this area.

Four surface mine reclamation scenarios are described in the paper.
One of these scenarios is no reclamation at all. This is included to
establish base values for the three other reclamation levels. The first
level of reclamation (level A) assumes that the mine operators will abide
by the Colorado Open Mining Land Reclamation Act of 1973. It is reasonable to assume that the operators will do so; therefore, this could be

*Research Associate, Lamont-Doherty Geological Observatory of
Columbia University, Palisades, New York.

This work was supported by the RANN Division of the National Science
Foundation, Grant No. ENV 76 16691. Lamont-Doherty Geological Observatory
Contribution No. 2634.

considered a most likely level of land reclamation. The most extensive
level (level C) consists of mulching the ground, fertilizing it, and
irrigating for three years. This treatment would produce a very low level
of additional waterborne sediments as opposed to the high level of waterborne sediments that could still occur under the first level of reclamation. However, it is unlikely that the mine operators will undertake this
level of land reclamation unless they are urged or coerced into doing so
because of one of the environmental laws, such as the Federal Water Pollution Control Act and Amendments of 1948 and 1972, respectively, the Endangered Species Act of 1973, or the Wild and Scenic Rivers Act of 1968.
For example, if it is shown that there are endangered species in the Yampa
River and that the strip mining could adversely affect them, then mine
operators could be required by law to engage in extensive land reclamation
procedures. Outside of some situation such as this, I do not think it is
reasonable to assume that the operators will engage in this level of land
reclamation.

Therefore, although the author states that his results should be
"used to illuminate data needs rather than to make predictions," based
on his current assumptions and analysis, the mine operators could comply
with the Colorado Open Mining Land Reclamation Act of 1973 and still destroy the trout fishery. This situation should lead to modification of the
act or an urgent need for more data and analysis, or both.

The author has used stochastic modeling techniques to generate
1,000 years of synthetic monthly streamflows and then divided these into
ten 100-year time series. These ten time series were then used to generate monthly base-line stream sediment concentrations. This is an appealing

technique in that it enables one to review likely stream sediment concentrations for various sequences of monthly streamflow. However, the author is remiss in just using historical streamflow data and not considering potential depletions in flow quantities themselves.

Consumptive Use and Streamflow

The Bureau of Land Management uses a production figure of 24 million tons of coal per year from surface mining in the Yampa Basin (U.S. Department of the Interior, Bureau of Land Management, 1976). The current tendency is to build power plants near the coal source. Based on the ratio of water use to coal use in energy developments in other areas of the Colorado River Basin, approximately 6 tons of water are totally consumed for every ton of coal consumed. Assuming this ratio to continue into the future, this production rate would mean that over 100,000 acre-feet of water per year will be used in converting the strip-mined coal to electrical energy.

Current consumptive uses of water in this area include irrigation waters diverted for 66,000 acres (U.S. Geological Survey, 1970). During the years of 1965 and 1966, approximately 300,000 acre-feet per year were diverted from the Yampa River for these irrigated acres (Hyatt and co-authors, 1970). Nineteen sixty-five was a high-flow year and based on the "run of the river" method of estimating irrigation diversion, perhaps this amount of water is unusually high, particularly for 1965. But this figure serves as an approximation of the relative magnitude of water diverted for irrigation.

The Four County Diversion Project which is proposed to divert water from the Yampa River to the Cheyenne-Laramie areas would divert 40,000 acre-feet per year out of the Yampa Basin (U.S. Department of the Interior, 1975). Thus, there is currently or planned a total consumptive use of over 400,000 acre-feet per year in this area (irrigation, export, and energy production combined).

The primary gaging station on the Yampa River in this region is at Maybell, Colorado. The average measured flow at this gaging station from 1916 to 1965, inclusive, is 1,125,000 acre-feet per year (U.S. Geological Survey, 1970). This can be used as an estimate of the surface water supply for this basin. However, this figure could be slightly high in that moisture-sensitive trees in the area have a mean standardized growth index (dimensionless) of approximately 1.4 during this time period in comparison to a long-term average growth index of 1. The only conclusion that should be drawn from these tree-ring data is that perhaps the measured flow produces a slightly high long-term mean. However, no definitive study has been made as yet of the Yampa River using tree-ring methods.

When all the depletions in the basin come into operation, they could total about one-third of the current surface water supply in the Yampa River at Maybell. It is important, therefore, to consider these depletions in estimating future streamflows in the basin. The sum total of these consumptive uses will certainly have a great impact on the flow in the lower main stem of the Yampa River, and could also have a sizable impact on the upper main stem and on some of the tributaries to the Yampa depending upon where the water is withdrawn.

In considering the consumptive use collectively, the average diversion could be as much as about 33,000 acre-feet per month, or 1,100 acre-feet per day. This approaches the monthly total flow for certain low-flow months. Therefore, in addition to considering the additional sediment load added to the river, one should also consider the scheduling of this consumptive use. In the paper by Walter Spofford in this volume, the author discusses the concept of critical periods with regard to the consumptive use of water and its effects on streamflow. This concept is particularly important in the Yampa River study in that it is the critical low-flow periods that are of most concern. There also may be critical reaches within the stream itself where volume and depth of flow could be detrimental to fish even though the flow and the depth may be quite adequate at certain gaging stations.

The suspended-sediment model constructed by Smith may not represent the realities of future streamflows and sediment concentrations in the Yampa River Basin at all since he has neglected to consider the depletion in streamflows which will result from the various developments both currently under construction in, and planned for, the Yampa River Basin. The author has developed a model which is very appealing and which appears to yield certain worthwhile results. However, by excluding some of the other realities of the river system being studied, the output of the model may not be as reliable as it might be.

The author states in one of his closing paragraphs that the purpose of the paper is to present a modeling technique for estimating the sensitivity of a fishery to hypothetical mining and reclamation scenarios. He has indeed presented a worthwhile technique, but it loses some of its value in not considering broader aspects. One of the problems with many of the environmental impact statements and analyses of the Colorado River Basin, and many other areas as well, is that analysts tend to focus on a single aspect or project that they are concerned with and not to consider this aspect or project within the broader context of development on a regional basis.

Reclamation Threshold Levels

One of the more positive aspects of this paper is that in comparing several of the plots of sediment concentration and adult population of trout, there appears to be a definite threshold level at which the population can continue at a quasi-equilibrium level. This threshold concept indicates that land reclamation to the most comprehensive level mentioned in this paper may not be needed, at least to protect the fisheries. According to the results, the biggest difference between impacts on the fishery is between reclamation levels A and B. Level B crosses some sort of threshold with regard to sediment yield per acre and if this threshold is surpassed in the reclamation procedures, the sediment yield is kept to an acceptable level and the trout population continues to flourish. Level B is expensive but there must be some less expensive level that could preserve the fishery. Possibly one year of irrigation added to level A would sufficiently control the sediment yield. This concept is worthy of more consideration and analysis in that reclamation level B is not reached by the Colorado Reclamation Act of 1973. One might envision laws being modified and some sort of monitoring system set up whereby reclamation is brought to the point where there is no appreciable damage to the fishery, and where reclamation does not become overly expensive and thus reach a point of diminishing returns. As the author states, this study points out the need for more data. I suggest that more research and data collection be undertaken to study the kinds of threshold effects presented by the author. This should be done to determine if these effects can be more rigorously defined in the context of expected streamflows rather than based on theoretical models built on historic streamflows without taking into consideration current and planned diversions and depletions.

Along with the development of surface coal mining in the Yampa River Basin will be an influx of people into the area and these people will be interested in preserving the same fish and wildlife values that their job-providing endeavours are in the process of depleting or destroying. This makes further study of the protection of fish and wildlife values in these areas even more important since the demands for these resources will increase at the very time that most of the damage is likely to occur.

Broader Implications for the Colorado River Basin

Potential problems in the Yampa River Basin are illustrative of some of the same problems that pervade the entire Colorado River Basin. There are plans for coal mining and other energy development activities which will consumptively use water and could contribute contaminants to both the surface and groundwater systems. There also will be increases in local population and, with the increase in the population nationwide, increased demands for fish and wildlife recreational values.

The Colorado River Basin, as many have observed, is a storehouse of fossil fuels and other substances that can be used for energy generation. It is also part of the so-called "sunbelt" of the United States where populations are increasing rapidly. This region is also part of the most water-deficient area in the United States. Thus, there are finite limits upon the renewable water resources of this area and certain proposed development plans could lead to approaching these finite limits in the not-too-distant future.

It has been stated by some of the other authors and discussants at this Forum that it is risky to make projections and predictions because of certain changes and innovations that may occur. This is true; however, there is a concept that is used in wagering on dog races called the principle of "last time out." This means that although there are unpredictable events that will occur in the future, the best method of evaluation is to use the performance the last time out. As previously noted, the existing power plants built in the Upper Colorado River Basin during the past few years use approximately 6 tons of water for every ton of coal. This water is consumptively used and does not return to the surface water system within the basin in any identifiable amount. There have been many new power plants proposed for the upper basin and, although there have been many postponements and even cancellations, some of these and other energy-related facilities are still proposed. And until new decisions are made, one should at least review the possibility of what might occur if these plans were to come to fruition. There are also new technologies, such as hybrid cooling towers, that can be used to conserve water. It is to be hoped that these new technologies will be employed in all future developments in the basin. However, until they are built and found to be economically feasible and reliably functional in their operation, one must work with the state of the art as it is now and has been the "last time out."

In an analysis of the available surface water supply of the Upper Colorado River Basin, the virgin flow was reconstructed for over 400 years (Stockton and Jacoby, 1976). This reconstruction helped place the measured flow data in a better historical context and revealed that the flow might be significantly lower than it was imagined to be during the early days of water resource development in the Colorado River Basin. These results support some of the figures used in the supportive documents for the Colorado River Basin Project Act of 1968 (House Report, 1968). Thus, there is a convergence of information leading to the possibility that water supplies will be somewhat limited or lower than originally envisioned.

In considering the overall picture, one can compare the estimated supply and the projected demand. It is to be emphasized that there are uncertainties in both of these estimates. Certain levels of demand were projected in the report Water for Energy in the Upper Colorado River Basin (U.S. Department of the Interior, Water for Energy Management Team, 1974). These projections included development of oil shale and the Kaiparowits generation station, as well as many other projected power plants. The Kaiparowits plant has been cancelled and the development of oil shale now appears to be farther in the future than estimated in this study. Other projections of use were made by a member of the Colorado River Board of California (Valentine, 1974). These may turn out to be more accurate in the final analysis. However, both these levels of projected use will, by the end of the century, reach the level of estimated renewable surface water supply. With this potential meeting in the near future of the supply of surface water and the demand for consumptive use in the Upper Colorado River Basin development of comprehensive water management plans for the region should be undertaken at this time. This is needed so that water can be allocated to its most beneficial consumptive uses. During the formation of such a management plan, it is necessary for those interested in, and responsible for, fish and wildlife values in this region to make themselves heard.

448

References

House Report No. 1312. 1968. 90 Cong. 2 sess., p. 24.

Hyatt, M. L., J. P. Riley, M. L. McKee, and E. K. Israelsen. 1970. Computer Simulation of the Hydrologic-Salinity Flow System Within the Upper Colorado River Basin (Logan, Ut., Utah State University) p. E-3.

Stockton, C. W., and G. C. Jacoby. 1976. Long-Term Surface Water Supply and Streamflow Trends in the Upper Colorado River Basin, Lake Powell Research Project Bulletin, No. 18, National Science Foundation. (Copies obtainable from: Institute of Geophysics and Planetary Physics, University of California, Los Angeles, California 90024.)

U.S. Department of the Interior. 1975. Quality of Water, Colorado River Basin Progress Report, No. 7 (Washington, D.C.) p. 56.

U.S. Department of the Interior, Bureau of Land Management. 1976. Final Environmental Statement--Northwest Colorado Coal (Washington, D.C.).

U.S. Department of the Interior, Water for Energy Management Team. 1974. Water for Energy in the Upper Colorado River Basin (Washington, D.C.).

U.S. Geological Survey. 1970. Surface Water Supply of the United States, 1961-1965, Part 9, Colorado River Basin, Vol. 1, Colorado River Basin Above Green River, U.S. Geological Survey Water-Supply Paper 1924 (Reston, Va.) p. 175.

Valentine, V. E. 1974. "Impacts of Colorado River Salinity," Journal of the Irrigation and Drainage Division, American Society of Civil Engineers, vol. 100, no. IR 4, pp. 495-510.

PREDICTING THE IMPACTS OF SURFACE COAL MINING
ON TROUT POPULATIONS IN THE YAMPA RIVER BASIN:
A DISCUSSION

Thomas H. Yorke*

Introduction

Richard Smith's paper on the potential impacts of surface coal mining on trout populations in the Yampa River Basin presents an interesting approach for assessing the impact of coal mining on fishery resources. The models assembled by the author use a combination of sediment data for the Yampa River Basin and laboratory and field studies of suspended sediment and trout survival studies to predict trout population following the initiation of coal mining activities. They are an excellent first step for assessing the total impact of sedimentation that will result from the extensive coal mining proposed for the western states.

The models provide a framework for evaluating the primary impact of increased suspended sediments, but they do not account for the secondary impacts of suspended sediments. The trout population model generates long-term populations based on egg and juvenile survival following prolonged exposure to high sediment concentrations. The impacts of high sediment concentrations on food sources and cover are not considered. More important, the model analyzes only suspended-sediment discharges and not bedload discharges. This discussion will concentrate on the impact of bedload and the secondary impact of suspended sediments to illustrate

*U.S. Fish and Wildlife Service, Columbia, Missouri.

the total impact on the fishery resource of surface mining for coal in the Upper Colorado River Basin.

Background

Since the Yampa River and its tributaries are some of the few clear water streams supporting trout populations in the Upper Colorado River Basin, a review of the physical setting and the sediment characteristics of the basin is appropriate. This will ensure that discussions of surface mining and the resultant sedimentation will be viewed from the proper perspective --its impact on a valuable and limited fishery resource.

The Yampa River is a tributary of the Green River and drains about 8,000 square miles in northwestern Colorado and south-central Wyoming. It flows westward from the west slope of the Rocky Mountains across rolling hills comprising the Yampa Plateau. Its course is marked by incised meanders and rugged canyons (Stokes, 1955). The headwater areas are underlain by precambrium granites and associated metamorphic rocks. The central part of the basin is underlain by sedimentary rocks of the Cretaceous and Tertiary Age. These include interbedded shales, sandstones, and limestones, and extensive coal deposits (U.S. Department of the Interior, Bureau of Reclamation, 1946).

The Yampa Basin is typical of all river basins in arid and semiarid regions. There is a delicate balance between vegetal cover and surface materials which is created by factors such as low precipitation, immature soils, and the relatively steep topography resulting from the diverse geologic formations. The low precipitation (10 to 20 inches per year) restricts the type and density of vegetal material. The immature soils and steep topography hold the basin on the threshold between stable surface conditions with low sediment yields and extensive erosion and gulleying with very high sediment yields.

Sediment yields in the Yampa River Basin are low compared to many of the other tributaries of the Green River. The average annual sediment yield near Maybell, Colorado, is 90 t/mi^2 (tons per square mile). The yield of the Little Snake River near Lilly, Colorado, the basin immediately north of the Yampa, is 295 t/mi^2. Yields of other basins range from 131 t/mi^2 for the White River at Buford, Colorado, to 2,600 t/mi^2 for the Price River at Woodside, Utah (Iorns and coauthors, 1965). Since the Yampa River Basin has a relatively high water yield compared to its sediment yield (the basin contributes 12 percent of the total streamflow and 1.5 percent of the total suspended-sediment discharge of the Upper Colorado River Basin), the differences between average suspended-sediment concentrations are even more dramatic. The average concentration for the Yampa River near Maybell is 196 mg/l (milligrams per litre) while the average concentrations for the Little Snake River and Price River are 1,790 and 34,000 mg/l, respectively.

The relatively low sediment yield of the Yampa River Basin in comparison to other basins in the Upper Colorado is not completely understood. However, because of similarities between the geology soils and the climate of the Yampa and other basins, any change in the vegetal cover or disruption of the surface materials will cause an increase in erosion and a sediment yield that will approach those of the other basins. Surface mining for coal will certainly provide the mechanism for increased erosion, particularly when the removal of 50 feet of overburden is required.

Increases in sediment discharges to the magnitude of adjacent basins will seriously alter the quality trout streams of the Yampa Basin.

Indirect Effects of Suspended Sediments

The author has illustrated the direct effects of increases in sediment concentration on trout survival. The discussion in this section will concentrate on the indirect effect of suspended-sediment increases associated with mining, principally the effect on food sources and cover.

The direct mortality of fish due to high suspended-sediment concentrations may significantly alter fish populations; however, suspended sediment and turbidity are usually more harmful to the nutritive richness of the water than directly to fish (Huet, 1965). Algae, a basic component of the food chain, are indicators of productivity. Sediment affects algae by physical settling, smothering attached algae, and reducing illumination necessary for photosynthesis. A study of a Georgia stream affected by silt pollution found only 126 plankton algal cells per ml (milliliter), representing two genera. Another stream unaffected by silt pollution had ten times as many cells and five times as many genera (Mackenthun, 1969). Cordone and Pennoyer (1960) also found that algae concentrations were destroyed by sediment in the Truckee River, California.

Benthic organisms, the principal food source for trout, are probably affected most by increases in suspended sediments. Attached algae and aquatic invertebrates thrive in the shelter and substrate provided by gravel and rubble. Sediments that cover this substrate and fill the interspaces can reduce benthos populations significantly. Cordone and Pennoyer (1960) found that silt from a gravel washing plant on Cold Creek, California, resulted in a 90 percent reduction in bottom organisms

immediately below the outfall and a 75 percent reduction 10 miles downstream. Cordone and Kelley (1961) reported that silt from a molybdenum
mining operation polluted Moore Creek in California. A total of 432 bottom
organisms were found in the clear water compared to thiry-two in the silted
area.

Another indirect impact of increases in suspended sediments is the
reduction in cover for fish, particularly the juvenile fish. Without the
cover of aquatic vegetation or the interspaces in gravel and rubble beds,
juvenile fish will be subjected to extensive predation. Sediments prevent
germination and retard growth of aquatics by reducing light penetration or
by directly smothering young plants. Fine sediment and sand also fill the
interspaces of gravel beds and completely bury other sources of cover in
streambeds.

Trout and other salmonids can survive losses in food sources and
cover; they can migrate to an unaffected section of stream. However,
the loss of food and cover represents a direct loss of quality habitat.
This type of habitat is already at a premium in the Upper Colorado River
Basin.

Effects of Bedload

Increases in bedload may have the most significant long-term effect
on trout populations in the Yampa River Basin. Sediments transported
from the mine sites to the stream channels may transform sections of the
Yampa River and its tributaries into sand bed streams. The shallow
residual soil and interbedded layers of shales and sandstones that lie
above the coal seams will provide a ready source of sand during mining
operations. These larger particles may not be eroded from the mining

sites during normal runoff events, but one large flood could introduce enough coarse sand into the channel system to provide a bedload source for many years.

A comparison of bed material samples in the Yampa River Basin and adjacent basins illustrates that a conversion to a sand bed stream is possible. Data presented by Iorns and coauthors (1964) show that the Yampa River and its tributaries have largely gravel and cobble beds while other tributaries have a high percentage of sand in the beds. The Yampa River near Oak Creek, the Yampa River at Steamboat Springs, and Williams Fork at Hamilton have bed material greater than 4.0 mm (millimeters). Only Fortification Creek in the Yampa Basin has bed material as fine as 0.250 mm. In comparison, Black Fork has fine sand (0.62 mm) in its bed, and the Price and San Rafael rivers have 20 to 40 percent of their bed material finer than sand (less than 0.062 mm). It is conceivable that the distribution of bed material of the Yampa River tributaries also could be dominated by smaller particle sizes if extensive erosion and sediment transport from the mine sites occur.

Large quantities of sand in the channels of the Yampa Basin will eliminate the spawning potential, food source, and desirable cover for trout and other fish. Sand will cover gravel bars or fill redds in the gravel beds before eggs have a chance to hatch. The shifting nature of sand beds will limit the ability of benthic organisms to survive in the channels, thereby eliminating a major component of a trout's diet. Gaufin and Tarzwell (1956) ranked substrates according to their ability to support macroinvertebrates, using a scale from one to 456. Shifting sand, which supported the fewest numbers, was ranked one, and gravel and rubble was ranked over 400. Sand deposits will also eliminate most of the cover available in the stream channels. Sand will fill the pools and small interspaces in the bed, and the shifting nature and abrasive action of the sand will limit the growth of aquatic plants.

The increase in bedload that can be expected because of mining activities in the Yampa River Basin may eliminate the affected channel reaches as trout habitat for many years. Large quantities of sediment will be eroded from mining sites during storms between the time the overburden is removed and the time the area is restabilized by various reclamation efforts. Intense summer storms will transport much of the sediment to the stream system. Most of the finer particles (clay and silt) associated with the shales will be transported through the entire Yampa system. In contrast, the heavier particles associated with sandstones probably will be deposited on the slopes between the mine and the channels or in the channels near the source. Subsequent storms will transport more fine particles downstream, also lifting sand particles and redepositing them further downstream. This process will continue adding more and more sediment to the stream channels until the mining sites are stabilized. Even after stabilization, the sands and gravels deposited in the basin will gradually move downstream during each spring thaw and summer thunderstorm, resulting in unstable streambeds and poor trout habitat.

Summary

These comments were intended to highlight the impact on trout habitat in the Yampa River Basin and the Upper Colorado River Basin of sedimentation resulting from strip mining. The author has shown that trout populations will not be directly affected by high suspended-sediment

concentrations if very strict reclamation efforts are employed to control
sediment discharge from mine sites. My point is that the major impact of
mining on trout will be the loss of food sources and cover, and not direct
mortality from suspended sediment. My comments only reinforce the author's
statement that the results of his modeling efforts are "to be used to
illuminate data needs rather than to make predictions."

References

Cordone, A. J., and D. W. Kelley. 1961. "The Influences of Inorganic Sediment on the Aquatic Life of Streams," California Fish and Game vol. 47, no. 2, pp. 189-228.

Cordone, A. J., and S. Pennoyer. 1960. "Notes on Silt Pollution in the Truckee River Drainage, Nevada and Placer Counties," (Sacramento, Calif., Department of Fish and Game, Inland Fisheries Administration Report 60-14).

Gaufin, A. R., and C. M. Tarzwell. 1956. "Aquatic Macroinvertebrate Communities as Indicators of Organic Pollution in Lytle Creek," Sewage and Industrial Wastes vol. 28, no. 7, pp. 906-924.

Huet, M. 1965. "Water Quality Criteria for Fish Life," in C. Tarzwell, ed., Biological Problems in Water Pollution (Washington, D.C., U.S. Department of Health, Education, and Welfare, Public Health Service Publication 999-WP-25) pp. 160-167.

Iorns, W. V., C. H. Hembree, and G. L. Oakland. 1965. "Water Resources of the Upper Colorado River Basin--Technical Report," U.S. Geological Survey Professional Paper 441 (Reston, Va., U.S. Geological Survey).

Iorns, W. V., C. H. Hembree, D. A. Phoenix, and G. L. Oakland. 1964. "Water Resources of the Upper Colorado River Basin--Basic Data," U.S. Geological Survey Professional Paper 442 (Reston, Va., U.S. Geological Survey).

Mackenthun, K. M. 1969. "Silts," in U.S. Department of the Interior, Federal Water Pollution Control Administration, The Practice of Water Pollution Biology (Washington, D.C., U.S. Department of the Interior) pp. 96-108.

Stokes, W. L. 1955. "Geomorphology of Northwestern Colorado," in Rocky Mountain Association of Geologists and Intermountain Association of Petroleum Geologists, Guidebook to the Geology of Northwest Colorado (Denver, Colo., and Salt Lake City, Ut.) pp. 56-59.

U.S. Department of the Interior, Bureau of Reclamation. 1946. The Colorado River--A Natural Menace Becomes a National Resource (Washington, D.C.).

Chapter 15

HABITAT EVALUATION PROCEDURES (HEP)

James M. Lutey[*]

Introduction

The Habitat Evaluation Procedures (HEP) were developed by the U.S. Fish and Wildlife Service, state fish and game agencies, and private conservation organizations to provide a uniform, nationwide method for assessing the impacts on fish and wildlife and their habitat resulting from water development projects. The system is currently being applied and refined by the Division of Ecological Services of the U.S. Fish and Wildlife Service in evaluating federal and federally assisted water resource development projects in accordance with the Water Resources Council's "Principles and Standards for Planning Water and Related Land Resources" (U.S. Water Resources Council, 1973). Directly related are activities currently underway by the Office of Biological Services of the U.S. Fish and Wildlife Service for a National Wetland Inventory,[1] and for a Habitat Classification and Analysis Project.[2] Among other important uses, these projects will serve as inputs to HEP.

---

[*]Area Supervisor, Division of Ecological Services, U.S. Fish and Wildlife Service, Kansas City, Missouri.

[1]The National Wetland Inventory will inventory and classify all wetlands in the United States. Wetlands will be displayed on maps according to a new classification system at a scale of 1:100,000. The program was begun in 1975 and is expected to be complete in 1980.

[2]The Habitat Classification and Analysis Project is the development of a system for classifying, inventorying, and evaluating existing fish and wildlife habitats. It is a basic step toward the evaluation, protection, and management of the nation's fish and wildlife. The project was initiated in January 1976 and is directed from Fort Collins, Colorado.

Background

Before discussing the present methodology for fish and wildlife impact assessment, it is important to understand what has happened in the past and why the procedures were developed.

The Fish and Wildlife Coordination Act, as amended in 1958, provides "...that wildlife conservation shall receive equal consideration and be coordinated with other features of water resource development programs." The act assumes the existence of an evaluation procedure; however, fish and wildlife agencies, unfortunately, did not strive toward a standardized methodology for assessing project impacts on fish and wildlife. Rather, project impacts were expressed in terms of loss or gain of resource use (man-days of hunting and fishing) and respective dollar values. Obviously, project evaluations based only on resource use do not include all the impacts of a project on fish and wildlife and their habitat.

In 1970, a National Coordinating Committee was formed. This committee consisted of representatives of state fish and game agencies, private conservation organizations, and the U.S. Fish and Wildlife Service. The committee was concerned about the fish and wildlife habitat that was being destroyed by the continuing development of water resource projects and the failure to receive adequate compensation for the losses of these habitats. The committee recommended opportunities to strengthen the consideration of fish and wildlife resources in water resource development programs. One of the recommendations was that the U.S. Fish and Wildlife Service move promptly to refine, establish and implement a system of habitat evaluation, based on nonmonetary measures of habitat values, to adequately display the beneficial and adverse effects of water development projects on fish and wildlife resources. This prompted the U.S. Fish and Wildlife Service, state fish and game agencies, and private conservation organizations to develop the Habitat Evaluation Procedures (HEP).

The Habitat Evaluation Procedures have a direct relationship to the National Environmental Policy Act (NEPA) of 1969 and to the Water Resources Council's Principles and Standards. Section 102(2)(B) of NEPA directs that "...all agencies of the Federal Government shall identify and develop methods and procedures...which will insure that presently unquantified environmental amenities and values may be given appropriate consideration in decisionmaking along with economic and technical considerations."

The U.S. Water Resources Council, established by the Water Resources Planning Act of 1965, established "Principles and Standards for Planning Water and Related Land Resources" that became effective in October 1973. Basically, the Principles and Standards set forth two national planning objectives: (1) National Economic Development (NED), and (2) Environmental Quality (EQ). Environmental quality objectives were put on an equal basis with economic development objectives in water resources planning. The NED objective, as defined, is enhanced by increasing the value of the nation's output of goods and services and by improving national economic efficiency. The EQ objective, as defined, is enhanced by the management conservation, preservation, creation, restoration, or improvement of the quality of certain natural and cultural resources and ecological systems. Each of these objectives has specific components which must be considered.

The EQ components can be set forth as follows:

A. Physical Land Resources

 1. Soil stability. As encompassed in the environmental

  quality objective, soil is valued as a basic national

  resource, rather than as a primary production factor,

  its more traditional role, contributing to increases

  in national output.

B. Air and Water Quality

 1. Air quality standards

 2. Water quality standards

C. Ecological Resources

 1. Terrestrial ecosystems

 2. Aquatic ecosystems

 3. Special ecosystem relationships and irreversible commitments

  of ecological resources

 4. Species threatened with extinction

D. Culturally Significant Resources

 1. Archeological resources

 2. Historical resources

 3. Areas of natural beauty

Item C, ecological resources, specifically deals with fish and wildlife impact assessments. Note that the four components listed under
ecological resources are all referred to as ecosystems. Early in the formulation stages of the new evaluation procedures, it was realized that
any system that did not identify the total range of impacts would be less

than adequate. The tendency in the past was to be concerned primarily with numbers of fish and wildlife, and then only with those species important to fishermen and hunters.

Habitat Evaluation Procedures

Because all fish and wildlife are dependent on habitat, it was determined that a methodology relating fish and wildlife to their habitat could best measure ecological impacts. Thus, the Habitat Evaluation Procedures developed recently are based on an assessment of the terrestrial and aquatic ecosystem habitat values for the fauna located in any given planning area.

The procedures consist of both a monetary and a nonmonetary evaluation. The monetary evaluation provides data on the supply and demand for fish and wildlife in the project area. It expresses the use of the resource (fishing, hunting, and nonconsumptive uses) in terms of man-use-days and equivalent dollar values. Monetary values are developed and displayed for the "without project" condition and for each alternative plan. This information is also used in the cost allocation process and in justifying fish and wildlife enhancement features.

Obviously, project evaluations in terms of resource use only cannot indicate the extent of the impacts of a project on fish and wildlife and their habitat. Although consideration has always been given to fish and wildlife habitats and maintenance of healthy ecosystems, these amenities have never been quantified. The nonmonetary portion of the evaluation procedures does this and is superior to the monetary portion for two reasons: (1) it provides a method for measuring the quality of habitat in a project-affected area for the full range of fish and wildlife present, using a rating scale of 0 to 10, 0 being poor and 10 being excellent; (2) it establishes a base for identifying measures needed to compensate for fish and wildlife losses.

A basic assumption of the habitat evaluation system is that the only meaningful way to mitigate project-caused losses to fish and wildlife is to increase the habitat carrying capacity of other lands to a level sufficient to offset habitat losses resulting from the project.

The first step in the evaluation process is to form an evaluation team and to make preliminary evaluations. It is important that representation include biologists from the state fish and game agency, the U.S. Fish and Wildlife Service, and the construction agency responsible for formulating the project. Representatives from private conservation agencies and environmental groups should also be welcomed. At least one member of the evaluation team should be familiar with the area and the ecosystem to be evaluated. The evaluation team should determine the area of project influence and develop baseline maps of the planning area with appropriate overlays that depict habitat types and proposed project segments and features.

The field evaluation sheet (figure 1) is used to rate the existing habitat for each species of fish and wildlife present. The evaluation form is used for both aquatic and terrestrial evaluations. The example presented in figure 1 is restricted to a single habitat type (riparian timber) of the terrestrial ecosystem. Representative species, or faunal groups, that are dependent to some degree on the habitat type being evaluated are listed across the top of the form. The objective is to consider the full range of animal life present. An adequate number of sample sites are evaluated for each habitat type present (forestland, cropland, pasture, etc.). The number

464

U.S. FISH AND WILDLIFE SERVICE
DIVISION OF ECOLOGICAL SERVICES

FISH AND WILDLIFE HABITAT FIELD EVALUATION SHEET

Page __1__ of __1__ pages

PROJECT NAME
Hypothetical Reservoir

DATE

HABITAT CODE 03	HABITAT TYPE Riparian Timber	ALTERNATIVE PLAN

EVALUATION ELEMENTS

SAMPLE SITE IDENTIFICATION NUMBER	W.T. Deer	Quail	Fox Squirrel	Raccoon	C. Rabbit	Raptors	Songbirds	Reptiles	Amphibians	Rodents	LINE TOTAL
1	6	7	5	7	5	5	8	7	8	5	63
2	5	4	6	5	4	6	6	5	4	5	50
3	6	8	5	6	7	6	8	7	8	6	67
TOTAL EVALUATION ELEMENT VALUES	17	19	16	18	16	17	22	19	20	16	180

Grand Total of All Evaluation Elements = 180 / Number of Sample Sites = 3 =	HABITAT TYPE UNIT VALUE 60	MANAGEMENT POTENTIAL UNIT VALUE (Wildlife habitat only) 40

INSTRUCTIONS

In order to evaluate the impact of the plan on the fish and wildlife habitat, it is necessary to know the value of the habitat itself. Here, each habitat type is assigned a value according to its worth for fish or wildlife. These resources are to be evaluated separately, and impacts and compensation needs are also computed separately. To determine this habitat type unit value, the evaluation team will complete a Form No. 3-1101 for each habitat type as follows:

1. Select ten representative species that are dependent to some degree on the habitat type being evaluated and which best express its diversity. These will be used in rating the sample sites. List them across the top of the chart at the left. The reasons for selecting these particular species should be noted and appended to this form. The objective is to consider the full range of animal life in assessing habitat quality. Normally ten species are selected, however the number of species used to evaluate a particular habitat type may vary. If another number is chosen, the rationale for this must be noted on the back of this form. These species, or evaluation elements, may not vary within a habitat type.

2. Select a number of sample sites agreeable to all members of the evaluation team. This number may vary with different habitat types.

3. Rate the capability of the habitat to meet the requirements of each of the evaluation elements on a scale of 1 through 10 at each sample site, the higher rating being given to the more desirable sites. All evaluation elements must be rated at each sample site.

4. The key criteria involved in making the above judgement should be recorded on the back of this form or on a separate sheet and attached.

5. Sum the values in each Evaluation Element column vertically, and write this number at the bottom of the column.

6. Sum each Sample Site line horizontally and sum the Total Evaluation Element Column. Write

SIGNATURE OF LEAD PLANNING AGENCY REPRESENTATIVE	LEAD PLANNING AGENCY
SIGNATURE OF STATE REPRESENTATIVE	STATE AGENCY
SIGNATURE OF FWS REPRESENTATIVE	ES FIELD OFFICE

Form No. 3-1101 (MAY 1976)

Figure 1. Fish and Wildlife Habitat Field Evaluation Sheet

of sample sites are tested statistically to determine the adequacy of the
sample size for each habitat type. If necessary, additional sample sites
are selected and evaluated. At each sample site, the capability of the
habitat to meet the requirements of a given variety of animals is rated
on a scale of 0 to 10 for each animal. A value of 0 refers to poor
habitat and 10 refers to excellent habitat. The rating is accomplished
by using a combination of biological judgment and key habitat criteria
for each species being evaluated. The key habitat criteria furnish a
common basis for each team member to rate the habitat. With such criteria,
biologists employ similar standards to evaluate each area resulting in more
uniform scoring. Such criteria include percent of vegetative cover,
percent of preferred food plant diversity, configuration of edge, browse
availability, and so on. The evaluation team should establish the
criteria that they will consider for each habitat type and each "evaluation
element" or species being evaluated. A basic need for the habitat evaluation
system is the development of key habitat criteria handbooks, on a regional
basis, to provide a uniform approach for translating descriptive data into
the numerical rating system. Examples of such key criteria have been
produced by Cooperative Fishery and Wildlife Research Units[3] and will
be coordinated with projects of the Office of Biological Services (U.S.
Fish and Wildlife Service) as input to the habitat evaluation procedures.

---

[3]Cooperative Fishery and Wildlife Research Units are associated with
various universities throughout the country. The Cooperative Units function
within the research management system of the U.S. Fish and Wildlife Service.
The purpose of the units is to involve graduate students in research activities and to gather data for specified research projects.

The values of the evaluation are summed vertically and horizontally. The total is divided by the number of sample sites. This yields an average habitat type unit value. This average is based on a scale of 0 to 100. If more or fewer than ten species are used, then the number obtained by this division must be prorated. For example, if only five species are used, the quotient must be multiplied by 10/5. If twelve species are used, the quotient must be multiplied by 10/12. Using professional judgment, the evaluation team then estimates the increase in wildlife habitat type unit value possible by proper management of the existing resources. This is the management potential unit value. Usually, this is the difference between the habitat type unit value and 100. This difference is the maximum management potential since 100 represents ideal conditions.

The evaluation system allows for the consideration of "interspersion" between habitat types. Parts of a given habitat type may be more valuable to certain wildlife because of their proximity to other habitat types. Interspersion can be considered in the base field evaluation, or it can be estimated separately and added to the base value. The procedure provides for a maximum of thirty additional habitat units to be added to the base value for interspersion.

The average habitat type unit value for each habitat type multiplied by the acreage of that type equals the total number of habitat units (table 1). This displays the "value" of the existing habitat "without" a project. Next, the project impacts must be analyzed for each alternative plan (table 2). The losses or gains in habitat units are calculated over the life of the project for a series of target years, usually 25, 50, 75, and 100 years. This furnishes an average annual habitat unit loss or gain

467

Table 1. Existing Habitat Values: An Illustration

Habitat type	Habitat area (acres)	Habitat unit value per acre (from figure 1)	Total habitat unit value
Riparian timber	200	60	12,000

Note: Simplified example using only one habitat type.

Table 2. Impact Analysis of Alternative Project Plan: An Illustration

Habitat type	Existing habitat (table 1)		New habitat			Habitat units lost or gained
	Area	Habitat units	% of area lost or gained	Remaining area	Habitat units	
Riparian timber	200	12,000	-20	160	9,600	-2,400

Note: Simplied example; changes over the period of analysis not included.

for a particular habitat type over the life of the project. The beneficial
and adverse effects are displayed for each alternative plan to indicate
the difference between them.

Finally, the compensation for habitat losses is calculated for each
habitat type (table 3). This computation determines how much intensively
managed land would be required to compensate for losses to wildlife habitat
resulting from the proposed construction for each alternative plan. It
indicates the full project impacts on wildlife and serves as a basis for
recommendations to mitigate fish and wildlife losses. The calculations
assume that land of like management potential (i.e., existing habitat
unit value equal to 60) would be provided. These lands would have to be
intensively managed for wildlife, thereby increasing their total habitat
value to 100, 40 more than the existing, (by improving the existing habitat
conditions and thereby increasing the carrying capacity) to offset habitat
losses caused by the project. If lands provided for mitigation are not
of like quality, then they must be evaluated and the new management
potential unit values used to determine habitat compensation requirements.

The goal is to compensate for habitat "in-kind"; that is, forestland
for forestland, grassland for grassland, and so on. Sometimes a trade-off
between habitat types is acceptable; however, some critical and unique
habitats cannot be compensated for, or adequately mitigated by, other
habitat types.

The results of the monetary and nonmonetary evaluations are summarized
in tables and with narratives, and incorporated into official fish and
wildlife impact reports to support recommendations for the preservation
and conservation of fish and wildlife resources.

Table 3. Compensation Determination: An Illustration

Habitat type	Management potential unit value per acre (from figure 1)	Habitat units lost or gained (from table 2)	Compensation requirement (acres)
Riparian timber	40	-2,400	60

Note: Calculations: $\dfrac{2,400 \text{ Habitat units lost}}{40 \text{ Habitat units/acre}} =$ 60 acres of intensive wildlife management needed to compensate for losses due to the project

The form and content of the procedures themselves will be refined as a result of field use and of input from other agencies. In addition, modification of the present procedures will be developed so that the system can be used in evaluating an expanded range of development projects. Eventually, the system should be applicable for evaluating the impacts of any development that would result in habitat changes.

Research Needs

This habitat evaluation (nonmonetary) system is superior to only a monetary evaluation system in that it is designed to quantify impacts on aquatic and terrestrial ecosystems. However, the development of regional key criteria handbooks to provide a data base useful for a uniform approach and input into the Habitat Evaluation Procedures is needed. The Habitat Evaluation Procedures have potential application to thousands of projects being planned or constructed by the U.S. Army Corps of Engineers, the Bureau of Reclamation, and the U.S. Soil Conservation Service throughout the United States. Regional handbooks would strengthen the conceptual basis for the evaluation procedures and would provide key habitat evaluation criteria to aid in implementation of these procedures. Written criteria would foster uniformity in habitat evaluation since all team members would employ similar standards to evaluate each area. With such criteria, biologists unfamiliar with the project area may have a better basis to work from. The development of these handbooks by the U.S. Fish and Wildlife Service is currently underway in various ecoregions of the United States.[4]

---

[4]See footnote 3, p. 8.

References

U.S. Water Resources Council. 1973. "Principles and Standards for Planning Water and Related Land Resources" (Washington, D.C., October).

HABITAT EVALUATION PROCEDURES
(HEP): A DISCUSSION

Ralph C. d'Arge*

There is perhaps no more important issue in the Rocky Mountain
states today than the direct and indirect impacts of energy development
on native ecosystems. Disturbance to flora and fauna from extraction,
processing, and transportation activities may be local and regional in
extent, but some of the more obvious direct impacts can be partially or
even completely ameliorated through land use planning and statutory requirements.[1] The indirect or "people" related effects due to demands
for outdoor experiences, low-density housing, convenient transportation,
and other "necessities" by a rapidly increasing and mobile population may
be much more severe, and ultimately more harmful. In order to assess
these effects, some type of predictive approach must be developed which
will allow, at minimum, reasonable assessment of whether substantial
negative impacts on species of an ecosystem will occur as a result of
energy construction and operation.[2] There are enough alternative locations for energy development, at least in the near future, that some sites
will be precluded from development because of potentially adverse ecosystem effects. In addition to siting alternatives, there is the related
issue of controls and management at the site to minimize environmental

---

*John S. Bugas Professor of Economics, University of Wyoming,
Laramie, Wyoming.

[1] An example is the Wyoming Environmental Quality Act of 1973.

[2] Predictive estimates are required in environmental impact statements although in practice, because of a lack of numerical models for
most species, these estimates are often nothing more than guesses.

disruption. There will undoubtedly be some important decisions on preservation or development decided by the perceived magnitude of both direct and indirect effects. Therefore, rational evaluation tools need to be developed and refined to assess effects and provide insight into relevant trade-offs.

A traditional tool for such assessments has been benefit-cost analysis; that is, the imputation of monetary values for each type of effect and a summing of these values to evaluate the efficiency gains or losses associated with a particular alternative.[3] While many difficulties have been identified with benefit-cost analysis, this tool does allow direct efficiency comparisons of various sites, and attributes of these sites. Its major shortcoming is in not being able to value nonmarket aspects of wildlife and ecosystems, most significantly values associated with existence, culture, heritage, and non-man oriented or innate values of ecosystems.[4]

---

[3]Traditional benefit-cost analysis defines a particular weighting system for the various categories of benefits and costs. The weight provided within time intervals equals 1 and the weight across time intervals is $\frac{1}{1+r}$. That is, benefits or costs accruing within a given time interval are weighted the same while benefits (costs) in time interval t+1 are worth $\frac{1}{1+r}$ what they are worth in time interval t, where r, a constant, denotes the societal rate of interest.

[4]Recently, a number of economists have attempted to conceptualize more clearly and to measure aesthetically and recreationally linked nonmarket values. See Krutilla, 1967; Krutilla and Fisher, 1975; and Brookshire, Ives, and Schulze, 1976.

The new technique described in this paper, called "Habitat Evaluation Procedures" or HEP, is one attempt to design an alternative evaluation system which would overcome some of the current nonmarket related deficiencies of benefit-cost analysis. The emphasis was to "refine, establish and implement a system of habitat evaluation, based on nonmonetary measures of habitat values." The U.S. Fish and Wildlife Service system, if I may simplify, proceeds as follows:

1. Selection of area to be impacted by the proposed project, and the subareas (sample sites) within this area.

2. Selection of ten "important" species.

3. Rank subarea habitats on a scale of 1 to 10 as to "quality" (using professional judgment) for each of the ten species.

4. Sum numerical scores over subareas and species.

5. Divide by number of subareas, or sample sites, to obtain "habitat type unit value" (HTUV).

6. Subtract HTUV from 100 to obtain maximum possible "management potential unit value."

7. Multiply the habitat type unit value (HTUV) by the total number of acres to equal a total number of habitat units.

8. By some arbitrary procedure, the loss (or gain) in number of habitat units is calculated given a project (I suspect in practice this loss is simply the number of acres taken up by the project), and these losses per year are summed over some arbitrary time period, say 100 years.

9. The amount of compensation for habitat losses is estimated, assuming that land "of like management potential would be provided." Again, the calculation appears to be in acres.

In my opinion, there are several fundamental problems with this "system" of evaluation that make it of little value for management purposes and perhaps even dangerous. First, all species' habitats are given equal weight! The quality of the habitat for moose or white tailed deer is assumed to be no more important than the quality of habitat for rodents or reptiles. A very low quality of habitat for any species, therefore, can be "compensated" for by a high value for any other species and the HTUV will not be affected. To take this point to its extreme, the ecosystem appears to be viewed as a homogeneous whole where all parts, regardless of relative importance for stability, resiliency, or nutrient flows, are valued equally. This is sort of a democratic rule gone mad or extreme "animalism" applied to ecosystems.

Second, because of the arbitrary 1 to 10 scaling of habitat "quality" and the selection of species, the HTUV cannot be compared across diverse ecological systems, i.e., how valuable is moose habitat compared with antelope habitat.

Third, there appears to be no way of interpreting the resulting calculations. What is meant by 11,000 HTUVs? How can it be compared with another site with only 9,000 HTUVs? Is the first site a better habitat on the average? This cannot be known without further information of a much more detailed type. The U.S. Fish and Wildlife Service has created a new unit of measurement; but, unlike Fahrenheit or Joule, they have not provided a qualitative or quantitative basis for comparison or understanding. Also, there is no reason to suspect that experience in its use (being hot or cold) will provide such an understanding.

Fourth, a rather subjective calculation is made of alternative lands to see whether they, if managed at a high level of intensity, would exactly, or nearly, substitute for the loss in "quality" of habitat due to the energy development or water resources project. Because the compensated average estimate is based on the previously calculated management unit potential value, however, even though it is in acres, there is no way of assessing where the acreage might come from or what characteristics the new land must have.

I would urge that HEP be abolished (or greatly modified), at least in its present form of development, and that a more general and intuitive approach to wildlife and ecosystem valuation be undertaken, at least until ecosystem modeling reaches a plateau of synthesis and application. To achieve the objectives of HEP as to compensated acreage, each project could be examined as to its reduction in a general type of habitat. A least cost substitute habitat could be identified where cost is defined and estimated for increased management requirements or for withdrawing land from other developments, or both. The measure of habitat loss could then be in terms of dollars, a convenient numeraire for comparative purposes. Note the least cost for replacing unique habitats with irreversible effects from development could approach infinity, an intuitively desirable result. In addition, sites of relatively high habitat value that are scarce could be expensive to compensate for, while sites that have relatively low habitat value and are easily substituted for will be cheap.

Mr. Lutey should be congratulated for a most topical and interesting presentation of HEP. Unfortunately, I cannot agree with his conclusion that HEP is superior to any approach without suggesting that it may have an equal probability of being inferior.

References

Brookshire, D., B. Ives, and W. Schulze. 1976. "The Valuation of Aesthetic Preferences," _Journal of Environmental Economics and Management_ vol. 3, no. 4.

Krutilla, John V. 1967. "Conservation Reconsidered," _American Economic Review_ vol. LVII, no. 4.

_____, and A. Fisher. 1975. _Economics of Natural Environments_ (Baltimore, Johns Hopkins University Press for Resources for the Future).

HABITAT EVALUATION PROCEDURES
(HEP): A DISCUSSION

Clarence Daniel*

The basis of the system presented by James Lutey of the U.S. Fish

and Wildlife Service is that all land has some value to wildlife, with

or without a water resource project. The system, based upon the profes-

sional evaluation of wildlife habitat, offers a rapid means of reducing

complex environmental factors to simple numbers that can be used to

compare development alternatives. Two major advantages of this system

are the rapidity and ease of application, and a numerical estimate of

environmental values.

The tendency in the past was to be concerned primarily with numbers

of fish and wildlife, and then only with those species important to fish-

ermen and hunters. Senate Document 97 (U.S. Senate, 1962) used the "man-

use-day" concept as a basis for evaluating the effects of water resource

projects.

Excerpts from a congressional report to the House Subcommittee on

Fisheries and Wildlife Conservation and the Environment, and to the House

Committee on Merchant Marine and Fisheries, entitled "Improved Federal

Efforts Needed to Equally Consider Wildlife Conservation with Other Fea-

tures of Water Resource Developments," compiled by the U.S. Comptroller

General's Office, dated 8 March 1974, clearly establishes previous

---

*Land Acquisition and Federal Liaison Supervisor, Wildlife Division, Missouri Department of Conservation, Jefferson City, Missouri.

problems with evaluating the effects of water resource projects on fisheries and wildlife solely from an economic viewpoint. The report covered

a review of twenty-eight water resource developments typical of those

which have had a major impact on wildlife in three geographic regions:

the Atchafalaya River Basin in Louisiana, the Cape Fear and adjoining

Neuse River Basins in North Carolina, and the Columbia River Basin in

the Pacific Northwest.

Chapter 2, page 8, of the report states that "generally, wildlife

conservation had not been considered equally with other features of the

28 water resource developments we reviewed, and the need for equal consid-

eration was demonstrated in each of the three river basins covered in our

review." Again on page 31, the report reads, "It is of primary importance

according to BSF&W [U.S. Fish and Wildlife Service] officials, to recog-

nize that the value of having a quality environment extends beyond the

value presently qualified in dollars, and that no dollar evaluation now

in use or in sight provides a reliable guide for deciding to destroy or

damage any given environment. They maintain that decisions to damage

any given habitat must not and cannot be made on the basis of only the

value of hunting and fishing use generated by that area." The "Principles

and Standards for Planning Water and Related Land Resources" (U.S. Water

Resources Council, 1973) that became effective in October 1973 attempt

to rectify past mistakes in the use of man-use-day figures for evaluating

water resource developments.

In Mr. Lutey's justification for a nonmonetary method for evaluating

projects, he states that "obviously, project evaluations based only on re-

source use do not include all the impacts of a project on fish and wildlife and their habitat." The evaluation system presented by Mr. Lutey is based on habitat values and has several advantages over an evaluation system based solely on the dollar value of man-days of use.

1. Habitat value represents an evaluation of the total resource. A balanced, diverse habitat providing food, cover, and water produces optimum populations of wildlife. Therefore, an evaluation of the habitat is actually an evaluation of the entire wildlife resource and not solely the part used by man.

2. Habitat values provide a numerical quantification of the fish or wildlife resource based on existing conditions within the boundaries of the project area.

3. The evaluation is derived through cooperative input from representatives of construction agencies and state and federal wildlife agencies.

4. Habitat values offer a means of assessing wildlife resource values with and without the project. With the numerical values, one can show: (a) the value of the resource without the project; (b) the losses within project by project segments (conservation pool, flood pool, and above flood pool); (c) the residual value (remaining value after deduction of losses) within the project by project segments; and (d) project benefits.

5. Habitat values offer a basis for determining when mitigation has been satisfied. Habitat units lost must be replaced with an equal number of habitat units.

6. Habitat values provide an approach for linking mitigation costs

to total project expense. It is the feeling of most states that the state should not have to bear the cost of development and maintenance of mitigation lands. Lands offered in mitigation have an existing value which cannot be used to offset losses because this value exists with or without a project. Compensation, therefore, can be achieved only through increasing the wildlife carrying capacity. This requires development, maintenance and management of the lands offered for mitigation. Since mitigation is not achieved until habitat development has taken place, the cost of development and maintenance should be included in the costs of the project.

7. The system also allows for recognition of project benefits. Project benefits are expressed in the same units of measure as project losses. The difference between habitat units of project benefits and project loss equals the mitigation needs.

Much of the criticism of the habitat evaluation system presented by Mr. Lutey has been aimed at the conversion of subjective values to a numerical index. As indicated in the last paragraph of Mr. Lutey's paper, there exists a need for "key criteria" handbooks. This author is directly involved with a research project financed by the U.S. Fish and Wildlife Service, and conducted by the Wildlife Unit of Missouri University, to develop a prototype handbook that can be adapted nationwide. I believe this handbook will remove some of the current criticism of HEP.

Success or failure of our nation to judge fairly and display accurately the effects of water resource developments on our natural resources will depend greatly on the acceptance of the professionalism of those involved in evaluations. The wildlife problems we now face, and will continue to face in the future, are too important for those of us who are involved to become discouraged by criticisms of our approach.

References

U.S. Congress, Senate. 1962. <u>Policies, Standards, and Procedures in the Formulation, Evaluation, and Review of Plans for Use and Development of Water and Related Land Resources</u>, Senate Document 97 (Washington, D.C.).

U.S. Water Resources Council. 1973. "Principles and Standards for Planning Water and Related Land Resources" (Washington, D.C., October).

PART IV. POLITICAL AND LEGAL INSTITUTIONS AFFECTING
THE USES OF THE LANDS AND WATERS IN THE
UPPER COLORADO RIVER BASIN

Chapter 16

INSTITUTIONAL ASPECTS OF WATER ALLOCATION
IN THE UPPER COLORADO RIVER BASIN:
IMPLICATIONS FOR FISH AND WILDLIFE

William B. Lord*

Introduction

Great changes will occur in water use in the Upper Colorado River Basin in the years immediately ahead. Energy development will surely be a major stimulus to these changes, and fish and wildlife resources will surely feel their impact. This Forum arises from a concern, in the U.S. Fish and Wild-life Service and elsewhere, that this rapidly changing situation be better understood and, by implication at least, more effectively controlled. The objective of that control is, at one level, to minimize the adverse effects which fish and wildlife resources may suffer as development proceeds. At another level, the objective is to achieve a socially acceptable (optimal may be too presumptuous a goal) pattern of resource use, in which the social gains outweigh the social costs of change.

The system which we are trying to understand and control is a complex one. Most of the contributions to this Forum address its biological and technological aspects. Clearly these must be understood, and in many re-spects better than they are currently understood, before efforts at control can achieve their objectives. But the social or institutional aspects of the system of western water use are the most strategic parts of the system from the point of view of controlling its behavior. It is only through

*Research Associate, Institute of Behavioral Science, University of Colorado, Boulder, Colorado.

human intervention that control is exercised, and that intervention takes place only through social institutions.

The definition of "institutions" employed throughout this discussion is a broad one, namely that an institution is a customary or generally accepted way of doing things. Thus, the U.S. Fish and Wildlife Service, an institution as narrowly defined, is an organization composed of and embedded in a context of many, many institutions from the perspective of my definition. Western water law is an expression standing for one or a set of several closely related institutions. The federal reclamation program is another set of closely related institutions, within which operate a number of organizations, including the U.S. Bureau of Reclamation.

For discussion purposes, I have separated water law from the political and administrative aspects, even though the separation is essentially dictated by our own disciplinary organization and not by boundaries evident in the social system. I have omitted any discussion of water quality considerations, partly because existing federal law states that there will be no water quality problem shortly, but perhaps more realistically because I could not add much to what has been said earlier in that respect (Lord and coauthors, 1975).

This paper is not intended to be a comprehensive discussion of social science research needs in the subject area, although it deals only with the social sciences. Economics, in particular, receives short shrift. Partly this is because economic matters are to be treated in other papers, partly it is because economic research is well developed, widely understood, fairly widely accepted, and in general able to make its own case. Mostly,

however, it is because economics, while studying the outcome of the functioning of social institutions, largely eschews the study of those institutions themselves.

The discussion, then, begins with a brief description of water law and
its implications for fish and wildlife in the Upper Colorado Basin, turns
next to a description of political, administrative, and other social institutions for water allocation, and concludes with some suggestions for future research in these areas.

Western Water Law

Water allocation was not a major problem when the constitutions of the
older eastern states were framed. In the absence of constitutional provision or subsequent legislation, the water law of the eastern states was
made by the courts and was based upon the riparian doctrine of the English
common law. Under this doctrine, rights to use surface waters inhere in the
owners of land immediately adjacent to those waters. Riparian water rights
are further limited to such use as will not diminish the rights of other
riparians.

The riparian doctrine served tolerably well wherever water supplies
were abundant and water usage was basically nonconsumptive. But by the time
statehood came to the arid western territories, it was clear that riparian
doctrine could not serve under the water-scarce conditions prevailing
there. Consequently, the laws of the western states provided for the
allocation of water according to the doctrine of prior appropriation. As
compared with the ambiguity inherent in riparian doctrine, appropriation
doctrine provides, in principle, a clear answer to whose claim on available
water supplies shall be honored, whatever might be the locational and

seasonal state of water availability. A clear answer to this question was a precondition to the economic development of the West, based as it was on consumptive water use.

Appropriation doctrine has two basic elements. The first is seniority--the embodiment of the notion of "first come, first served," or "first in time, first in right." In other words, the date of establishment of a particular water right is crucial, for scarce water supplies go to satisfy completely the entitlements of those with the rights of longest duration. If there is not enough to meet all claims, those whose claims are most junior go dry.

The second basic element of appropriation doctrine is beneficial use-- the notion that a water right cannot be acquired by claim alone, but only through making productive use of water. The beneficial use notion is clearly both reasonable and necessary, but it is also the source of difficulties. State constitutions commonly enumerate and even order beneficial uses. In Colorado, for example, beneficial uses are defined constitutionally as domestic, agricultural, and manufacturing, in that order of preference (Colorado Constitution, Article XVI, Section b). The courts have held that this order of preference is not self-executing. It must be exercised by condemnation and the payment of compensation by the preferred user-condemnor to the condemnee whose right, though not preferred, may nevertheless be senior in priority (Dewsnut and coauthors, 1973).

Another important aspect of the beneficial use notion in appropriation doctrine is the widespread linking of beneficial use and diversion. Historically, water rights were established when water was diverted from its natural course and put to a beneficial use. Constitutional and statutory language reflected this tradition, and continues to do so in many western states. Water uses which do not involve diversion encounter great difficulty in qualifying as bases for the acquisition of water rights, although recent legislative and judicial decisions have relaxed this requirement.

Nowhere may one find appropriation doctrine in its pure form governing water allocation in a western state. The California situation is the most complex, embodying a mixture of appropriation doctrine, riparian doctrine, pueblo doctrine (from the Spanish laws in effect before the Treaty of Guadalupe Hildago), and a permit system (Ostrom, 1971).

In all but one of the states of the Upper Colorado River Basin the doctrine of prior appropriation is implemented through use of a permit system. Colorado, the earliest and "purest" appropriation state, does not employ the permit system, although well-drilling permits are required there. A permit system provides that water rights are acquired through application to a state water agency. The conditions which must be met in order to qualify for a permit may or may not be identical to those necessary for the acquisition of a water right by appropriation (diversion, beneficial use, existence of unallocated water, etc.). However, a records system is established and water conflicts are greatly reduced. Priority of use, or acquisition of use permit, still rules.

The basic thrust of riparian doctrine was to protect the public interest; to safeguard the quantity and quality of surface waters, with the reasons for doing so left largely unstated. The contrasting thrust of appropriation doctrine was to facilitate and encourage full beneficial use of surface waters, with beneficial use specifically defined to consist of

economically-oriented consumptive uses. In the East, riparian doctrine
has been modified progressively to recognize and tolerate consumptive water
uses. In the West, appropriation doctrine has been modified progressively
to recognize and protect nonconsumptive water use.

Maintenance of natural ecosystems, including their fish and wildlife
components, was neither contemplated nor provided for in the early development of western water law. In fact, it would scarcely be possible to de-
liberately design a legal system more inimical to such values than is pure
appropriation doctrine. However, modifications of that doctrine over the
past century, and particularly during the past decade, have ameliorated
its adverse effects upon fish and wildlife. The practical effect of that
amelioration has been modest to date. Its prospective effect could be
much greater. The extent to which this actually occurs will depend upon
the climate of public opinion and values within which judicial decision
and legislative action take place.

The first major breakthrough in the modification of appropriation
doctrine occurred with the adoption of permit systems. Each of the states
of Arizona, New Mexico, Utah, and Wyoming directs the state official approv-
ing water use applications to consider the public interest in so doing. The
directive may be broad and nonspecific, as in Arizona, New Mexico, and
Wyoming, or it may specify that water use detrimental to the natural stream
environment may constitute grounds for denying an application (Utah Code Ann.
sec. 73-2-8). In these four permit states, the way seems open to a gradual
evolution of the public interest concept toward increasing recognition of
fish and wildlife values, through court decision if not through legislative
enactment.

The situation in Colorado is more difficult. There exist no permit system and no public interest criterion for limiting new appropriations. In fact, the Colorado Constitution states that "the right to divert the unappropriated waters of any natural stream to beneficial uses shall never be denied" (Colorado Constitution, Article XVI, Section 9). Barring constitutional amendment, it would appear that water can be assured for fish and wildlife purposes only through establishing an appropriation (a valid water right). But this conclusion only leads to other difficulties. The Colorado Supreme Court in 1965 held that:

> The first essential of an appropriation is the actual diversion of the water.... Water can actually be diverted only by taking it from the stream.... There is no support in the law of this state for the proposition that a minimum flow of water may be "appropriated" in a natural stream for piscatorial purposes without diversion of any portion of the water "appropriated" from the natural course of the stream. (158 Colorado 331-35)

The Colorado legislature in 1973 enacted a statute declaring (1) that fish and wildlife purposes are beneficial uses, (2) that the state may appropriate minimum flows between specified points in order to preserve the natural environment to a reasonable degree, and (3) that controlling water in its natural course constitutes a diversion (Colorado Revised Statutes 37-92-103). It remains to be seen how this 1973 law will fare when subjected to constitutional test. The layman, not possessing the gift of legal ingenuity in interpreting our common language, may be excused for wondering how a diversion can be accomplished through explicitly forbidding diversions. He may wonder how additional beneficial uses can be proclaimed by statute when the constitution is specific and limiting in defining beneficial use. All that is sure is that Colorado, like the other basin states, will strengthen its protection of fish and wildlife values by limiting the rights of private water users in the years ahead. Whether this will be accomplished by amending an outdated constitution, by enacting new laws which blaze new etymological paths, or by calling upon its courts to mediate between precedent and current perceptions of social need and public interest cannot be foreseen. Most likely, all three approaches will be employed.

All five basin states are moving to recognize fish and wildlife values and their stake in the process of allocating as yet unappropriated water. But for many basin ecosystems this is too little and too late. Some of these systems are already altered beyond the point of no return. Others are not, but they are threatened by the prospect of continued and more extensive exploitation of existing water rights. The obvious solution, yet the one which few wish to mention, let alone propose, is the condemnation of existing water rights in order to protect and even to restore natural environments. We are well accustomed to condemning private land for public purposes, mainly in order to achieve orderly economic growth, but also to accomplish environmental purposes. We are accustomed as well to the condemnation of private water rights in order to serve such preferred uses as domestic water supply, even when the need is not immediate but simply based upon the projection or aspiration of future population growth. But we too readily assume that the environmental concerns of an increasingly urban population must take a back seat to traditional attitudes toward water allocation. The beating recently taken by the Denver Water Board when it opposed congressional creation of an enlarged Eagles Nest Wilderness Area suggests that public opinion is shifting and that the old taboos which assert the sacred nature of water rights are no longer beyond challenge.

No discussion of western water law would be complete without mention of federal reserved rights. A long succession of judicial decisions has established the doctrine that when parts of the public domain were withdrawn from entry and reserved for public purposes a water right was established implicitly at that same time. The right is to as much of the water originating on or flowing across the reserved land as may be necessary to accomplish the purpose for which the reservation of land was made.

A large share of western water supplies originates on or crosses federal reserved lands (national forests, national parks, wildlife refuges, and many other categories), and the dates of reservation of these lands are generally early enough to make them senior to most private water rights. Furthermore, since federal reserved rights are not limited by the quantity of water actually diverted and put to beneficial use at some past time, as are private water rights, they are open-ended and presently unquantified.

Indian reserved water rights are similar to the water rights appurtenant to other reserved lands, except that they are even more secure, since Congress may neither limit nor disclaim Indian rights.

The import of the reserved rights doctrine for fish and wildlife resources would seem to be three-fold. First, eventual quantification of reserved rights will strongly shape water use, and thus development in general, in the West. The resulting patterns of water use and development will largely determine the character of the environment within which fish and wildlife will exist. Second, whatever may be the nature of other water use, reserved rights will dictate the quantity of water presently unappropriated, and thus available for appropriation or reservation under state water laws for preserving minimum streamflows and otherwise directly safeguarding fish and wildlife resources. Finally, it is possible that federal reserved rights may be asserted and recognized for the purpose of protecting fish and wildlife resources in the first instance. Such reserved rights would be pretty much beyond dispute in the case of lands explicitly reserved for fish and wildlife protection. Whether they might be implicit in the case of the national forests, for example, is another question. The issue may not be a major one with respect to fish and wildlife resources on the federal lands in question. However, the eastern national forests were acquired under the authority of the Weeks Law of 1911, and that law was predicated upon the effect of forests in maintaining (minimum) streamflows (downstream), which was in turn held to be a matter of federal concern under the Commerce Clause of the U.S. Constitution.[1] Similar reasoning could be extended to reserved national forests in the West, with the result being the assertion of a federal reservation of minimum streamflows below such forests.

Political and Administrative Aspects

Political and administrative behavior powerfully affects water allocation at international, national, state, and local levels of government. In the Upper Colorado River Basin, the force of appropriation doctrine is especially strong in dictating the limits within which such political and administrative behavior must operate. But within these limits, the history of water use decisions in the first half of this century has constituted a

---

[1] The Commerce Clause (clause 3, section 8) in the first article of the U.S. Constitution assigns Congress the power to regulate interstate commerce.

remarkably faithful example of what Lowi has called distributive politics (Lowi, 1972).

In distributive politics, a congeries of local interests puts together a "Christmas tree" package of projects from which each stands to gain and none stands to lose, arranges that the lion's share of the costs of the package will fall upon some larger entity (most usually the federal treasury), and agrees to stand together in support of the package as an inviolable whole. The package must be large enough to generate sufficient political support to ensure its adoption. It must generate overwhelming local support for each of its constituent projects (Ingram, 1971). Any consideration of broad national objectives, whether for the program area itself (in this case, water policy), or for related areas (such as environmental quality), or for economic efficiency or equity, constitutes a potential threat and is to be avoided if possible or obfuscated, at a minimum. The discerning reader will note that pure appropriation doctrine is itself remarkably congruent with the distributive mode of political decision making.

The basic instrument of distributive water politics in the Colorado River Basin has been the federal reclamation program. It became operative in 1902 in response to the growing capital requirements for making irrigation water available under then-existing technology, to the perception of western states that the national concern with developing the West could be turned into a strong rationale for distributive politics, and to a national strain of Jeffersonian agricultural fundamentalism (Caulfield, 1975). Over the intervening decades, the federal taxpayer has subsidized a complex program of western water development which has shifted the production of some crops from the humid East or South to the arid West; has created a system of storage reservoirs and associated water delivery systems whose effect upon western ecosystems, both aquatic and terrestrial, has been profound (profoundly good to some, profoundly bad to others); has established public energy production as a reality which it could not become east of the Mississippi, save in the Tennessee Valley; and which, with increased fishing pressure and stocking of nonnative species, has led to the radical changes in the composition of Upper Colorado fisheries described elsewhere in this volume.

Social scientists have described and analyzed the process through which the waters of the Colorado River were divided up, first between the upper and lower basin states in the 1922 Colorado River Compact and later among the upper basin states in the Upper Colorado River Basin Compact of 1948 (Mann and coauthors, 1974). In fact, if all of the claimants were to fully utilize their allotted shares, in nine years out of ten there would be no water left for fish and wildlife, for our Mexican neighbors, or even for fully meeting the claimants' entitlements themselves. The latest chapter in this saga has been marked by federal assumption of the responsibility for alleviating the adverse water quantity and water quality effects upon Mexico produced by the usage which the basin states have made and will make of the waters of the Colorado (Mann and coauthors, 1975).

Attempts have been made to restrain or to redirect, or both, the classically distributive U.S. water resource programs, including that for reclamation. A few Presidents (albeit half-heartedly), economists both within and outside the government, a handful of congressmen with high principles buttressed by little stake in water programs, many militant conservation organizations, and a number of water planners within the water agencies themselves have been responsible for benefit-cost analyses, the National Environmental Policy Act (NEPA), the new Principles and Standards of the Water Resources Council, and specific changes such as the deletion of Echo Park Dam from the Colorado River Storage Project Act of 1965.

Observers differ in the extent to which they see change occurring in the classical pattern of distributive politics in western water development. After analyzing the decision rules for distributive water politics, one observer finds significant change in the operation of each, all in the direction of making water resource development in the classical pattern less feasible (Ingram, 1972). Another observer also notes changes, mostly in the way of introducing some elements of regulatory politics (marked by bargaining and compromise, concern with policy issues, greater openness, and broadened decision making), but concludes that the politics of western water development remains predominantly distributive and is likely to continue in this pattern (Mann, 1974). A third observer notes the complicated mixture of both modes in practice, and laments the modest progress of attempts to reinforce regulatory politics (attempts in which he personally has played a major role) as opposed to distributive politics in federal water programs (Caulfield, 1975). One thing is clear; social scientists differ greatly on many water resource issues but they stand united in condemning the distributive political process which has so powerfully shaped western water allocation (Fox, 1965).

Earlier writers were sharply critical of the federal water agencies (Maass, 1951; Hart, 1957; Marshall, 1965), although the Congress and its committees shared this criticism. Another, not altogether separate, strain of criticism has been directed at the engineering profession, with its alleged "construction ethos." But this criticism fails to account for the fact that the federal and state fish and wildlife agencies, whose employees are biologists, not engineers, also have joined the game of distributive politics. They may not have played as successfully or as enthusiastically as the water agencies (would greater success have engendered greater enthusiasm?), but their repeated acquiescence to environmentally damaging water projects for the price of a fish hatchery here or a wildlife refuge there, while called mitigation, is classical distributive politics (Nienaber and coauthors, 1976).

It is always popular to blame objectionable political decisions on the powerful vested interests, and indeed local business interests have been intimately involved in distributive water politics. But the pervasive role of interest groups in American political institutions has been long evident, and the positive aspects of that role have been emphasized as well (Bentley, 1935; Truman, 1951). One must ask whether environmental interest groups have opposed the conduct of distributive politics. The answer seems to be that by and large they have not. Instead, they, too, have joined the game, as so well illustrated in the conservation lobby's acquiescence to the Colorado River Storage Compact in 1955, for the price of deletion of Echo Park Dam (Mann and coauthors, 1974).

Lowi (1972), in addition to elucidating the paradigms for distributive, regulatory, and other political modes, asserts that the particular political mode employed in a specific policy context will be determined by the perceived nature of that policy. This point of view seems promising, for it directs attention to all of the mutually reinforcing elements in the policy context rather than finding fault with a particular agency or group. While it is true that politicians like to be reelected, that bureaucrats like to manage growing agencies, that engineers like to build things, and that local businessmen prefer to do more business rather than less, the behaviors which follow from these ubiquitous motivations differ immensely within the context in which they occur. It seems likely that western water politics will remain predominantly distributive for just so long as the public perceives water issues to be basically local ones and for so long as water planners can successfully design non-zero-sum alternatives which offer more benefits than costs to all important local interests, whatever may be the burden on the federal treasury.

The decision-making process for federal water resource development is composed of three stages. The first stage, from the initial conception of a project up to the point of authorization by the Congress and the President, may be called planning. From a technical point of view, this is the period during which the nature of the project is refined and its technical feasibility determined. From an economic point of view, it is the period during which the project's potential benefits and costs are examined and its economic desirability determined. From a political point of view, it is the period during which solid local support is generated, and acquiescence of federal and state agencies is achieved. It is during this time that the potential project must clear the many hurdles which are required for it to become accepted currency for the distributive political process within the Congress. Academic research into the decision-making process has focused mostly on this first stage, and it is reasonably well understood and described, insofar as the behavior of bureaucrats and politicians is concerned. The evolution of the public attitudes and beliefs which support and condition the behavior of bureaucrats and politicians has not been much investigated and is not well understood.

One aspect of the planning stage has received much research attention. This is the matter of project evaluation. Economists, in particular, have been contending with water agency planners and with each other for a quarter century over the proper conduct of benefit-cost analysis and the desirability of such analysis as compared with broader multiple objective approaches. Most of the substantial literature in this area is outside the scope of this discussion because it avoids institutional considerations almost completely. However, evaluation procedures are used within an institutional setting, and their use becomes an important aspect of that institutional setting.

A favorable benefit-cost ratio has been almost a precondition for project authorization. In a regulatory political decision process there would be great interest in seeing that the analysis was performed correctly because the goal of the process would be to select those projects which were economically most desirable (or most desirable by such other criteria as might be included in a multiple objective evaluation scheme). In a fundamentally distributive mode, however, the goal is to provide a rationale to support and justify a project chosen on completely different grounds, grounds which cannot themselves be stated explicitly. From this point of view, the interesting and important research questions are not how to improve or perfect evaluation techniques, or how to enforce the proper use of existing evaluation techniques. Instead, they are concerned with how to move from distributive to regulatory politics and how to employ information, such as that provided by evaluation procedures, to accelerate that movement.

The second stage of the federal water resource decision process begins with project authorization and ends with completion of the project. This is the appropriations process, and it is the heart of distributive politics. Each water resource agency has a backlog of authorized but unfunded projects, each of which has successfully met the several tests embodied in the planning process. If and when such a project becomes reality now depends upon the political clout possessed by its congressional sponsor and his ability to build coalitions with his colleagues in support of a package of "new starts" (many authorized projects are never constructed). Chairmen of the appropriations committees and subcommittees of the Congress are key figures in this process and derive great power from their ability to include or exclude specific projects. The appropriation process has been studied by political scientists and is reasonably well understood by them, although other social scientists are inclined to overlook its importance, perhaps because there is so little opportunity for their participation at this stage, as compared with the more technical and more open planning process.

One aspect of this stage of the decision process deserves greater research attention. As soon as a project receives initial funding, it moves into the engineering design, or preconstruction planning, phase. It is common for projects to change substantially in character, scale, and cost during this phase. The project which is finally constructed may be quite different from that which met the various tests leading up to congressional authorization. The courts have repeatedly upheld the authority of the water agencies to make these post-authorization changes, even though the character

of the project may be changed substantially, without returning to the
Congress for a new authorization (customary practice is to clear such
changes with the appropriate congressional committee, however). Post-
authorization changes result from many considerations, some of them the result of engineering studies, some the result of changing economic conditions,
and many the result of a bargaining process between water planners and local
interests over cost sharing and other political considerations. Given the
magnitude of these changes, a more intensive study of this process than
has yet occurred seems appropriate. The informal and covert nature of the
process, and the limited number of participants (local beneficiaries, water
agency personnel, and congressional committee chairmen), makes it a congen-
ial process for distributive politics and a hostile one for regulatory
politics.

Much of the same can be said for the third stage of the federal water
development decision process. This is the period of project operation,
from its completion to the end of its useful life. Very little research
attention has been focused on this stage. The level of congressional interest drops greatly once the project is completed, and many of its opponents accept the fait accompli, moving on to other conflicts where they
may be more successful. Yet major decisions continue to be made, for example, in the execution of contracts for water deliveries and the determination of water release schedules, and such decisions can have profound implications for fish and wildlife resources. There are no statutory requirements for inter-agency review and other tools of regulatory politics, except for the section 102 environmental impact statement requirements of
NEPA. It may be that substantial opportunities for influencing fish and

wildlife impacts of federal water resource developments exist and can be rather easily grasped at this stage of the decision process, although our knowledge is so limited that such a judgment cannot be made with confidence.

Lowi (1972) has suggested that the type of decision making employed with respect to a particular policy issue will be primarily a function of the perceived nature of that issue. I would emphasize the word "perceived," for it is only through the screen of human perceptions that institutional behavior is shaped. This much seems obvious, but our knowledge of how this shaping occurs or how perceptions are formed and altered is rudimentary at best.

In the area of water allocation and environmental protection there has been considerable research in human attitudes. We know that the historically predominant attitude towards public action in western water resource allocation has been that such action is undertaken to remove the water-related constraints to private economic development. This is the essence of appropriation doctrine and is stated repeatedly by legislators and others whose views are recorded in newspapers and other media. The existence of a different attitude, that public action in water resource allocation should be undertaken to maximize the economic well-being of society as a whole, is obvious in the writings and public statements of economists and some others. A third attitude has also been apparent in some quarters, that public water resource development can be destructive of environmental values and should not be undertaken in those cases. But each of these attitudes is typical of only a small number of people, those whose societal roles make their attitudes a matter of public record. Until recently, we have had almost no information on wider public attitudes, beliefs, and values specific to this area of concern.

Over the past decade this situation has changed. Sociologists and other social scientists have conducted a number of studies, using survey research techniques, which have revealed public attitudes towards water resource development or environmental protection, or both. Substantial descriptive material now exists, and some studies have been undertaken to correlate attitudes with demographic, economic, and other social variables. However, very little has been done to explore attitude change or the relationship of attitude to behavior in the area of public participation in decision-making institutions. (This latter topic is most difficult, of course, for it will involve interdisciplinary research by social psychologists and political sociologists.) Until such research is undertaken our ability to reshape the distributive water resource decision-making process will be sharply limited.

Research Needs

The general area of water resource institutions has received proportionally more attention from social scientists than most other areas of public policy (Kneese, 1965). That attention has been highly selective, however. There has been much economic research on evaluation techniques and upon economic impacts of water development. There has been substantial political science research on decision-making institutions, particularly at the congressional and agency levels. There has been a fair amount of social science research in various disciplines on the planning process, and upon public involvement within it. There has been some sociological research on perceptions, opinions, attitudes, and beliefs. Most of this research has produced useful results, but it has been fragmented and specialized, and little of it has concerned the fish and wildlife aspects of water allocation.

Ten years ago, Smith called for more integration of social science research in water resources (Smith, 1965). He noted the impressive research output on economic evaluation, together with the difficulty of applying it in decision making because it is largely unrelated to the institutional context within which decisions are made. The result was satisfactory to neither economists nor decision makers--perceived misuse of technical procedures on the one hand and perceived obstruction of established decision-making procedures on the other (Smith, 1965).

Many things have happened since Smith described this problem. Much additional research has further improved benefit-cost analysis in principle, including recent progress in deriving quasi-market values for incommensurables such as fish and wildlife values, and incorporating irreversibility as an explicit and quantitative consideration in the analysis. (See Krutilla, 1972, for several notable examples.) On the decision-making side, we have seen major development in handling incommensurables in a totally different way, as embodied in the NEPA environmental impact statements and the multiple objectives and accounts of the Water Resources Council's Procedures and Standards. For all of its problems, the multi-objective approach seems more likely to focus attention on issues and facilitate bargaining between competing groups and objectives than does the benefit-cost notion of finding the analytically proper social welfare optimum. It is exactly such things as issue orientation and meaningful bargaining which characterize the regulatory mode of political decision making as opposed to the distributive mode. Of course, we do not know that evaluation procedures, however well designed and executed, will be effective stimuli to a shift toward a more regulatory mode, but research into the effects of NEPA and the Principles and Standards seems appropriate to shed light on this important question.

Considerations such as these caused Smith to emphasize the need for interdisciplinary research built upon the idea of understanding the behavior of the social decision-making system for water resources allocation. During the intervening decade we have created many more of the building blocks for such a conceptual structure, but we have seen only modest progress in putting those blocks together to form the structure which is so badly needed. Research of the kind needed to do this is difficult and risky in a research community and funding structure which rewards specialization through its organization into disciplines, departments, and journals. Nor is the record of interdisciplinary research such as to inspire confidence in a successful outcome. Nonetheless, the potential payoff is great enough, compared to the risks involved, to make this type of research the highest priority, in my estimation.

Of course, many of the pieces which must contribute to a model (loosely defined) of the social decision-making system for water resources exist only in tentative and imperfect form as well. Perhaps the most important of these is the formation and change of values, attitudes, and beliefs concerning water resource allocation and its environmental consequences, together with the effects of such social variables upon forms of participation in decision making. Our knowledge is scanty in this area, and a better

understanting is essential to the redesign of decision-making institutions.

Research techniques exist, from prior work in other areas, to permit a more

vigorous research effort, and high priority should be given to doing so.

Determinants of human behavior deserve more research attention, not

only with respect to participation in public water resource decision making

but also with respect to a variety of individual and group behaviors which

affect water use and, consequently, the environments of fish and wildlife.

Some work is now under way on factors affecting the adoption of energy-

saving behavior and technology in the residential and transportation areas.

Little is being done concerning water-saving behaviors, yet there would

seem to be great opportunities for increasing water availability at little

cost to fish and wildlife resources through so doing. Economists have long

studied the use of economic incentives to influence behavior, and sociolo-

gists have more recently investigated a range of factors which influence

the adoption of new agricultural technology. Parallel studies in the area

of water use might pay substantial dividends.

A number of more specific areas of high priority also may be identified.

Each grows out of an existing practical problem and is directed to the so-

lution of that problem. None depends upon achieving a better understanding

of the entire social decision-making system, and successful research and

problem resolution in these areas may not do much to change that system

fundamentally. They are of the nature of short-run modifications to make

existing institutions function somewhat more effectively.

The traditional distributive mode of water politics has been characterized by the avoidance of conflict. Such progress as has been made in shifting to a more regulatory mode has produced increasing levels of conflict.
This has been viewed with alarm by those accustomed to the rules of distributive politics. However, conflict need not be seen in a completely, or
even predominantly, negative light. It is both a necessary and useful characteristic of regulatory politics. Nonetheless, too much time and energy
aan be absorbed in conflict, and conflict resolution techniques should be
established features of water resource decision-making systems. Research
is indicated to explore the extension of conflict resolution techniques developed elsewhere to the water resources sphere, where they have been un-
needed and unused in the past. Such research may require the classification
of water resource and other conflict situations in such a way as to indicate the type of conflict resolution technique likely to be most effective.
Interdisciplinary research by psychologists, sociologists, political scientists, and possibly other social scientists would seem necessary to achieve
sufficient breadth of concept and utilization of the full range of conflict
resolution techniques.

One of the major problems in making water allocation decisions more
sensitive to fish and wildlife impacts is our lack of ability to predict
those impacts. This inability in turn is due in part to gaps in scientific
knowledge, but it is also due to a lack of systematic information about fish
and wildlife resources and the natural environments upon which they depend.
Inventory information of this sort is available for most kinds of inani-
mate resources, e.g., topographic, geologic, and hydrologic maps and inven-
tories, and for human and cultural resources, e.g., census reports. It is

not available comprehensively in any degree of detail for biological resources. In itself, of course, this is not an institutional problem, and
the purpose here is not to suggest that additional biological research,
data gathering, or information dissemination is called for. That is taken
for granted. The institutional problem, rather, is why this situation
should exist, after the U.S. Fish and Wildlife Service and its precursor
agencies have been in operation for so many years. What institutional
changes are called for to ensure that the information base exists to permit the consideration of fish and wildlife values in water resource decision making in the event that a commitment develops to do so?

One area of current institutional innovation is the modification of
appropriation-based water law to protect some remaining fish and wildlife
resources. Most of this effort has been directed to maintaining minimum
streamflows, certainly the most critical problem for aquatic ecosystems.
Several western states have recently enacted laws permitting state authorities to reserve or establish rights to minimum flows, in the interest
of safeguarding instream uses which include fishery maintenance. Approaches taken differ somewhat among states. A totally different approach is to employ the public interest criterion in deciding upon whether
new water use permits should be issued, although by precedent, the public
interest is usually given an economic rather than an environmental interpretation.

There has now been sufficient experience in the application of minimum
streamflow preservation institutions to justify a study and evaluation of
that experience. The results would be useful to states which do not now

possess such institutions but which may shortly adopt them. They may also
suggest revisions in existing institutions and may even offer some insights
into the design of institutions going beyond the protection of minimum
streamflows alone.

References

Bentley, Arthur F. 1935. The Process of Government: A Study of Social Pressures (Evanston, Ill., Principia Press of Illinois).

Caulfield, Henry P., Jr. 1975, "The Politics of Multiple Objective Planning." Paper delivered at the Multiple Objective Planning and Decision-Making Conference of the Water Resources Research Institute (Boise, Idaho, University of Idaho).

Dewsnut, Richard L., Dallin W. Jensen, and Robert W. Swenson, eds. 1973. A Summary - Digest of State Water Laws (Arlington, Va., National Water Commission).

Fox, Irving K. 1965. "Policy Problems in the Field of Water Resources," in Allen V. Kneese and Stephen C. Smith, eds., Water Research (Baltimore, Johns Hopkins University Press).

Hart, Henry C. 1957. The Dark Missouri (Madison, Wis., University of Wisconsin).

Ingram, Helen. 1971. "Patterns of Politics in Water Resources Development," Natural Resources Journal vol. 11, No. 1, pp. 102-118.

_____. 1972. "The Changing Decision Rules in the Politics of Water Development," Water Resources Bulletin vol. 8, pp. 1177-1188.

Kneese, Allen V. 1965. "Introduction: New Directions in Water Resources Research," in Allen V. Kneese and Stephen C. Smith, eds., Water Research (Baltimore, Johns Hopkins University Press).

Krutilla, John V. 1972. Natural Environments: Studies in Theoretical and Applied Analysis (Baltimore, Johns Hopkins University Press).

Lord, Wm. B., Susan K. Tubbesing, and Craig Althen. 1975. Fish and Wildlife Implications of Upper Missouri Basin Water Allocation (Boulder, Colo., University of Colorado Institute of Behavioral Science).

Lowi, Theodore J. 1972. "Four Systems of Policy, Politics, and Choice," Public Administration Review vol. 32, p. 298.

Maass, Arthur B. 1951. Muddy Waters: The Army Engineers and the Nation's Rivers (Cambridge, Mass., Harvard University Press).

Mann, Dean E. 1974. "Political Incentives in U.S. Water Policy: Relationships Between Distributive and Regulatory Politics," in Matthew Holden, Jr. and Dennis L. Dresang, eds., What Government Does. Vol. 1, Sage Yearbooks on Politics and Public Policy (Beverly Hills, Calif., Sage Publications).

Mann, Dean E., Gary D. Weatherford, and Phillip Nichols. 1974. Legal-Political History of Water Resource Development in the Upper Colorado River Basin. Lake Powell Research Bulletin, No. 4, National Science Foundation. Copies obtainable from: Institute of Geophysics and Planetary Physics, University of California, Los Angeles, California 90024.

_____. 1975. "Politics in the United States and the Salinity Problem of the Colorado River," Natural Resources Journal vol. 15, pp. 113-128.

Marshall, Hubert. 1965. "Politics and Efficiency in Water Development," in Allen V. Kneese and Stephen C. Smith, eds., Water Research (Baltimore, Johns Hopkins University Press).

Nienaber, Jeanne, Helen Ingram, and Daniel McCool. 1976. "'The Rich Get Richer' Phenomenon: Comparing Innovation in Six Federal Agencies." Paper presented at the 1976 Annual Meeting of the Mid-West Political Science Association, Chicago, Illinois.

Ostrom, Vincent. 1971. Institutional Arrangements for Water Resource Development (Arlington, Va., National Water Commission).

Smith, Stephen C. 1965. "Major Research Problems in the Social Sciences," in Allen V. Kneese and Stephen C. Smith, eds., Water Research (Baltimore, Johns Hopkins University Press).

Truman, David B. 1951. The Government Process: Political Interests and Public Opinion (New York, Knopf).

INSTITUTIONAL ASPECTS OF WATER ALLOCATION
IN THE UPPER COLORADO RIVER BASIN--
IMPLICATIONS FOR FISH AND WILDLIFE:
A DISCUSSION

Henry P. Caulfield, Jr.[*]

Introduction

The major objective of the Forum, which included the paper by William
Lord under discussion here, was to identify specific fish- and wildlife-
related problems and to delineate research needs. The identification of
problems and delineation of research needs was not to advance science
per se, but to apply science to a cause: the betterment of the nation's
fish and wildlife in the interests of those concerned.

Lord's paper is a good general overview of his subject, leading to
statement of research proposals which certainly are worthy of further definition and consideration. One could raise and discuss specific questions
about his analysis of western water law, political and administrative behavior, and research needs. Instead, I choose to discuss certain general
matters that I see involved in the predicament of fish and wildlife in our
society. I believe that this clarification of approach is essential to
looking more clearly at problems of fish and wildlife in the institutional
setting set forth by Lord. Hopefully, it will provide useful analytical
understanding of the process of policy change and implementation, and a
guide to needed research.

---

[*]Professor, Department of Political Science, Colorado State University,
Fort Collins, Colorado.

In what follows, I will discuss: (a) economics and politics generally; (b) Lord's application of political science concepts to natural resources environmental problems; (c) application in combination of distributive, regulative and redistributive politics; and (d) a suggested political strategy for enhancing fish and wildlife values generally and for guiding selection of applied research projects.

Economics and Politics

Both economics and politics deal with the allocation of value in a society. Neither tries to explain value preferences per se (e.g., one's like or dislike of fishing, hiking, or spinach). But both are concerned with the allocation of resources in a society in terms of value.

Economics as an academic and professional discipline is both positive (i.e., scientific) and normative. Its normative paradigm of welfare economics is taken to be a norm of what ought to occur in the allocation of resources (Pigou, 1920). Individual decisions within self-regulating markets result in an allocation of resources that is taken to be optimum. Individual decisions bringing about this good result are collectively viewed as "consumer sovereignty," which is taken to be a fundamental good (Krutilla and Eckstein, 1958).

Lord is right when he asserts that economic analysis, both positive and normative, as applied to problems of natural resources/environment is overdone compared to analysis from the perspectives of other disciplines that are involved or could become so. Without disparaging the need for research by these other disciplines (or for true, rigorous interdisciplinary research), I choose to confine my perspective here to the discipline of political science.

"Politics," for the purpose of this analysis, can best be defined as the process by which a society makes authoritative decisions about the allocation of values (Easton, 1965a). It is a "process" that is structured and occurs over time; it is "societal" in that it is a collective process in which individual decisions play a role; and it is "authoritative" in the sense that outputs of the process are generally accepted and have the force of law.

The study of politics (although now generally labeled "political science") has always had positive and normative approaches, and admixtures thereof. In the last twenty-five years or so, the positive approach calling for strict use of scientific method in studying political behavior has been emphasized. There appears to be little concensus on a normative paradigm, except in the gross sense of a preference for "democracy" over "totalitarianism." Certainly, there is little concensus as compared to that among economists.

The output of politics can be said to be "policy" plus the value-significant results of its application by decision makers responsible for implementing policy. "Policy," in the abstract, can be said to be the criteria by which a decision maker decides what to do or what not to do in a given situation (Friedrich, 1960).[1] More concretely, policy is the criteria contained in constitutions, laws and legislative history, judicial interpretations, regulations and executive orders, and so on, followed by decision makers in the implementation process.

The process by which political outputs occur can be viewed as involving public problem identification, solution formulation, solution

---

[1] Somewhat similarly, policy is defined by David Easton (1965b) as "decision rules adopted by authorities as a guide to behavior...."

legitimation, implementation, evaluation and feedback. This process once-through involves incremental policy change. It can be viewed as cyclical within a longer time frame (Jones, 1970).

The foregoing brief conceptual framework from political science can be viewed as an appropriate context (particularly the parts dealing with "formulation" and "legitimation" in both the legislative and executive branches) within which to discuss the types and processes of decision identified by Theodore Lowi (1972) and discussed by Lord:

Type of decision[2]   Basic process of decision

1. Distributive   vote trading[3]

2. Regulative   bargaining, resulting in compromise

3. Redistributive   ideological adherence

Among these three types of decision, welfare economics is redistributive.[4] Its process of decision involves ideological adherence to a normative paradigm. In effect, the private economy is taken to be the appropriate norm for public decision making: prices (including discount rate) used in calculating benefits and costs largely simulate the private economy

---

[2]Lowi's fourth type of decision, "constituent," is not discussed here because it is not immediately relevant. This writer's perception and use of Lowi's concepts does not necessarily coincide with that of Lowi. Probably there are some important, but now undefined, differences.

[3]Popularly termed within this country, the politics of "pork barrel" and "logrolling," particularly as applied to water resource development projects.

[4]Welfare economics, in essence, comes today under other academic labels: modern political economy, public choice theory and the rational comprehensive approach. Its practical manifestations in government are benefit cost analysis and the major analytical components of the former PPBS system and the current zero-based budgeting system.

and are taken to be objective expressions of value. That alternate proposed action having maximum net benefits is accepted as being most efficient, the right decision.[5]

Welfare economics and its derivative ideologies have contemporary relevance in political decision making in this country. However, they should not be viewed as having such normative power as to rightfully pre-empt the field. Welfare economics is not some "super truth" before which all else should give way.

Certain normative ecological principles are now being viewed by many environmentalists as "super truths." Their implementation is viewed as necessary to establish a sustainable relationship for man in nature for the longest possible time. The basic presuppositions of the Water Pollution Control Act Amendments of 1972 (PL 92-500) might be said largely to reflect such normative ecological principles.

The Principles and Standards of the Water Resources Council, however, accord only equal status to National Economic Development (NED) and Environmental Quality (EQ) (Federal Register, 1973). They presuppose superordinate ideological status of neither of these objectives. Trade-offs in practical terms between these two ideological positions, NED and EQ, as well as others, no doubt, will continue for some time to come.

---

[5]The author recognizes, of course, that economists have long recognized "external" benefits and costs, other market failures and their inability to value all relevant values in monetary terms (i.e., in commensurable terms). To present to the world at large the normatively "right answers," economists appear to be desperately attempting to overcome certain of those difficulties, particularly to commensurate the incommensurable.

Application of Political Science Concepts
to Natural Resources/Environmental Problems

Lord recognizes that distributive politics is a fact, but he considers it bad. He derisively refers to "congeries of local interests putting together a 'Christmas tree' package" of water resource development projects for passage through the Congress, with the "lion's share" of the costs of the package to be borne by the general taxpayer. "Any consideration of broad national objectives," he says "...constitutes a potential threat and is to be avoided...."

Lord finds western water law "remarkably congruent with the distributive mode of political decision making." He mentions the trend in western water law to enable action providing minimum low flows for fish. He fails to mention that this development fosters conflict in water use, bargaining and compromise--an example of increasing use of regulative politics.

Lord chides federal and state fish and wildlife agencies for their "repeated acquiescence to environmentally damaging water projects." The bargaining, in the process of going along with the project, to achieve mitigation of damage (or, since 1958, he could have mentioned "enhancement" features of a project) he views as essentially distributive politics. This bargaining and compromise, of course, represent regulative politics, not "vote trading."

Finally, to clinch his argument against distributive politics, Lord states: "One thing is clear; social scientists differ greatly on many water resource issues but they stand united in condemning the distributive political process...."

Lord notes, apparently approvingly, that regulative politics is making some inroads on distributive politics. By implication, in referring to my efforts when in government to foster regulative politics, he has in mind that a political decision for action within the context of multiple objective planning involves bargaining and compromise. Thus, it is a manifestation of regulative politics (Caulfield, 1975).

The great ideological support for public participation in water resource planning in recent years might also be viewed as supportive of regulatory politics. What is being sought is public conflict between groups having widely differing values. In the past, there has been well organized public participation (i.e., by local groups affiliated with the old Rivers and Harbors Congress and the National Reclamation Association), but this participation involved people who were largely like-minded and intensely interested. However, today, with wider public interests and their participation, a sufficient incentive does not seem to exist to bargain and compromise by either developmentalists or environmentalists (Caulfield, 1975, 1976a). And, if and when a bargain is struck, there is not the continuing organization of the coalition of interests involved in the bargain to see that action eventually materializes.[6]

Redistributive politics, Lowi's third decision concept, is not mentioned by Lord. Normatively, Lowi would appear to rank his three types of

---

[6]I am indebted to J. F. Mangan, Planning Officer, Regional Office, Bureau of Reclamation, Boise, Idaho, for this critically important observation.

politics, as follows (Lowi, 1972):[7]

Distributive politics	bad
Regulative politics	better
Redistributive politics	good

Normatively, Lord proceeds in the same direction. Possibly, in his effort
to get intellectually outside the discipline of economics, he is unwilling
to endorse welfare economics; but, as yet, he has not found a satisfactory
substitute paradigm applicable to water resource development and fish and
wildlife.

Application in Combination of Distributive,
Regulative, and Redistributive Politics

Normative stances are important to maintenance of individual integrity
as well as to social and political well-being. But, to the extent they in-
terfere with clear understanding of reality (which to some degree they do),
they tend to be a disservice both to the individual and to society. Political reality, I will argue here, usually involves application in combination
of distributive, regulative, and redistributive politics.

The longstanding revulsion of eastern intellectuals against water
resource development projects as pork barrel and the product of logrolling,

---

[7]My ranking is perceived as implicit in Lowi's discussion; it is not
explicit. He sees distributive politics as aimed at achieving private ends
by public means, by implication, wrong. Bargaining between different interests in regulative politics is better in terms of the public interest.
Redistributive politics he sees as approaching the public interest, because
he sees all citizens as affected by it.

and the spread of this sentiment among intellectuals and others across the country, is clear.[8] Lowi's first example of distributive politics is federal water resource development projects. He cites a case study of passage of the Rivers and Harbors Act of 1952 as his evidence (Bailey and Samuel, 1952).

On the basis of the political reality revealed by that case, Lowi's conclusion appears correct. However, if one were to refer to the ideological origins for political support of navigation projects in Henry Clay's proposed "American system" for national economic development (i.e., protective tariffs for eastern manufacturers and financing of federal navigation projects east of the Mississippi River), one can see that redistributive politics has been involved too in political decisions on federal water development projects. This political position was later adopted in the founding platform (1856) of the Republican Party (Porter and Johnson, 1970). It was refortified by ideology of the Conservation Movement led by Theodore Roosevelt and Gifford Pinchot. And further analysis would reveal the major role that redistributive politics has played in development of policy over many years (e.g., benefit-cost analysis) that has constrained the acceptability of the projects which find their way by distributive politics into annual authorizing or appropriation acts, such as the Rivers and Harbors Act of 1952.

Lowi's perspective on the application of his three concepts leads him to see each act of the Congress as largely determined by only one of his types of politics. On the other hand, my research as well as my experience in the

---

[8] I seem to recall from research some twenty years ago that Albert Bushnell Hart, professor of history at Harvard University, in the latter 19th century was the writer who first utilized the words "pork barrel" and "logrolling" to express antipathy for navigation projects of the Army Corps of Engineers.

processes of policy formulation and legitimation have led me to the conclusion that all three types of politics are usually, if not always, involved
in the policy-making process.[9]

The Wild and Scenic River Act of 1968 is a good case in point. The
idea of a federally authorized and administered wild river stems, as far as
I am aware, from the report of the Comprehensive Inter-Agency Study of the
Arkansas-White-Red River Basins, Inter-Agency River Basins Committee (1955)
and inclusion of this proposal was largely due to the professional leadership of Irving Fox supported by planners in the National Park Service.
However, the genesis of the idea of wild rivers, per se, apparently stems
from the writings of the revered environmentalist, Aldo Leopold. By the
late 1950s, the ideas of the Arkansas-White-Red River Basins study mater-
ialized within the National Park Service as the Ozark National Rivers bill.
Efforts to gain political support for it by the National Park Service were
underway when the Kennedy Administration began.

Early in his tenure, Secretary of the Interior Udall, reflecting upon
the Ozark bill and his experience as a congressman, generated the idea that
what was needed was a bill authorizing a "national wild river system." The
bill should provide, as he envisioned it, that certain rivers would be
"instantly" authorized and others would be authorized for intensive study.
A joint Forest Service-Bureau of Outdoor Recreation study was then undertaken to review rivers, or river segments, that might qualify under certain

---

[9]The author was assistant director and then director, Resources Program
Staff, Office of the Secretary, Department of the Interior, February 1961 to
April 1966; and executive director, U.S. Water Resources Council, 1966-1969.
He was closely associated with Secretary of the Interior Stewart L. Udall when
he provided the high level political leadership that initiated the environmental
movement in the 1960s that continues today.

criteria as worthy of being national wild rivers. The list of worthy candidates, well spread out geographically, served as a basis for initial recommendation and later adjustment of those which should be authorized by Congress
instantly and which should be authorized for further study. The conscious
political strategy throughout the formulation and legitimation process of
what came to be the Wild and Scenic River Act of 1968 was to include only
political "sure things" in the instant authorization list and less sure, but
politically viable for study, rivers in the study list. During the process
of majority building, rivers shifted from the former to the latter category
and were added to and dropped from the bill, depending upon the political
attitudes of the senators and congressmen most concerned. The act as passed
contained eight "instant rivers" and twenty-seven "study rivers."

Clearly, distributive politics was important to legislative success.
But regulative politics was involved, too. The idea of only "wild rivers"
had to be compromised so that "scenic rivers" and "recreational rivers"
(i.e., rivers touched in varying degrees by the hand of man) could also be
included in the national system. The original idea of administration only
by the National Park Service or the Forest Service had to be compromised to
permit state administration within the national system. And, for certain
important legislators, the most critical compromise of all was this: that
the act require that reports to Congress, recommending future rivers or segments for inclusion in the system, should inform the Congress, in effect, of
the developmental benefits that would be foregone. The Congress could then
weigh the two alternatives.

Lastly, and maybe most importantly, redistributive politics was involved.
By 1968 the environmental movement, to which Secretary Udall provided

ideological leadership (especially as regards aesthetic and ethical concerns
and the beginning of normative ecological ideas) was well advanced. In 1963
he published a widely-read book, The Quiet Crisis, to promote ideological
adherence (Udall, 1963). By 1968 the movement had already been highly productive of much environmental legislation (Caulfield, 1976b).

Somewhat similar analyses, using all three of Lowi's concepts in combination, could be made for the Wilderness Act of 1964 and the Land and Water
Conservation Fund Act of 1965, as well as the Fish and Wildlife Service autho-
rizations that long predate the 1960s; for example, the Duck Stamp and Wetlands purchase programs to provide nesting and resting areas for ducks and
geese. Similar analyses, but largely within the ideological context of the
traditional conservation movement of the turn of the century, could be made
for the Water Resources Research Act of 1964 and the Water Resources Planning
Act of 1965.

Political Strategy for Enhancing
Fish and Wildlife Values Generally

Application in combination of Lowi's three concepts will now be taken
as an empirically established practical political approach to policy change.
But this empirical analysis does not argue its normative propriety, particularly inclusion of distributive politics. Many people today would boggle at
attaching the epithets of pork barrel and logrolling to political decisions
on wild and scenic rivers. Nevertheless, they could properly be so attached.
Vote trading clearly was involved in building the majority in favor of the
Wild and Scenic River Act.

Can the problem of normatively justifying vote trading be solved within
an acceptable ideological context of democratic government? This problem is

important and intriguing, and a good problem for basic political research. Its solution cannot be attempted here. Instead, it will be assumed that this normative problem can be acceptably solved. The following strategy for enhancing fish and wildlife values is taken to be justifiable, both positively and normatively.

When public problems with respect to fish and wildlife are identified, the fullest possible range of alternative technically-feasible solutions should be formulated. Their political efficacy should then be appraised in terms of the three types of decisions that potentially could be involved.

Distributive Politics. Does the policy proposal call for or imply a number of geographically widespread local projects, with local benefits greater than local costs? Can it be modified to provide this local interest? If this is done, means should be sought to apprise local interests of this potentiality with the expectation of their active political support. Established national fish and wildlife and environmental interest groups can provide this service. Historically, the fish and wildlife program has utilized distributive politics very effectively. The Pittman-Robinson (1937) and Dingell-Johnson (1950) acts which provide for the collection of federal taxes on guns, ammunition and fishing tackle, and the distribution of the proceeds by formula to all states (which, of course, strongly support these federal taxes) are major examples.

Regulative Politics. As fish and wildlife concerns developed, conflict with other public and private activities occurred. The present Fish and Wildlife Coordination Act, the first version of which was enacted in 1934, is aimed at providing a confrontation between water resource development interests and fish and wildlife interests, involving mitigation of damage to fish and wildlife (e.g., fish ladders in the Columbia River) as well as their enhancement (e.g., variable release of water from dams to establish and maintain downstream cold-water fishery) within water resource development projects. The Water Resources Planning Act of 1965 also encourages such confrontation. Its passage was strongly supported by leaders of the Fish and Wildlife Service, but only reluctantly supported by water resources development interests.

Solution of new public problems will increasingly involve confrontation and conflict. Bargaining and compromise (trade-offs) will necessarily be involved. Fish, wildlife and other ecological values are not seen politically, as yet at least, as superordinate. Political planning to get the best possible compromise should be the name of the game. Widespread support, beyond support from those gaining directly from the proposed policy, should be sought. If possible, such support should be so great and evident that the other side in confrontation will "blink first."

Redistributive Politics. Widespread support comes from ideological adherence. Today, the cutting edge of ideology relevant here is normative ecology, backed up by scientific ecology. Aesthetic and ethical ideological concerns, however, should not be neglected. Fish and wildlife policies should be capable of being rationalized as supportive of these ideologies. To make them so may well involve policy changes within the fish and wildlife area itself; for example, reconciliation of policy differences between the fishermen and hunters on the one hand and Audubon Society members and general ecological ideologists on the other.[10]

---

[10]For a brief, but more substantive, fish and wildlife policy critique, (with suggestion of policy directions for the future), see Caulfield (1971).

Looking at alternative technically feasible policy proposals in terms of these three types of decisions will not just require good off-the-top-of-the-head political judgment. Careful planning and studies based upon the existing state of knowledge will be highly desirable. But improving the knowledge base through applied research could also prove very helpful. Specific topics could be deduced from consideration generally of the political strategy suggested here as well as induced from experience in utilizing this strategy.

References

Arkansas-White-Red River Basins Inter-Agency River Basins Committee. 1955.
    "Arkansas-White-Red River Basins--A Report on the Conservation and
    Development of the Water and Land Resources." Mimeographed. (Wash-
    ington, D.C., U.S. Water Resources Council) Part I, pp. 217-218;
    Part II, pp. 175-176.

Bailey, Stephen K., and Howard Samuel. 1952. Congress at Work (New York,
    Hold).

Caulfield, Henry P., Jr. 1971. "Political Considerations," in Richard D.
    Teague, ed., A Manual of Wildlife Conservation (Washington, D.C.,
    The Wildlife Society) pp. 9, 10.

_____. 1975. "Politics of Multiple Objective Planning," in E. L.
    Michalson, E. A. Englebert, and W. Andrews, eds., Multiple Objective
    Planning Water Resources Vol. 2 (Moscow, Idaho, Idaho Research Founda-
    tion, Inc.) pp. 82-94.

_____. 1976a. "Let's Dismantle (Largely but not Fully) the Federal
    Water Resource Development Establishment, or The Apostasy of a Longstanding
    Water Development Federalist," Denver Journal of International Law and
    Policy vol. 6 (Special Issue) pp. 395-402.

_____. 1976b. "Perspectives on Instream Flow Needs," in J. F. Orsborn
    and C. H. Allman, eds., Instream Flow Needs (Washington, D.C., American
    Fisheries Society) pp. 8-14.

Easton, David. 1965a. A Framework for Political Analysis (Englewood Cliffs,
    N.J., Prentice-Hall) pp. 50, 96, 97.

_____. 1965b. A Systems Analysis of Political Life (New York, John
    Wiley and Sons) p. 358.

Federal Register. 1973. Vol. 38, no. 174 (10 September) Part III.

Friedrich, Carl J. 1960. Constitutional Government and Democracy (Boston,
    Ginn and Co.) p. 362.

Jones, Charles. 1970. An Introduction to the Study of Public Policy (Belmont,
    Calif., Wadsworth Publishing Co.).

Krutilla, John V., and Otto Eckstein. 1958. Multiple Purpose River Develop-
    ment--Studies in Applied Economic Analysis (Baltimore, Johns Hopkins
    University Press for Resources for the Future) pp. 18, 43, 91, 92.

Lowi, Theodore J. 1972. "Four Systems of Policy, Politics and Choice,"
    Public Administration Review vol. 32 (July-August) pp. 298-310.

Pigou, Arthur Cecil. 1920. The Economics of Welfare (London, McMillan).

Porter, Kirk H., and Donald B. Johnson, comp. 1970. National Party Platforms, 1840-1968 (Urbana, Ill., University of Illinois Press) p. 28.

Udall, Stewart L. 1963. The Quiet Crisis (New York, Holt, Rinehart and Winston).

INSTITUTIONAL ASPECTS OF WATER ALLOCATION
IN THE UPPER COLORADO RIVER BASIN--
IMPLICATIONS FOR FISH AND WILDLIFE:
A DISCUSSION

Harvey R. Doerksen*

Introduction

A common assumption expressed by practitioners in fish and wildlife
agencies goes something like this: "When we get the data, we can prove we
are right." Raleigh (1977) has observed that, by contrast to political
decision makers, biologists as a profession tend to adhere to the belief
that there is an ultimate truth, knowledge of which is obtainable through
the acquisition of objective facts. Further, biologists tend to feel
threatened by conflict, for, in an arena of ultimate truth, conflict can
only mean that someone is wrong.

In placing such faith in objective, biological data, the biological
community often has neglected the assertion made by Bill Lord that "the
social or institutional aspects...are the most strategic parts of the system from the point of view of controlling its behavior." Fish and wildlife
biologists have not had overwhelming success in competing for limited resources
due in part to the lack of attention paid to the decision-making processes.

Western Water Law

The author presents a reasonable overview of western water law. However,
I have some concerns about certain conclusions and future directions stated

*Research Manager, Western Water Allocation Project, Western Energy
and Land Use Team, U.S. Fish and Wildlife Service, Fort Collins, Colorado.

in the paper. First, the author asserts that state water right permit systems may open the way to gradually recognizing fish and wildlife. While a number of western states have long had available the mechanism for denying permits as a means of preserving flows, this mechanism has not been used with great frequency. In many western states, in fact, a permit is granted almost automatically for any purpose recognized by the state constitution as beneficial.

Use of the permit system would require only a change in decision maker behavior and not a change in the law. Even if such a behavioral change is made, it has distinct disadvantages from the standpoint of fish and wildlife resources. The most crucial problem is that discretionary denial of permits is not the same as recognition that fish and wildlife maintenance constitutes a beneficial use and, more importantly, it does not positively assert a water right for fish and wildlife purposes with a priority date. Time is the critical element in western water allocation. While use of permit systems for occasional denial "evolves," all permits not denied would constitute water demands with a higher claim than fish and wildlife, since fish and wildlife rights would never be positively established.

Another avenue of change would be in the law itself, and in most cases, in the state constitution. Such a change has been made in Colorado and in at least three other western states. Yet even where a constitutional amendment provides that water uses for fish and wildlife are beneficial, the priority date still remains a significant problem.

According to the author, the "obvious solution" to continued exploitation of existing water rights is condemnation of existing rights in order to protect and restore natural environments. Certainly, condemnation of private property for public uses has considerable precedent. In the area of water resources, it would appear especially likely that this mechanism could be used, because western waters are considered to be a public good appropriated to private uses by the states. Return to public good status would seem to follow.

There are other mechanisms which also seem obvious. These include condemning portions of all existing water rights rather than the totality of certain water rights. Under the present system, the state can continue to grant appropriations far in excess of existing water supplies simply because the mechanism exists to protect prior rights. Some kind of proportional reduction of the rights of all users for such reasons as seasonally low flows or establishment of flow reservations for fish and wildlife purposes may, among other things, result in pressures placed on the state by present users to limit future appropriations. This mechanism, combined with the public interest doctrine and selective condemnation, could create the flexibility necessary to recognize changing social demands. Such modifications also would create bloodshed of no small measure.

Another possible solution involves incentives for water conservation. State water allocation laws generally discourage conservative use because maintenance of a water right generally requires active exercising of that right. Some form of water conservation could be implemented through (a) requiring rigid documentation of water needs in support of permit applications; (b) initiating positive and negative incentives; and (c) monitoring water use.

Another mechanism for providing flexibility in water allocation systems to be responsive to changing public demands might be the issuance of term permits. For example, the state of Washington is attempting to initiate a system whereby the large water permits (e.g., for irrigating large, corporate farms) are issued for a period of fifty years after which time the water is available for reallocation by the state.

The federal reserved rights doctrine may, as the author suggests, provide the opportunity to reserve flows for fish and wildlife purposes. Certainly, the recent court decision on the desert pupfish in Nevada, which placed a limit on the pumping of groundwater in the vicinity, gives promise to this argument (Cappaert vs. United States, 1976). On the other hand, the recent decision by Sandy White, referee in the "Eagle County Case" in Colorado (United States vs. District Court, 1971), tends to refute the author's statement that dates of reservation on federal lands are generally early enough to make them senior to most private water rights. White placed the priority data for fish and wildlife purposes at 1960, when the multiple use concept was recognized in national law.[1] In Colorado, and in many other western states, 1960 is not a very desirable priority date.

Whether or not the Forest Service obtains an early priority date, it still can effectively use its authority to grant easements for diversion ditches and other rights-of-way. Judicious use of this authority may

---

[1] President Truman specifically reserved Devil's Hole National Monument for the desert pupfish by Presidential Proclamation in 1952. Thus, there was a federal reserved right as of that date which precluded irrigation rights which were perfected later. With regard to the Forest Service, the act creating the national forests specified the purposes of watershet management and tree propagation. Therefore, these purposes have priority coincident with establishment of the forest. Other uses were given a priority date of 1960.

result in the maintenance of flow reservations at least within the forest
boundaries.

The Eastern example of establishing forests under the Weeks Act to
maintain downstream flows for transportation under the commerce clause may
not apply directly to the West. In the West, water navigation simply is
not an important water resource consideration. A more promising assertion
of federal rights for preserving minimum streamflows would be related to
federal hydroelectric power dams. Hydroelectric power is a nonconsumptive
use of water, but hydroelectric power dams require large amounts of water to
be delivered at a particular location. Whether or not a legal right to such
water for delivery at that point is obtained from the state, the existence
of the dam and its attendant benefits is, in itself, a strong argument for
the provision of adequate flows from upstream areas.

Political and Administrative Aspects

The argument is adequately supported that water use decision making follows Lowi's (1966) concept of distributive politics. Further, the extent to
which this decision arena will continue to reflect distributive politics,
as opposed to the regulatory model, is highly uncertain.

As is pointed out by the author, a number of people have attempted to
make water resources politics more responsive to the broader public interest,
as those observers perceive the public interest. More than one President
has taken a hard look, for example, at the relative independence of the Corps
of Engineers, contemplated some form of reorganization, and finally concluded
that, in political terms, "life's too short." While Helen Ingram (1972) is undoubtedly correct in observing changes in the decision rules, partially out of

disruption of tradition by the environmental movement, I tend to adopt the stance taken by Dean Mann (1974). It is my belief that the "system" displayed a certain amount of adaptive behavior by including a broader set of interests into the distributive mode with a subsequent retreat into tradition when the pressure was off.

I am less convinced than the author that fish and wildlife agencies have joined the game of distributive politics. I think it is more accurate to say that these agencies have been involved in the game, but they have been playing by the wrong rules!

As described by the author, distributive politics exhibits certain characteristics. In distributive politics, there is a "Christmas tree" full of goodies. However, fish and wildlife agencies know only what they don't want. In distributive politics, there is something for everyone. However, fish and wildlife agencies perceive every change as a loss. In distributive politics, the cost is borne by the federal taxpayer. However, fish and wildlife agencies assume the loss to be theirs. In distributive politics, there is support of the entire program by all participants. However, fish and wildlife agencies more often oppose the entire program.

The few instances in which the fish and wildlife agencies have played the game reasonably well should serve as guides for future behavior. One example is Topock Marsh in the Havasu National Wildlife Refuge on the Colorado River in Arizona. The Bureau of Reclamation participated with the Fish and Wildlife Service to dredge and dike a portion of the marsh boundary in order to stabilize water levels, to more effectively use the water allotment, and to improve water circulation. If this kind of cooperative effort were the rule rather than the exception, it could clearly be said that fish and wildlife agencies have joined the game.

If the assumption is correct that water politics in the near term will more closely approximate distributive politics than regulatory politics, then fish and wildlife agencies should articulate ways in which the traditional water development agencies can create Christmas packages for them as well. In the past, fish and wildlife agencies knew precisely what they did not want, but did not know what they did want. Distributive politics requires articulation of packages to be placed under the Christmas tree, and opposition to all of the other agencies' packages cannot itself be easily packaged.

Now, more than any previous time, construction agencies are responsive to positive habitat improvement suggestions. For example, one regional director of the Bureau of Reclamation recently invited fish and wildlife agencies to submit a list of projects they would like to have constructed. While this offer was made in the mode of classical distributive politics, it requires fish and wildlife agencies to break with their own traditions in order to benefit fish and wildlife resources.

I agree with the author that the post-authorization phase of water development project planning, which offers substantial opportunities for inputs by fish and wildlife agencies, is an important one. The point is well made that opponents of projects (generally including fish and wildlife agencies) lose interest toward the end of the project development process. Followup studies and monitoring activities to identify fish and wildlife values in existing projects are almost nonexistent. Yet, it appears to me that active advocacy of fish and wildlife projects at the planning stage,

539

coalition-building in the authorization process, informal bargaining at the post-authorization stage, and operation monitoring following project completion could also yield substantial payoffs for fish and wildlife agencies.

I disagree with the author in his preference for moving to a regulatory mode. In the foreseeable future, substantially more could be gained for fish and wildlife resources if the advocate agencies learned to play the distributive game by the correct rules.

There is no assurance that fish and wildlife interests would be better represented under a regulatory decision-making mode. The same cast of characters would still be present and the same inequities of budget, personnel, and political support would still exist. It is still highly likely that the agencies presently strong would continue to get the lion's share of the benefits. The fact remains that the societal "best" typically has been based on economic criteria and there is no assurance that noneconomic criteria will be any more important under regulatory politics than they are presently. Further, it is a common observation that regulatory agencies follow the "regulatory ideal" only for a brief time after which they mature as bureaucracies and begin to serve the constituencies that they were created to regulate. From this standpoint, I am impatient with some of the research needs suggested by the author which are oriented to finding ways in which we can move from a distributive system to a regulatory system.

I suggest the following research opportunities, some of which are articulated by the author:

1. Information gathering about fish and wildlife resources in the natural evironments upon which they depend should, and will, continue into the future. However, I would go further and recommend that the stress be

equally divided between development of biological data on the one hand, and development of appropriate information formats and timing on the other, so as to be maximally useful to other agencies participating in the distributive game. In other words, we should identify ways in which biological data can be related to the political process.

2. An area of research alluded to indirectly by the author would relate to ways in which traditional development agencies could be utilized to provide benefits for fish and wildlife agencies. Such research would be intended to change the traditional way of thinking from opposition in the context of open warfare, to positive advocacy of fish and wildlife resource development.

3. An evaluation should be made of the experience of those western states which have enacted minimum flow laws. Such evaluations could be useful to other states considering such laws.

4. Attention should be given to how fish and wildlife resource development components can be introduced at the post-authorization stages of water projects.

References

Cappaert et al. vs. United States et al., 7 June 1976.

Ingram, Helen. 1972. "The Changing Decision Rules in the Politics of Water Development," Water Resources Bulletin vol. 8, pp. 1177-1188.

Lowi, Theodore J. 1966. "Distribution, Regulation, Redistribution: The Functions of Government," in R. Ripley, ed., Public Policies and Their Politics (New York, Norton).

Mann, Dean E. 1974. "Political Incentives in U.S. Water Policy: Relationships Between Distributive and Regulatory Politics," in Matthew Holden, Jr. and Dennis L. Dresang, eds., What Government Does vol. 1, Sage Yearbooks on Politics and Public Policy (Beverly Hills, Calif., Sage Publications).

Raleigh, Robert. 1977. "Western Energy Scenario," Fisheries (January-February).

United States vs. District Court, County of Eagle, Colorado, 401 U.S. 520, 91 S. Ct. 998 (1971).